AF601567

Birkhäuser

Progress in Computer Science and Applied Logic

Volume 29

More information about this series at http://www.springer.com/series/4814

Patrick Schultz • David I. Spivak

Temporal Type Theory

A Topos-Theoretic Approach to Systems and Behavior

Patrick Schultz
Massachusetts Institute of Technology
Cambridge, MA, USA

David I. Spivak
Massachusetts Institute of Technology
Cambridge, MA, USA

ISSN 2297-0576 ISSN 2297-0584 (electronic)
Progress in Computer Science and Applied Logic
ISBN 978-3-030-00703-4 ISBN 978-3-030-00704-1 (eBook)
https://doi.org/10.1007/978-3-030-00704-1

Library of Congress Control Number: 2018962871

Mathematics Subject Classification: 03B44, 03B45, 03G30, 18B25, 18F20, 93A30

This book is published under the trade name Birkhäuser, www.birkhauser-science.com by the registered company Springer Nature Switzerland AG.
The registered company address is: Gewerbestrasse 11, 6330 Cham, Switzerland

Acknowledgments

Many of the ideas presented here began to take form in 2015–2016 while working closely with Christina Vasilakopoulou, a postdoc in our group at the time. We benefitted greatly from—and thoroughly enjoyed—this collaboration.

We have learned a great deal from our collaborators Kevin Schweiker, Srivatsan Varadarajan, and especially Alberto Speranzon at Honeywell Labs as well as from our sponsors at NASA, including Alwyn Goodloe, Anthony Narkawicz, and Paul Minor. We also want to thank Aaron Ames, Brendan Fong, Tobias Fritz, Ben Sherman, and Rémy Tuyéras for helpful discussions. Thanks also to Ben Sherman and Rémy Tuyéras for comments on an early draft of this book.

The work presented here was supported by NASA grant NNH13ZEA001N as well as by AFOSR grants FA9550–14–1–0031 and FA9550–17–1–0058.

Contents

Chapter 1
Introduction

1.1 Overview

In this book we provide a new mathematical formalism for proving properties about the behavior of systems. A system is a collection of interacting components, each of which may have some internal implementation that is reflected in some external behavior. This external behavior is what other neighboring systems interact with, through a shared environment. Properties of a behavior can be established over a given duration (sometimes called frame or window) of time, and we propose a mathematical language for working with these behavioral properties.

1.1.1 Behavior Types

A *behavior type* B is the information of a set of "different things that can occur" over any length of time. For example, a movie is a behavior type: given any duration of time, there is a set of possible snippets of the movie that have said duration, and the snippets of the movie are what we would call its behaviors. In fact, "all possible movies" is another a behavior type, because to any duration of time, we could associate the set of all 24-frame/second sequences of photographs together with an overlay of sound. All possible music, all possible fights or boxing matches, all behaviors that an airplane or set of airplanes are capable of, etc.—each of these can be modeled as a behavior type.

We give a category-theoretic description of behavior types, using the language of sheaves. That is, to every time window, say of length ℓ, a behavior type B is responsible for providing a set $B(\ell)$ of all behaviors, called *length-ℓ sections of B*, that can possibly occur over this time window. And for every inclusion of one time window into another, the behavior type B is responsible for providing a *restriction map* that restricts each long behavior to the shorter time window.

P. Schultz, D. I. Spivak, *Temporal Type Theory*, Progress in Computer Science and Applied Logic 29, https://doi.org/10.1007/978-3-030-00704-1_1

We say a behavior type *has composable sections* when, for any two overlapping time windows, behaviors over the first that match behaviors over the second can be *glued* to form a behavior over the union time window.

Although many behavior types have composable sections, not all do. For example, the sheaf of monotonic functions to $\mathbb{R}$ has composable sections, because if for every small interval $[t, t+\epsilon]$, we have $f(t) \leq f(t+\epsilon)$ then for every interval $[t_1, t_2]$ whatsoever, we also have $f(t_1) \leq f(t_2)$. In contrast, the following sort of "roughly monotonic" behaviors are not composable:

$$\forall t_1 \forall t_2. (t_1 + 5 \leq t_2) \Rightarrow f(t_1) \leq f(t_2). \tag{1.1}$$

This formula says "if you wait at least 5 seconds between taking samples, you will find f to be increasing." Such a property cannot be determined except on intervals of length at least 5. Any behavior — if tested over a very short time window—will comply with this property, but if one glues two compliant behaviors that agree along an overlap, the result may not be compliant. This example has the property that it satisfies a composition property for pairs of intervals whose overlap is large enough. An example of a behavior type without any sort of composability is what we might call "functions of bounded difference," e.g., satisfying:

$$\forall t_1 \forall t_2. |f(t_1) - f(t_2)| \leq 5. \tag{1.2}$$

Composable sections or not, however, all the behavior types we consider will be *sheaves* in an appropriate setting, and morphisms of sheaves allow us to connect one behavior type to another. For example, there is a morphism from the sheaf corresponding to a movie to the sheaf corresponding to its sound-track. One might formalize the system–environment notion as a morphism $S \to E$ and say two systems S_1 and S_2 share an environment when given morphisms $S_1 \to E \leftarrow S_2$. Thus, it is through sheaf morphisms that different behaviors can interact. Two singers may be hearing the same drum beat, or two airplanes may share the same communication channel.

The behavior types and their morphisms form a category $\mathcal{B}$ of sheaves, and as such, a *(Grothendieck) topos*. Toposes are particularly nice categories, in that they enjoy many properties that make them in some sense similar to the category of sets. One way to make this precise is to say that toposes satisfy the Giraud axioms [AGV71]. But to more clearly draw the analogy with **Set**, let us suffice to say that toposes are regular categories, and that they have limits and colimits, effective equivalence relations, exponential objects, and a subobject classifier Ω.

The subobject classifier in particular is an important object when thinking about properties, e.g., properties of behavior. For **Set**, the subobject classifier is $\Omega_{\mathbf{Set}} = \{\text{True}, \text{False}\}$. Many properties take the form of "yes/no" questions; such questions distinguish elements of a set according to a given property. That is, for any set S and property P, there is a corresponding yes/no question $S \to \{\text{True}, \text{False}\}$ that classifies the subset of P-satisfying elements in S. For example, the property of an integer being odd is classified by a function `is_odd`: $\mathbb{Z} \to \{\text{True}, \text{False}\}$.

An analogous fact holds for behavior types. There is a behavior type $\Omega_{\mathcal{B}}$ which classifies behavior subtypes. Rather than "yes/no" questions, properties of behavior types are "compliant when" questions. For example, the property of a continuous function $f:\mathbb{R}\to\mathbb{R}$ being greater than 0 is classified in terms of the largest open subset $\{x\in\mathbb{R}\mid f(x)>0\}$ on which it is compliant. Similarly, the property of f satisfying Eq. (1.2) is classified by the collection of all open intervals over which f has upper and lower bounds less than 5 apart.

Every topos has an associated internal language and higher-order logic. The symbols of the logic are the typical ones, $\top$, $\perp$, $\wedge$, $\vee$, $\Rightarrow$, $\Leftrightarrow$, $\neg$, $\forall$, and $\exists$, and the sort of reasoning steps allowed are also typical constructive logic. For example, given P and $P\Rightarrow(Q\wedge Q')$, one can derive Q. However, our topos $\mathcal{B}$ is not Boolean, so $P\vee\neg P$ does not hold in general, nor does $(\neg\forall x.\,Px)\Rightarrow\exists x.\,\neg Px$. Again, the reasoning is *constructive*, meaning that a proof of a statement gives a witness to its truth: to prove an existential, one must provide a witness for it and similarly for a disjunction.

One need not imagine behavior types as sheaves when doing this logic—it is all purely formal—and yet anything that can be proven within this logic reflects a specific truth about behavior types as sheaves. The separate-but-connected relationship between the logic and its semantics allows us to work with a highly abstract and complex topos using standard constructive logic, reasoning as though with sets and their elements. Thus for example, one could prove sound theorems about $\mathcal{B}$ in an undergraduate course on formal logic, without ever discussing categories, let alone toposes.

1.1.2 Goal: To Prove Properties of Systems

Our goal is to understand what can possibly occur when multiple components—each with their own type of behavior, interact in any given way. If we know something about the behavior of each component, if each component has a *behavior contract* or guarantee that specifies something about how it will behave among all its possibilities, then we may be able to guarantee something about how the entire system will behave. This is relevant to industries in which different suppliers provide different parts, each with its own behavioral guarantee. In such a setting, one still wants to draw conclusions about the system formed by arranging these components so that they interact in some specified way.

For example, a thermostat, a furnace, and a room comprise three components, each of which may be guaranteed to satisfy a certain behavior contract: the thermostat promises to sense temperature of the room and promises to send a signal to the furnace if the temperature is too low. The furnace promises to heat the room air at a certain rate when given the signal, and the room promises to lose heat at some maximum rate. We may want to prove a behavior contract for the whole thermostat–furnace–room system, e.g., that the temperature will remain within certain bounds.

In this book, we provide a formal system—a temporal type theory and a higher-order temporal logic—in which such proofs can be carried out.

In other words, we will provide a language for proving properties of interconnected dynamical systems, broadly construed. Dynamical systems are generally considered to come in three flavors: continuous, discrete, and hybrid, according to how time is being modeled. In the above example, the temperature of the room could be modeled by a continuous dynamical system, the thermostat by a discrete dynamical system, and the whole setup is a hybrid system. However, our notion of behavior type is much more general, serving as a sort of "big tent" into which other conceivable notions of behavior can be translated and subsequently compared. For example, there is no differential equation whose solution set is the roughly monotonic curves from Eq. (1.1), because two such trajectories f can "cross," but these trajectories constitute a perfectly good behavior type in $\mathcal{B}$. Behavior types also include infinite-dimensional systems, and as such may be a good language for considering adaptive control [ÅW13], though we do not pursue that here.

There are already many different temporal logics—including linear temporal, (metric, Halpern–Shoham) interval temporal, and signal temporal [RU12, AFH96, MN04, HS91]—for describing behavior. Some of these generalize others, but none is most general. While we do not prove it here, we believe and will give evidence for the assertion that our temporal logic can serve as a "big tent," in which all other such logics can embed.

Some of these temporal logics have very powerful model checkers that can produce a proof or counterexample in finite (though often exponential or doubly exponential) time, and we make no such claim: our logic is wildly undecidable. However, as a big tent, our formalism can embed these proofs from other logics and integrate them as a last step in a process.[1] Our work may also be useful to those who simply want a topos-theoretic and category-theory-friendly approach to understanding behavior.

1.2 Behavior Types as Sheaves

While the introduction has been informal so far, the work in this book is fairly technical. We assume that the reader has a good understanding of category theory, including basic familiarity with Grothendieck toposes and internal languages; see [Joh02, MM92]. Some familiarity with domain theory [Gie+03], and of frames and locales [PP12] would be useful but is not necessary.

Consider the usual topological space of real numbers, $\mathbb{R}$, which has a basis consisting of open intervals (r, s). One might be tempted to model a behavior type as a sheaf S on $\mathbb{R}$, but we did not make this choice for several reasons that we will

[1] It may be objected that these temporal logics are often Boolean whereas our topos $\mathcal{B}$ is not; in this case, one would simply embed the statements into the Boolean subtopos $\mathcal{B}_{\neg\neg} \subseteq \mathcal{B}$.

soon explain. Nonetheless, sheaves on $\mathbb{R}$ will be important for our work, and they give a good starting point for the discussion, so let's briefly consider the structures and properties of a sheaf B on $\mathbb{R}$.

For any open interval $(r, s) \subseteq \mathbb{R}$, there is a set $B(r, s)$ whose elements we call behaviors of type B over the interval. If $r \leq r' \leq s' \leq s$, there is a restriction map $B(r, s) \to B(r', s')$ expressing how a system behavior over the longer interval restricts to one over the shorter interval. Given two overlapping open intervals, say (r_1, r_3) and (r_2, r_4) where $r_1 < r_2 < r_3 < r_4$, and given behaviors $b' \in B(r_1, r_3)$ and $b'' \in B(r_2, r_4)$ such that $b'|_{(r_2,r_3)} = b''|_{(r_2,r_3)}$, there exists a unique behavior $b \in M(r_1, r_4)$ extending both: $b|_{(r_1,r_3)} = b'$ and $b|_{(r_2,r_4)} = b''$. This sort of "composition"-gluing condition can be succinctly written as a finite limit:

$$B(r_1, r_4) \cong B(r_1, r_3) \times_{B(r_2,r_3)} B(r_2, r_4) \tag{1.3}$$

A gluing condition—expressed as the limit of a certain diagram—holds more generally for any open covering of an open interval by other open intervals in $\mathbb{R}$. However, it suffices to add just one more kind, which we might call the "continuity" gluing condition; namely, those for the following sort of telescoping inclusion:

$$B(r, s) \cong \lim_{r<r'<s'<s} B(r', s'). \tag{1.4}$$

These two gluing conditions—composition gluing and continuity gluing—on a functor B allow us to identify it with a sheaf on $\mathbb{R}$. However, we do not take sheaves on $\mathbb{R}$ as our model of behavior types for two reasons: non-composability of behaviors and translation invariance of behavior types. We explain these next.

1.2.1 Non-composable Behaviors

One usually imagines behaviors as composable: given two behaviors that match on an overlapping subinterval, they can be joined to form one long behavior. This was expressed in condition (1.3), but it is not always what we want. For example, if we glue together two roughly monotonic curves as in Eq. (1.1), the result may fail to be roughly monotonic.

Another way to see this is in terms of the internal logic and its semantics. In any sheaf topos—though we will speak in terms of topological spaces—when a proposition is true over some open set, it must be true over every open subset. Such a property in a temporal setting is often called a *safety property*: if a system is to be compliant over an interval of time, it must be compliant over every subinterval. The semantics of safety properties is thus that of *falsifiability*: a system satisfies a proposition over an interval if and only if there is no subinterval on which the proposition is false. For example, a property like "if event a happens, then event b will happen 10 seconds later" is impossible to falsify on intervals of length less than

10 s, so we must consider it to be vacuously satisfied on such short intervals. Again, if an interval is too short to falsify a proposition, the proposition is deemed true on that interval.

Now we can more clearly see why the composition gluing condition creates a problem. It says that if a proposition is true on each element of an open cover, then it is true on the union. In $\mathbb{R}$, we have $(0, 4) \cup (2, 6) = (0, 6)$, so if a proposition is unfalsifiable on $(0, 4)$ and on $(2, 6)$ we must call it true on $(0, 6)$. Again, consider the formula Eq. (1.1). It gives an example of a proposition that is unfalsifiable on intervals whose length is strictly less than 5, and hence would be true on $(0, 4)$ and $(2, 6)$. Thus, it would be forced by the composition-gluing condition to be true on $(0, 6)$—and, by induction, on an interval of any length—which is not the semantics we want.

The way we solve the above problem is by enlarging our topological space. Consider the (non-Hausdorff) topological space $\mathbb{IR}$, called the *interval domain*, whose points are compact intervals $[d, u] \subseteq \mathbb{R}$. As with $\mathbb{R}$, the space $\mathbb{IR}$ has a basis of open sets indexed by pairs of real numbers $r < s$. Namely, for any $r < s$, the corresponding basic open is $(r, s) := \{[d, u] \mid r < d \leq u < s\} \subseteq \mathbb{IR}$. Then, a subset $U \subseteq \mathbb{IR}$ is open if and only if it can be expressed as a union of these basic opens.

Note that $(0, 4) \cup (2, 6) \neq (0, 6)$ in this topology, because the left-hand side does not contain the point (compact interval) $[1, 5]$. We will see that $\mathbb{IR}$ is a domain, called the *interval domain*, meaning that it is a topological space—the topology defined in the previous paragraph is called the Scott topology for the domain—and its points have a natural poset structure that completely determines its frame of open sets. Likewise, $\mathbb{IR}$ is a sober topological space, so its frame of open sets completely determines its poset of points.

We use $\mathbb{IR}$ rather than $\mathbb{R}$ as our main space of interest because it allows us to capture non-composable behaviors.

1.2.2 Translation-Invariant Behavior Types

The second reason for not using $\mathbb{R}$ persists even when we pass to $\mathbb{IR}$, but is simpler to explain. Namely, both $\mathbb{R}$ and $\mathbb{IR}$ come equipped with a specific reference point in time, the origin. An important assumption in science is that experiments run today are still valid tomorrow, all other things being equal. The concepts being tested may depend on durations of time, but they are independent of the "date," say, the number of years since the big bang or the birth of an influential person. Some behaviors, such as holidays, are dependent on date, but we regard such time dependence as an additional feature—something to made explicit—rather than as the norm.

Another way to understand what we might call "the translation problem" is that there is no connection in $\mathbb{IR}$ between the interval $(0, 1)$ and the interval $(5, 6)$, unless one somehow builds in the translation action of $\mathbb{R}$. But, one would be tempted to have $\mathbb{R}$ also act on the objects of the topos: to every sheaf X and real number $r \in \mathbb{R}$, there is a translated sheaf, say $X \rhd r$, whose (a, b)-behaviors are the $(a + r, b + r)$-behaviors of X. This creates an interdependence between the real numbers—as an object in the topos—and the rest of the objects in the topos. We believe that this would have made the logical system much more complex.

To remedy the translation problem, we work in a translation-invariant setting: A behavior that can occur over one interval could also occur over any other. In this setting, the time-line itself becomes a behavior type, we call `Time`. Over a given interval of length ℓ, a behavior of type `Time` can be regarded as the behavior of a perfect clock: it starts at some time t_0 and ends at $t_0 + \ell$. If a specific behavior b is dependent on a choice of clock time, we make that dependence explicit in the sense that $t : \mathtt{Time}$ will occur in the formula for b.

To get a bit more technical, our base topos $\mathcal{B}$ is a quotient of $\mathsf{Shv}(\mathbb{IR})$, i.e., there is a geometric surjection $p_* \colon \mathsf{Shv}(\mathbb{IR}) \to \mathcal{B}$. One characterization of geometric surjections is that the inverse image part, in this case $p^* \colon \mathcal{B} \to \mathsf{Shv}(\mathbb{IR})$, is faithful. Another is that $\mathcal{B}$ is the category of coalgebras for the left-exact comonad $p^* p_*$ on $\mathsf{Shv}(\mathbb{IR})$. Intuitively, objects in $\mathcal{B}$ can be thought of as sheaves $X \in \mathsf{Shv}(\mathbb{IR})$ that are *translation-invariant*, i.e., for which one has coherent isomorphisms $X(a, b) \cong X(a + r, b + r)$ for every open interval (a, b) and real $r \in \mathbb{R}$. This is the translation action of $\mathbb{R}$ on $\mathbb{IR}$, mentioned above, and we denote by $\mathbb{IR}_{/\rhd}$ the localization of $\mathbb{IR}$ at the collection of translation maps. $\mathbb{IR}_{/\rhd}$ is no longer a space—more precisely, it is a category and not a poset—but it has a natural site structure and $\mathcal{B} := \mathsf{Shv}(\mathbb{IR}_{/\rhd})$ is the corresponding topos of sheaves.

In fact, $\mathcal{B}$ is an étendue, meaning that there is a specific object, namely `Time`, such that the slice topos $\mathcal{B}/\mathtt{Time} \cong \mathsf{Shv}(\mathbb{IR})$ is localic.

$$\mathsf{Shv}(\mathbb{IR}) \cong \mathcal{B}/\mathtt{Time} \underset{p_!}{\overset{p_*}{\underset{\longleftarrow p^* \longrightarrow}{\longrightarrow}}} \mathcal{B} \cong \mathsf{Shv}(\mathbb{IR}_{/\rhd})$$

The word étendue means "extent" and indeed objects in $\mathbb{IR}_{/\rhd}$ are extents—or durations—of time, over which behaviors can occur.

1.2.3 Four Relevant Toposes

All four toposes in the square below play a role in this work:

$$\begin{array}{ccc} \mathsf{Shv}(\mathbb{R}) & \rightarrowtail & \mathsf{Shv}(\mathbb{IR}) \\ \downarrow\!\!\downarrow & & \downarrow\!\!\downarrow {\scriptstyle p_*} \\ \mathcal{B}_\pi & \rightarrowtail & \mathcal{B} \end{array}$$

The toposes in the top row comprise temporal sheaves—those over a specified time-line—whereas those in the bottom row comprise behaviors that are translation-invariant. Sheaves in the left column have composable behaviors, whereas sheaves in the right column can include more general, non-composable behaviors. We discussed $\mathsf{Shv}(\mathbb{R})$, $\mathsf{Shv}(\mathbb{IR})$, and $\mathcal{B} = \mathsf{Shv}(\mathbb{IR}_{/\rhd})$ above.

The horizontal maps are geometric embeddings, i.e., they correspond to modalities, and the vertical maps are geometric surjections. The topos $\mathcal{B}_\pi$ and the left-hand and bottom maps are uniquely determined from the top and right-hand maps, by way of the surjection-embedding factorization systems on toposes. The π stands for "pointwise," which means that properties of sheaves in $\mathcal{B}_\pi$ can all be determined in neighborhoods of points in $\mathbb{R}$.

1.3 Temporal Type Theory

The main subject of this book is the definition of a type theory—and its associated logic—that has its semantics in the topos $\mathcal{B} = \mathsf{Shv}(\mathbb{IR}_{/\rhd})$ discussed above.

The logic we present is higher-order logic, plus subtypes and quotient types. Higher-order logic—as well as its strong connection with topos theory—has been very well-studied [LS88, Fou77, BJ81, Awo16, Joh02], which is part of our motivation for using it. In fact, toposes support not just higher-order logic but also dependent type theories. This means that one can use an automated proof-assistant based on dependent type theory—such as Coq or Lean [CH88, Mou+15]—to validate proofs.

The primary goal of the type theory and logic we present is to support defining and reasoning about behavior types. As such, the logic is capable of expressing statements from standard temporal logics, such as linear temporal logic (LTL). The primary "temporal operator" from LTL is called *until*. The meaning of until—written $\mathcal{U}$ in the logic—is often presented by a formula such as:

$$(\phi_1 \,\mathcal{U}\, \phi_2)(t) := \exists(r : \mathtt{Time}).\left[(t < r) \wedge \phi_2(r) \wedge \forall u.\,(t < u < r \Rightarrow \phi_1(u))\right]. \tag{1.5}$$

Here, the temporal propositions ϕ_1 and ϕ_2 are being represented as propositions that have an explicit dependence on time, i.e., as functions $\mathtt{Time} \to \mathtt{Prop}$. As one can attempt to read off from Eq. (1.5), $\phi_1 \,\mathcal{U}\, \phi_2$ constructs a new temporal proposition which is true if both the following hold: ϕ_2 is true sometime in the future, and ϕ_1 is true from now until that point.

In fact, Eq. (1.5) is an example of a formula in our logic. The type `Prop` is what makes the logic higher order, and the type `Time`—which is definable from the one atomic term of our theory, and whose semantics is the behavior type `Time` discussed in Sect. 1.2.2—allows for temporal statements like $\phi_1 \,\mathcal{U}\, \phi_2$ to be expressed. It is from this perspective that we find it reasonable to refer to our system as *temporal type theory*. See Sect. 8.6 for a more in-depth discussion of how our type theory and logic relates to existing temporal logics, in particular to LTL and metric temporal logic.

Every property we can discuss is a safety property, in the sense described in Sect. 1.2.1, so all of our connectives and quantifiers take safety properties to safety properties. An example of the expressive power of the temporal type theory: we are able to internally define real-valued functions of time, as well as their derivatives, and prove that the derivative satisfies the Leibniz rule.

As a first test that our formal system is strong enough to be useful in practice, we use a simplified version of the safe separation problem for airplanes in the US National Airspace System (NAS). There one wants to avoid situations in which airplanes get too close to one another. To achieve this, the current system consists of an interaction between radar, a traffic collision avoidance system (or TCAS), pilot decision-making, and actuators and thrusters on the surface of the airplane. The position of each airplane roughly follows a differential equation, with time-varying parameters such as "climb at rate $r(t)$," supplied by the pilot. The pilot's decisions take into account the commands from air traffic control and the advice of the TCAS, which alerts the pilot to urgent situations and suggests corrective maneuvers. In turn, this information is determined by the relative position of the airplanes, completing the loop.

The NAS thus requires continuous interaction between many different types of behavior. Some of the components are modeled continuously, such as the motion of the plane, while others are modeled discretely, such as the TCAS alerts and suggestions. The pilot generally takes the advice given and carries it out after some delay. It is the combination of continuous, discrete, and delay behaviors that we believe is the essence of the safe separation problem. In Chap. 8, we will prove a version of the safe separation property—a version that is greatly simplified but that still involves the above essential elements—in which there is only one airplane that must obtain safe separation from the ground.

1.4 Related Work

Spivak et al. [SVS16] presents a topos similar to that described above, in Sect. 1.2. The authors (ourselves plus C. Vasilakopoulou) consider sheaves and presheaves on a certain category **Int** of intervals: sheaves for behavior types and presheaves for behavior contracts. Sheaves on **Int** can be identified with discrete Conduché fibrations—or unique factorization liftings—over the monoid $(\mathbb{R}_{\geq 0}, 0, +)$ of non-negative real numbers. This may have interesting connections with decomposition spaces [GKT15], but more closely related are applications to dynamical systems, pursued in [Law86, BF00, Fio00].

This book considers instead a subtopos $\mathcal{B} \subseteq \mathsf{Shv}(\mathbf{Int})$, obtained by sheafifying **Int**-sheaves or **Int**-presheaves with respect to the "continuity" gluing condition (1.4), and hence building in continuity at the ground floor. This is a natural step, as the category $\mathbf{Int}^{\mathrm{op}}$ is already a continuous category in the sense of Johnstone and Joyal, and $\mathcal{B}$ is equivalent to the topos of continuous functors $\mathbf{Int}^{\mathrm{op}} \to \mathbf{Set}$; see [JJ82] or Appendix B. The present work develops a type theory with a semantics in the sheaf topos $\mathcal{B}$, but which can be used independently of the sheaf semantics.

There are also connections with [JNW96] and following works, where dynamics is considered in terms of morphisms from a given category of paths. For example in [HTP03], an object in the path category is an interval "modeling a clock running on [an open interval] at unit rate," a perspective quite similar to our own.

Our general approach to dynamical systems also has something in common with that of the early cyberneticists, such as Ashby or Weiner; for example, our notion of behavior type is roughly what Ashby [Ash13] calls the "field" of a system. However, our work is more closely aligned with the relatively recent "behavioral approach," as advocated in [Wil07]. There, a dynamical system is defined to be a triple $(\mathbb{T}, \mathbb{W}, \mathcal{B})$, where $\mathbb{T}$ represents time, and which we fix to be $\mathbb{T} = \mathbb{R}$, where $\mathbb{W}$ is an arbitrary set of "signal values," and where $\mathcal{B} \subseteq \mathbb{W}^{\mathbb{T}}$ is the set of possible behaviors as a subset of all functions $\mathbb{T} \to \mathbb{W}$.

An object of the category $\mathcal{B}$ is closely related to such a dynamical system. The primary difference is that, instead of specifying only those possible "infinitely extended" behaviors $\mathbb{T} \to \mathbb{W}$, an object $B \in \mathcal{B}$ must specify a subset $B(a, b) \subseteq \mathbb{W}^{(a,b)}$ for every open interval $(a, b) \subseteq \mathbb{R}$, subject to the conditions that for any $f : (a, b) \to \mathbb{W}$:

- if $f \in B(a, b)$, then $a \leq a' < b' \leq b$ implies $f|_{(a',b')} \in B(a', b')$;
- if $f|_{(a',b')} \in B(a', b')$ for any $a < a' < b' < b$, then $f \in B(a, b)$;
- if $f \in B(a, b)$, then $[t \mapsto f(t - r)] \in B(a + r, b + r)$ for any $r \in \mathbb{R}$.

We find this to be more natural from a modeling perspective, since an infinitely extended behavior is by definition unobservable.

Behavior types $B \in \mathcal{B}$ are also more general, in that we can consider behavior types in which there exist behaviors $B(0, 5)$ which admit no extension to $B(0, 10)$, say. As an example of this phenomenon, we could define $B(a, b)$ to be the set of all differentiable functions $f : (a, b) \to \mathbb{R}$ satisfying the differential equation $f' = 1 + f^2$. Then, $\tan(t): (-\frac{\pi}{2}, \frac{\pi}{2}) \to \mathbb{R}$ defines a behavior in $B(-\frac{\pi}{2}, \frac{\pi}{2})$ for which there is no extension to a behavior on any larger interval.

As another example, consider the behavior type B of possible trajectories of population levels of wolves and deer in some ecosystem. For the sake of analysis, we may want to consider the subobject $B' \hookrightarrow B$ consisting of only the desired behaviors, say where neither the wolves nor the deer go extinct. It is certainly possible for the ecosystem to be in a state in which, according to the model being used, there does not exist a future in which neither species goes extinct.

Finally, as mentioned earlier in this introduction, one of the goals of this work is to serve as a "big tent," so that many approaches to models of systems can be embedded into our category $\mathcal{B}$, and many temporal logics invented to analyze those models can be embedded into our axiomatics. In that sense, our goal is not to compete with other bodies of work, but to help weave them all together.

1.5 Notation, Conventions, and Background

Here, we lay out some of the basic notation and conventions used throughout the book. We then give a lightning review of sites, sheaves, and toposes.

1.5.1 Notation and Conventions for Categories

Notation 1.1 We say that a set S is *inhabited* if there exists an element $s \in S$. This is the constructive way to say that S is nonempty. Similarly, we say that a category is inhabited if it has an object.

For any $n \in \mathbb{N}$, we sometimes write n to denote the linearly ordered category $\bullet^1 \to \cdots \to \bullet^n$ with n objects.

All categories in this book are 1-categories unless otherwise stated. We denote composition of morphisms $A \xrightarrow{f} B \xrightarrow{g} C$ in the classical "Leibniz" order $g \circ f \colon A \to C$, and we denote identity morphisms as id. If C is a category, we may denote the hom-set between objects $c, c' \in C$ either by $C(c, c')$ or $\mathrm{Hom}_C(c, c')$, or simply $[c, c']$ if C is known from context.

1.5.2 Definition of Sites, Sheaves, and Toposes

Whenever we speak of a topos in this book, we always mean a Grothendieck topos, which is the category of sheaves on a site. We quickly remind the reader of the definition of site and topos. Readers for whom the definition seems overly abstract may simply skim it on a first reading. The following is taken from [Joh02].

Definition 1.2 (Coverage, Site, Sheaf, Topos, and Geometric Morphism) Let C be a category. For any object $U \in C$, a *family over* U is a subset of $\mathrm{Hom}_C(-, U)$. Explicitly, a family consists of an object U, a set I, and for each $i \in I$ a morphism $f_i \colon U_i \to U$ for some $U_i \in C$. Let $\Phi(C)$ denote the set of families in C.

A *coverage* χ consists of a set T and a function $F \colon T \to \Phi(C)$ satisfying the following condition. Suppose given an object $U \in C$, a morphism $g \colon V \to U$, and a $t \in T$, such that $F(t) = (f_i \colon U_i \to U \mid i \in I)$. Then, there exists a $t' \in T$ such that $F(t') = (h_j \colon V_j \to V \mid j \in J)$ and such that for each $j \in J$ there exists some $i \in I$ and some $g_{j,i} \colon V_j \to U_i$ making the following square commute:

$$\begin{array}{ccc} V_j & \xrightarrow{h_j} & V \\ {\scriptstyle g_{j,i}}\downarrow & & \downarrow{\scriptstyle g} \\ U_i & \xrightarrow[f_i]{} & U \end{array}$$

A *site* is a category C equipped with a coverage $\chi = (T, F)$. For any $t \in T$, we say $F(t) = (U, I, f)$ is a *χ-covering family over U*.

Suppose given a functor $X: C^{\text{op}} \to \textbf{Set}$, a family $(f_i : U_i \to U \mid i \in I)$, and an element $x_i \in X(U_i)$ for each $i \in I$. We say that the x_i are *compatible* with respect to the family if, for every pair of elements $i, i' \in I$, object $W \in C$, and commutative square:

$$\begin{array}{ccc} W & \xrightarrow{g_i} & U_i \\ {\scriptstyle g_{i'}}\downarrow & & \downarrow{\scriptstyle f_i} \\ U_{i'} & \xrightarrow[f_{i'}]{} & U \end{array}$$

the equation $X(g_i)(x_i) = X(g_{i'})(x_{i'})$ holds in $X(W)$.

A *sheaf* on a site $S = (C, \chi)$ is a functor $B: C^{\text{op}} \to \textbf{Set}$ such that for each χ-covering family $(f_i : U_i \to U \mid i \in I)$ and compatible family $\{b_i \mid i \in I\}$, there exists a unique $b \in B(U)$ such that $B(f_i)(b) = b_i$ for each $i \in I$. A *morphism of sheaves* $B \to B'$ is simply a natural transformation of functors, and this defines the category of sheaves on (C, χ). The category of sheaves on the site S is denoted $\mathsf{Shv}(S)$.

A *topos* is any category $\mathcal{E}$ for which there exists a site S and an equivalence of categories $\mathcal{E} \cong \mathsf{Shv}(S)$. If $\mathcal{E}'$ is another topos, a *geometric morphism* $f: \mathcal{E} \to \mathcal{E}'$ is a functor which has a left adjoint, such that the left adjoint preserves finite limits.

1.6 What to Expect from the Book

We hope the reader can learn useful things by reading this book. Of course, what is learned depends on the reader: her background, interests, level of effort, etc. Before giving an outline of the chapters, we will discuss some potential items of interest.

1.6.1 What the Reader Can Hope to Learn

Unfortunately, this book could not be written for a general audience. For example, the authors are well-aware that anyone unfamiliar with toposes probably had difficulty knowing how to think about Definition 1.2. There is certainly material in this book that is amenable to such readers, including Sect. 2.4, Chaps. 4, 5, 7 and 8.

However, as mentioned above in Sect. 1.2, the book is mainly written for readers who have seen toposes before. We would like to think that students of all levels can learn something about toposes by reading this book. Indeed, it can be considered as an extended example of a single topos $\mathcal{B}$ as well as its slices and subtoposes. Chapter 4 is a stand-alone chapter that is meant to introduce readers to type theory and logic, especially as they relate to toposes. For example, we provide detailed type-theoretic accounts of real numbers and other numeric objects that have semantics in an arbitrary topos.

While some knowledge about Scott domains would certainly be useful at times, it is not necessary. Again, we hope that this book will provide insight into the subject of domain theory by offering an extended example. Semantically, we will work in a quotient of the domain $\mathbb{IR}$, but we will also have occasion to consider domains internal to a topos. For example, the derivative of a continuous function is an interval-valued function, and it is defined in a domain-theoretic style (see Definition 7.25).

Finally, the reader will also learn about temporal logic. We spend a bit of space to try and convey how existing temporal logics fit into our theory. But mostly, we hope that the reader will have interest in our own particular *higher-order temporal logic*, as it motivates everything in the book.

1.6.2 Contributions

Our main contribution is to offer a new temporal type theory (the first of its kind as far as we know), together with a novel topos-theoretic semantics. We believe that it can mediate between several existing formal systems for dealing with time, e.g., for describing cyber-physical systems.

Along the way to understanding Dedekind real numbers—and generalizations like proper and improper intervals—in our topos-theoretic semantics, we were led to a number of results which we believe might be of independent interest. When using the temporal type theory to reason about behaviors which are represented by continuous real-valued functions, we found the need to make use of Dedekind real number objects in various subtoposes as well as the standard Dedekind real number object. In the type theory, these appear as objects defined using versions of the standard Dedekind axioms which have been modified by modal operators. Studying these "modal Dedekind real number objects" was greatly simplified by considering generalizations which remove some of the standard Dedekind axioms. The reason is that these generalized numeric objects form *domains*, which one can think of as a particularly nice class of topological spaces which are intimately connected to order theory. We found that for particular kinds of subtopos—in particular closed, quasi-closed, and dense proper subtoposes—and for particular kinds of domains which are presented by sufficiently nice bases, we could give strong comparisons between domains in a subtopos and domains in the enclosing topos.

These general domain-theoretic results are collected in Appendix A—a section we tried to make readable, independently of the rest of the book—while the special cases about generalized Dedekind numeric objects are collected in Chap. 7. As an example application, the theory of differentiation that we develop in the temporal type theory in Sect. 7.3 makes essential use of generalized Dedekind numeric objects in several different subtoposes.

In addition, we contribute a few other new ideas to the literature. In Sect. 2.5, we prove a result about dense morphisms of posites. While this formally follows from the results in [Shu12], we have provided a new direct proof of the simpler posite case. In Sect. 4.1, we give an informal introduction to type theory and higher-order logic, a subject for which informal accounts seem to be lacking in the literature.

In Chap. 8, we generalize the usual notion of hybrid system from the control theory literature. We also show how to integrate several different temporal logics into our own.

Finally in Appendix B.2, we prove the existence and various properties—e.g., regarding interaction with the Ind-completion—of the (connected, discrete bifribration) orthogonal factorization system on **Cat**. Although this material is well-known, it seemed difficult to locate in the literature.

1.6.3 Chapter Outline

This book has eight chapters and two appendices. Chapter 1 has hopefully summarized and explained the goals of this work: to study temporal properties of a very general class of behavior types, using the language and tools of toposes. We said that each behavior type is modeled as a set of possible behaviors over each interval of time, namely as sheaves on some sort of time-line. However, we made two choices that keep this from being as simple as one might be tempted to expect. The first, and more important, is that matching behaviors on overlapping intervals need not be composable (Sect. 1.2.1); the second is that the set of possible behaviors over an interval should not depend on its position in the time-line (Sect. 1.2.2).

We formalize these two ideas in the following two chapters. In Chap. 2, we define the time-line we will use, which is called the interval domain and denoted $\mathbb{IR}$. The real line sits inside it as the length-0 intervals $\mathbb{R} \subseteq \mathbb{IR}$. We give a continuous bijection between $\mathbb{IR}$ and the upper half-plane in $\mathbb{R}^2$, which not only allows the reader to visualize $\mathbb{IR}$ but also proves quite useful for semantic purposes throughout the book. In this chapter, we also review the definition of posites, (0, 1)-sheaves, Scott domains, and give four equivalent definitions of the time-line as a posite.

In Chap. 3, we deal with translation invariance, roughly by taking the quotient of $\mathbb{IR}$ with respect to the translation action of the group $\mathbb{R}$. The result is a category $\mathbb{IR}_{/\rhd}$, which is no longer a Scott domain, but instead a continuous category, in the sense of Johnstone and Joyal. The topos of sheaves on the corresponding site is $\mathcal{B}$, our main topic of study throughout the book.

We transition from the external viewpoint to the internal viewpoint in Chap. 4. This chapter is fairly independent of the rest of the book, standing as a review of some connections between toposes, type theory, and higher-order logic. In particular, we spend the first half of the chapter reviewing these notions at a high-enough level not to get bogged down in specifics, but allowing the book to be fairly self-contained in terms of its type theory and logic. We then provide a short section recalling the notion of modalities j—also known as Lawvere–Tierney topologies—and their relationship to subtoposes. We spend the remainder of the chapter discussing real numbers and related numeric objects in subtoposes. In particular, we explain j-local arithmetic and inequalities.

The technical heart of the book is in Chaps. 5 and 6. In Chap. 5, we axiomatize the higher-order logic of our temporal type theory, which includes one atomic predicate—defining `Time`—and several axioms. We also discuss a few modalities that correspond to important subtoposes of $\mathcal{B}$. In Chap. 6, we prove the soundness of our axioms in $\mathcal{B}$. This requires explaining the semantics of the various numeric objects, such as the real numbers, as well as the semantics of the modalities.

In Chap. 7, we work with numeric objects relative to various modalities j. For example, using what we call the "point-wise" modality, we have access to the sheaf of real-valued functions on the usual real line, inside of $\mathcal{B}$. In particular, we compare these numeric types internally for differing modalities in Sect. 7.1 and give their $\mathcal{B}$-semantics in Sect. 7.2. This section relies on technical work from Appendix A. Perhaps most interestingly, in Sect. 7.3 we internally define the derivative of such a real-valued function with respect to t : `Time` and prove that this definition is linear and satisfies the Leibniz rule. We also prove externally that its $\mathcal{B}$-semantics is that of derivatives in the usual sense.

The main body of the book concludes with Chap. 8, where we discuss several applications of the work. For example, we give an embedding of discrete, continuous, and hybrid dynamical systems into our temporal type theory. We also explain delays, and our general perspective on behavior contracts for interconnected systems. One of our main inspirations for this work was a case study involving safe separation in the National Airspace System; this is discussed in Sect. 8.5. Finally in Sect. 8.6, we discuss the relationship between our higher-order temporal logic and some of the better known temporal logics from the literature.

The book also has two appendices. In Appendix A, we define a technical tool—which we call predomains—by which to reduce the complexity of domains and the morphisms between them. We explain how our constructions work relative to arbitrary modalities, i.e., within arbitrary subtoposes. Finally, in Appendix B we prove that $\mathbb{IR}_{/\rhd}$ is a continuous category in the sense of Johnstone and Joyal.

Chapter 2
The Interval Domain

In this chapter, we will introduce the *interval domain* $\mathbb{IR}$, which is a topological space that represents the line of time in our work to come. The points of this space can be thought of as compact intervals $[a, b]$ in $\mathbb{R}$. The specialization order on points gives $\mathbb{IR}$ a non-trivial poset structure—in fact it is a domain—and as such it is far from Hausdorff.

The topos of sheaves on the space $\mathbb{IR}$ will play a very important role throughout the book, so we begin in Sect. 2.1 with a review of sheaves on topological spaces, or more generally on posets equipped with a coverage. In Sect. 2.2, we review the theory of domains, and we define $\mathbb{IR}$ in Sect. 2.3. In Sect. 2.4, we show how to view $\mathbb{IR}$ in terms of the usual Euclidean upper half-plane. In Section 2.5 we discuss Grothendieck posites, which allow us to prove the equivalence between four different formulations of the topos of sheaves on $\mathbb{IR}$.

There are a few places in this chapter where we refer to predomains, which are fairly technical and are the subject of Appendix A. However, none of that material is necessary to understand the present chapter. It will become more important for the technical results about arithmetic between real number objects in various subtoposes, which we will discuss later in Sect. 4.3. For the time being, we suggest the reader look briefly at the predomain material if the interest arises, but otherwise feel free to skim it or take it on faith.

2.1 Review of Posites and (0, 1)-Sheaves

The material in this section is largely taken from [nLabb]. For any poset $(P, \leq)$ and $p \in P$, the down-set $\downarrow p := \{p' \in P \mid p' \leq p\}$ will play a fundamental role throughout this section. For a set S, let $P(S) := \{V \mid V \subseteq S\}$ denote its power set. Given $V \subseteq S$, we write $\downarrow V$ to denote the set $\bigcup_{v \in V} \downarrow v$.

P. Schultz, D. I. Spivak, *Temporal Type Theory*, Progress in Computer Science and Applied Logic 29, https://doi.org/10.1007/978-3-030-00704-1_2

A posite is a site—i.e., a category equipped with a coverage (see Definition 1.2)—for which the category is a poset. Any topological space has an underlying posite: the poset is that of open subsets, and covering families are collections of open sets whose union is another. The definition of a coverage can be simplified slightly on categories that are posets.

Definition 2.1 (Coverage on Poset) A *coverage* on a poset S is a relation $\rhd \subseteq P(S) \times S$ such that:

- If $V \rhd u$, then $V \subseteq {\downarrow} u$.
- If $V \rhd u$ and $u' \leq u$, then there exists a $V' \subseteq ({\downarrow} V \cap {\downarrow} u')$ such that $V' \rhd u'$,

If $V \rhd u$, we say that V is a *basic cover* of u. A *posite* is a poset equipped with a coverage.

Example 2.2 If B is a basis for a topological space X, then B ordered by inclusion is a posite. For any $b \in B$ and $V \subseteq B$, we have $V \rhd b$ iff $\bigcup V = b$ (where the union is taken as subsets of X, which is not necessarily the join in B).

As a posite is in particular a site, the usual definition of sheaves of sets applies, constructing a (localic) topos from any posite; see Definition 1.2 or [Joh02, Theorem C.1.4.7]. However, for any posite S we can also consider "sheaves of truth values," or (0, 1)-sheaves, thus constructing a frame $\mathsf{Shv}_0(S)$. If B is a basis for a topological space X, then $\mathsf{Shv}_0(B)$ will simply be the frame of opens in X.

Definition 2.3 ((0, 1)-Sheaf) Let S be a posite. A (0, 1)-*sheaf*[1] is a subset $I \subseteq S$ such that:

Down-closure: If $u \leq v$ and $v \in I$, then $u \in I$.
Sheaf-condition: If $V \rhd u$ and $V \subseteq I$, then $u \in I$.

The poset of (0, 1)-sheaves on S is denoted $\mathsf{Shv}_0(S)$.

A coverage on a category C is called *subcanonical* if the representable functor $C(-, c)$ is in fact a sheaf, for every $c \in C$. The coverage is called *canonical* if it is the largest subcanonical coverage.

Example 2.4 A frame is a poset with infinite joins, finite meets, and which satisfies the infinite distributive law. Any frame S has a *canonical coverage*, with $V \rhd u$ iff $\sup V = u$. For any topological space X, $\mathsf{Shv}(X) \cong \mathsf{Shv}(\Omega(X))$, where $\Omega(X)$ is the frame of opens of X with its canonical coverage.

[1] (0, 1)-sheaves are also sometimes called ideals, but this terminology clashes with two other notions we use in this book that are also called ideals—namely, ideals in a poset and rounded ideals in a predomain (see Definitions 2.5 and A.6)—thus we call them (0, 1)-sheaves to avoid confusion.

2.2 Domains and Posites

In this section, we will quickly review domains in general; see [Gie+03] for a thorough treatment. We also show how to endow an arbitrary domain with a natural posite structure.

2.2.1 Review of Domains (Continuous Posets)

Domains come at the intersection of order theory and topology. They can be thought of as spaces for which the specialization order on points determines the topology, or as posets with "two extra left adjoints," as we will see in Definition 2.10. In this section and throughout the rest of the book, we often use the notation $\sqsubseteq$ for the order relation on domains.

Definition 2.5 (Directed Sets, Dcpos, Ideals, and Down-Closure) Let $(S, \sqsubseteq)$ be an arbitrary poset. A subset $D \subseteq S$ is called *directed* (or *up-directed*) if D is inhabited and, for any $u_1, u_2 \in D$, there exists a $u' \in D$ with $u_1 \sqsubseteq u'$ and $u_2 \sqsubseteq u'$. A *directed-complete poset (dcpo)* is a poset S for which every directed subset $D \subseteq S$ has a supremum $\sup D$. A subset $I \subseteq S$ is called an *ideal* if it is directed and down-closed: $i \in I$ and $j \sqsubseteq i$ implies $j \in I$. The collection $\mathrm{Id}(S)$ of all ideals in S is a poset ordered by inclusion. Given any element $u \in S$, its *down-closure* $\downarrow u := \{ v \in S \mid v \sqsubseteq u \}$ is an ideal, and this map defines a monotonic map $\downarrow : S \to \mathrm{Id}(S)$.

Proposition 2.6 *Let S be a poset. The down-closure $\downarrow : S \to \mathrm{Id}(S)$ has a left adjoint if and only if S is directed-complete. The left adjoint* $\sup : \mathrm{Id}(S) \to S$ *sends any ideal $I \subseteq S$ to its supremum* $\sup I$.

Definition 2.7 (Way Below) Let $(S, \sqsubseteq)$ be a directed-complete poset and let $u, v \in S$ be elements. One says that u is *way below* v, denoted $u \ll v$, if the following is satisfied for each ideal $I \subseteq S$: if $v \sqsubseteq \sup I$, then $u \in I$.

Remark 2.8 If $u \ll v$, then $u \sqsubseteq v$ since $v = \sup \downarrow v$. From the definition, it is easy to check that if $u' \sqsubseteq u$ and $u \ll v$ and $v \sqsubseteq v'$, then $u' \ll v'$. It follows that $\ll$ is transitive.

Proposition 2.9 *Let S be a directed-complete poset. The following are equivalent:*

- *For every $u \in S$, the set $\twoheaddownarrow u := \{ v \in S \mid v \ll u \}$ is an ideal, and $u = \sup \twoheaddownarrow u$.*
- *The functor* $\sup : \mathrm{Id}(S) \to S$ *has a further left adjoint.*

The left adjoint to $\sup : \mathrm{Id}(S) \to S$ sends an element $u \in S$ to $\twoheaddownarrow u$.

Definition 2.10 (Domain) A directed-complete poset $(S, \sqsubseteq)$ satisfying one of the equivalent conditions of Proposition 2.9 is often called a *continuous poset*; we will usually call it a *domain*.[2] The order $\sqsubseteq$ is called the *specialization order*.

Example 2.11 The category 2 (see Notation 1.1) is posetal: $0 \sqsubseteq 1$. Every nonempty subset of it is directed. It is directed-complete and its ideals are $\{0\}$ and $\{0, 1\}$. Its way-below relation coincides with its order $a \ll b \Leftrightarrow a \sqsubseteq b$, and it is a domain.

Proposition 2.12 *Let S be a directed-complete poset and $U \subseteq S$ an up-closed subset. Then, the following are equivalent:*

1. *For any directed set $D \subseteq S$, if $\sup D \in U$ then there exists $s \in S$ with $s \in D \cap U$.*
2. *For any ideal $I \subseteq S$, if $\sup I \in U$ then there exists $s \in S$ with $s \in I \cap U$.*

If S is a domain, then the above are also equivalent to the following:

3. *For all $u \in U$, there exists some $u' \in U$ with $u' \ll u$.*

Proof This is [Gie+03, Lemma II-1.2] and [Gie+03, Proposition II-1.10], but it can also be checked directly. For example, $2 \Rightarrow 1$ because for directed D, the down-closure $I := \downarrow D$ is an ideal, and given $s \in I \cap U$ we have $s \in U$ and $s \leq d$ for some $d \in D$. Since U is up-closed, $d \in D \cap U$. □

We next define Scott-open sets in any directed-complete partial order, even though all the dcpos we consider in this paper will be domains.

Definition 2.13 (Scott-Open Set) A subset $U \subseteq S$ in a dcpo is called *Scott-open* if it is up-closed and satisfies any of the equivalent conditions of Proposition 2.12.

We prove a few properties of domains that may be useful; others can be found in [Gie+03, Section II-1].

Proposition 2.14 *Let $(S, \sqsubseteq)$ be a domain.*

1. *The way-below relation is interpolative: $s_1 \ll s_2$ implies $\exists s$ with $s_1 \ll s \ll s_2$.*
2. *For any element $u \in S$, the set $\twoheaduparrow u$ is Scott-open.*
3. *A set U is Scott-open iff $\uparrow u \subseteq U$ and $\twoheaddownarrow u \cap U \neq \varnothing$ for each $u \in U$.*
4. *The Scott-opens constitute a topology on the set S, called the* Scott topology.
5. *The set $\{ \twoheaduparrow u \mid u \in S \}$ is a basis for the Scott topology.*
6. *The order $\sqsubseteq$ is the specialization order on points in the Scott topology.*
7. *A set C is closed in the Scott topology iff $\downarrow c \subseteq C$ and $\sup I \in C$ for every ideal $I \subseteq C$.*
8. *The closure of a point $s \in S$ is its down-closure $\downarrow s$.*

Proof Suppose $s_1 \ll s_2$. Then, one can check that $I := \bigcup_{s \ll s_2} \twoheaddownarrow s$ is directed, and by Remark 2.8 it is down-closed. For all $s \ll s_2$, we have $\twoheaddownarrow s \subseteq I$, which implies $s = \sup \twoheaddownarrow s \sqsubseteq \sup I$. It follows that $s_2 = \sup \twoheaddownarrow s_2 \sqsubseteq \sup I$, and this implies $s_1 \in I$.

[2] This nomenclature is the same as that in [Gie+03, Definition I-1.6]. Continuous posets, and their generalizations to continuous categories, are also discussed in [Joh02, C.4.2] and [JJ82].

This proves 1., and 2. and 3. follow directly. Using 3. the only non-trivial part of 4. is showing that the intersection of two open sets is open, but this follows from the fact that $\twoheaddownarrow u$ is directed. For 5., each $\twoheaduparrow u$ is open by the interpolative property, and 3. implies that these form a basis. For 6., we need to show $u \sqsubseteq u'$ iff $(u \in U \Rightarrow u' \in U)$ holds for all open U. It suffices to consider $U = \twoheaduparrow u_0$, in which case the forward direction is obvious and the backward direction is $u = \sup \twoheaddownarrow u \sqsubseteq \sup \twoheaddownarrow u' = u'$. For 7., use that C is closed iff its complement is up-closed and satisfies Proposition 2.12. For 8., $\downarrow s$ satisfies the properties from 7. and is the smallest such set containing s. □

Example 2.15 For any set X, the power-set $S := 2^X$, ordered by inclusion $u \sqsubseteq v$ iff $u \subseteq v$, is a domain. One can check that $u \ll v$ iff u corresponds to a finite subset of v.

We define a function between domains to be *continuous* if it satisfies any of the equivalent conditions in Proposition 2.16; see [Gie+03, Proposition II-2.1] for proof.

Proposition 2.16 *Let S, T be domains, and let $f\colon S \to T$ be a function between their underlying sets. The following are equivalent:*

- *f is continuous with respect to the Scott topology,*
- *f is order-preserving and $f(\sup(I)) = \sup(f(I))$ for all ideals I,*
- *$f(s) = \sup\{f(s') \mid s' \ll s\}$ for all $s \in S$.*

Remark 2.17 The inclusion functor from domains to topological spaces is fully faithful, and in fact, domains are particularly nice topological spaces. On the one hand, the frame of open sets in a domain can be recovered from—in fact is defined in terms of—its poset of points under specialization. On the other hand, any domain is a sober space (see [Gie+03, Proposition II-1.11(ii)]), meaning that its poset of points can be recovered from its frame of open sets.

2.2.2 Predomains and Their Corresponding Posites

We give a couple of technical results which involve material on predomains from Appendix A.1, and which will be used in the next section. One may choose to skip the proofs on a first reading.

A predomain is like a basis for a domain. Indeed, a predomain B determines a domain denoted $\mathrm{RId}(B)$, called the domain of *rounded ideals*, but B is typically more convenient to work with because it involves much less data. We now show that for every predomain B, its upper specialization order $(B, \preccurlyeq)$ has a natural posite structure, and $(0, 1)$-sheaves on this posite can be identified with Scott-opens in $\mathrm{RId}(B)$. Thus, the set B can directly encode the frame of opens of the space $\mathrm{RId}(B)$, and the topos of sheaves on $\mathrm{RId}(B)$, in addition to its points. Since every domain is naturally a predomain (see Proposition A.12), the following results also hold when B is a domain.

Proposition 2.18 *Let* $(B, \prec)$ *be a predomain and* $\lessdot$ *its upper specialization order. Then, there is a posite* S_B *whose underlying poset is* $(B, \lessdot)^{\mathrm{op}}$, *and whose coverage is defined by* $V \rhd b$ *iff* $V = \Uparrow b$.

Proof We only need to prove that $\rhd$ is a coverage, i.e., the two conditions of Definition 2.1. For the first condition, if $v \in \Uparrow b$, then $b \lessdot v$. For the second, suppose $b \lessdot b'$. The condition follows from the fact that $\Uparrow b' \subseteq \Uparrow b$. □

Proposition 2.19 *Let* $(B, \prec)$ *be a predomain,* $\lessdot$ *its upper specialization order, and consider the posite* S_B *as defined in Proposition 2.18. The maps* $I \mapsto \{b \in B \mid \exists b' \in I.\, b' \prec b\}$ *and* $U \mapsto \{b \in B \mid \Uparrow b \subseteq U\}$ *define an isomorphism of frames,* $\mathsf{Shv}_0(S_B) \cong \Omega(B)$.

Proof The well-definedness of the maps and the fact that they are mutually inverse can be checked directly from Definitions 2.3, A.1, and A.4. □

Proposition 2.20 *Suppose that for a predomain* $(B, \prec)$, *lower specialization implies upper specialization; that is, for all* $b, b' \in B$, *if* $b \leqslant b'$ *then* $b \lessdot b'$. *Then, there is a posite* S'_B *with underlying poset* $(B, \leqslant)^{\mathrm{op}}$, *and with coverage defined by* $V \rhd b$ *iff* $V = \Uparrow b$.

Proof The proof that $\rhd$ defines a coverage is essentially the same as in Proposition 2.18. □

Remark 2.21 We show in Corollary 2.53 that, in the setting of Proposition 2.20, there is an isomorphism:

$$\mathsf{Shv}_0(S_B) \cong \mathsf{Shv}_0(S'_B).$$

2.3 The Interval Domain and Its Associated Topos

In Sect. 2.3.1, we define one of the central objects of this book: the interval domain $\mathbb{IR}$. If this is the reader's first encounter with the interval domain, its order may appear opposite to what is expected. Namely, $\sqsubseteq$ refers to the specialization order on intervals rather than to the inclusion order.

Next in Sect. 2.3.2, there is a brief section on the relationship between $\mathbb{IR}$, its subspace $\mathbb{R}$, and the notion of Dedekind cuts, which will also be a central theme throughout the book. In Theorem 2.27, we state four equivalent definitions of the topos $\mathsf{Shv}(\mathbb{IR})$.

2.3.1 Definition of the Interval Domain

Definition 2.22 (The Interval Domain) The *interval domain* $\mathbb{IR}$ is the set $\{(d, u) \in \mathbb{R} \times \mathbb{R} \mid d \leq u\}$, an element of which we denote by $[d, u]$, with order given by $[d, u] \sqsubseteq [d', u']$ iff $d \leq d' \leq u' \leq u$.[3]

Conceptually, the interval domain is the set of bounded closed intervals in $\mathbb{R}$, ordered by reverse inclusion. Thus, the maximal elements of $\mathbb{IR}$ are the zero-length intervals, which we can identify with elements of $\mathbb{R}$. For intuition about the order, regard intervals as *approximations of real numbers*, so that an increase in domain order corresponds to an increase in precision.

We have the following characterization of the way-below relation (Definition 2.7) in $\mathbb{IR}$.

Proposition 2.23 *Every ideal $I \subseteq \mathbb{IR}$ has a supremum. The way-below relation in $\mathbb{IR}$ is given by:*

$$[d', u'] \ll [d, u] \quad \textit{iff} \quad d' < d \textit{ and } u < u',$$

and $[d, u] = \sup \downarrow\!\!\downarrow[d, u]$. Thus, $\mathbb{IR}$ is indeed a domain in the sense of Definition 2.10.

Proof Define $\bar{d} = \sup_{[d,u]\in I} d$ and $\bar{u} = \inf_{[d,u]\in I} u$. We will show that $[\bar{d}, \bar{u}]$ is the supremum of I in $\mathbb{IR}$. First, we must show that $\bar{d} \leq \bar{u}$, or equivalently, that for any $[d_1, u_1], [d_2, u_2] \in I$ we have $d_1 \leq u_2$. But I is directed, so there exists a $[d', u'] \in I$ such that $[d_1, u_1] \sqsubseteq [d', u']$ and $[d_2, u_2] \sqsubseteq [d', u']$, and it follows that $d_1 \leq d' \leq u' \leq u_2$. Finally, it is clear that $[\bar{d}, \bar{u}]$ is the least upper bound of I.

Now, suppose $d' < d$ and $u < u'$; we want to show $[d', u'] \ll [d, u]$. Suppose I is an ideal with $\sup I = [\bar{d}, \bar{u}]$, such that $[d, u] \sqsubseteq [\bar{d}, \bar{u}]$. Then, $d' < \bar{d}$ and $\bar{u} < u'$, which implies that there exist $[d_1, u_1], [d_2, u_2] \in I$ with $d' \leq d_1$ and $u_2 \leq u'$. Then by directedness, there exists a $[d'', u''] \in I$ with $[d_1, u_1] \sqsubseteq [d'', u'']$ and $[d_2, u_2] \sqsubseteq [d'', u'']$. It follows that $d' \leq d_1 \leq d''$ and $u'' \leq u_2 \leq u'$, so $[d', u'] \sqsubseteq [d'', u''] \in I$ shows $[d', u'] \in I$.

Conversely, suppose $[d', u'] \ll [d, u]$. Define $I := \{[d'', u''] \mid d'' < d \leq u < u''\}$. It is easy to see that I is an ideal, and that $\sup I = [d, u]$. Hence, $[d', u'] \in I$, showing that $d' < d$ and $u < u'$. □

We often make use of the following notation:

$$\downarrow\!\!\downarrow[d, u] = \{\, [d', u'] \mid d' \leq d \leq u \leq u' \,\} \quad \text{and} \quad \uparrow\!\!\uparrow[d, u] = \{\, [d', u'] \mid d < d' \leq u' < u \,\}.$$

[3] Throughout this book, we often use d or δ to stand for "down" and u or υ (upsilon) to stand for "up."

A set U is called *filtered* if it is nonempty and down-directed; it is an open filter if it is additionally open. Here is a proposition that gives a few facts about $\mathbb{IR}$ that will be useful later.

Proposition 2.24

1. *Sets of the form $\Uparrow[d, u]$, for $d \leq u$ form a basis for the topology on $\mathbb{IR}$.*
2. *A subset U is open iff it is up-closed and $\Downarrow[d, u] \cap U \neq \varnothing$ for every $[d, u] \in U$.*
3. *$\mathbb{IR}$ has binary meets.*
4. *$\mathbb{IR}$ has conditional joins: if $[d_1, u_1]$ and $[d_2, u_2]$ have an upper bound, then they have a join.*
5. *For any extended real numbers $r_1, r_2 \in \mathbb{R} \cup \{-\infty, \infty\}$ with $r_1 < r_2$, the open set $U_{r_1,r_2} := \{[d, u] \in \mathbb{IR} \mid r_1 < d \leq u < r_2\}$ is filtered.*
6. *Every filtered open set U is of the form U_{r_1,r_2}, for some $r_1 < r_2$ in $\mathbb{R} \cup \{-\infty, \infty\}$.*

Proof 1 and 2 are shown in Proposition 2.14. Let $[d_1, u_1]$ and $[d_2, u_2]$ be arbitrary points. It is easy to check that their meet is $[\min(d_1, d_2), \max(u_1, u_2)$, showing 3. For 4, if there exists $[d, u]$ above each, then $d_1 \leq d$ and $d_2 \leq d$, so $\max(d_1, d_2) \leq d$, and similarly $\min(u_1, u_2) \geq u$, so their join is $[\max(d_1, d_2), \min(u_1, u_2)$.

For 5, clearly $U = U_{r_1,r_2}$ is nonempty and it is open by condition 2. If $[d_1, u_1] \in U$ and $[d_2, u_2] \in U$, then $[\min(d_1, d_2), \max(u_1, u_2)] \in U$, so U is down-directed.

For 6, let U be an open filter, and let $r_1 := \inf_{[d,u]\in U} d$ and $r_2 := \sup_{[d,u]\in U} u$. Clearly, $U \subseteq U_{r_1,r_2}$, so take any $[d, u]$ such that $r_1 \leq d \leq u < r_2$; we must show $[d, u] \in U$. We know that there exist $[d_1, u_1]$ and $[d_2, u_2]$ in U such that $d_1 \leq d$ and $u \leq u_2$. But then, their meet is in U, and it contains $[d_1, u_2]$ and hence $[d, u]$. □

2.3.2 Discussion of $\mathbb{IR}$, $\mathbb{R}$, and Dedekind Cuts

The logic and type theory we present in this book is constructive. As the interval domain plays an important role within the type theory, not just in defining its semantics, we make a brief digression to discuss how to define $\mathbb{IR}$ constructively, and its relation to standard constructive definitions of the real numbers.

Each element of $\mathbb{IR}$ can be identified with a pair of subsets $\delta, \upsilon \subseteq \mathbb{Q}$ satisfying the following conditions:

1. δ and υ are inhabited,
2. $d \in \delta \Leftrightarrow \exists(d' \in \delta).\, d < d'$,
3. $u \in \upsilon \Leftrightarrow \exists(u' \in \upsilon).\, u' < u$, and
4. $\delta \cap \upsilon = \varnothing$.

The element $[d, u] \in \mathbb{IR}$ corresponds to (δ, υ) where $q \in \delta$ iff $q < d$, and $q \in \upsilon$ iff $u < q$.

When the pair of subsets (δ, υ) satisfies the above four conditions, we call it a *disjoint pair of cuts*, in the sense of Dedekind cuts. This is part of a larger story, including the domain structure for this formulation of $\mathbb{IR}$, which is fleshed out in Proposition 4.30 and Appendix A.1; see in particular Example A.9.

Classically, it is easy to show that maximal elements of $\mathbb{IR}$ correspond precisely to real numbers. However, constructively, maximality is slightly too weak to capture the Dedekind reals. Indeed, seeing each element of $\mathbb{IR}$ as a disjoint pair of cuts, it is a simple exercise to show that $(\delta, \upsilon) \in \mathbb{IR}$ is maximal iff it is a *MacNeille real*, i.e., if it satisfies the following axioms for any $d, u : \mathbb{Q}$

$$[(d < u) \wedge \neg(u \in \upsilon)] \Rightarrow d \in \delta$$

$$[(d < u) \wedge \neg(d \in \delta)] \Rightarrow u \in \upsilon.$$

For most purposes, the *Dedekind reals* $\mathbb{R}$ are constructively better behaved. Rather than maximal, a Dedekind real is a disjoint pair of cuts (δ, υ) that is also *located*, i.e., such that:

$$d < u \Rightarrow (d \in \delta \vee u \in \upsilon).$$

Every Dedekind real is a MacNeille real; the converse holds classically but not constructively. Throughout this book, we will be focused on Dedekind reals.

2.3.3 Five Equivalent Definitions of the Topos $\mathbf{Shv}(\mathbb{IR})$

We first define two different posites, though we will see in Theorem 2.27 that they define the same topos.

Definition 2.25 (The Interval Posite $S_{\mathbb{IR}}$) The *interval posite*, which we denote $S_{\mathbb{IR}}$, is what we called $S'_{\mathbb{IR}}$ in Proposition 2.20. Explicitly:

- The underlying set is $\mathbb{IR} = \{\, [d, u] \in \mathbb{R} \times \mathbb{R} \mid d \leq u \,\}$ as in Definition 2.22,
- The order is given by inclusion, $[d, u] \leq [d', u']$ iff $d' \leq d \leq u \leq u'$, and
- The coverage is given by $V \rhd [d, u]$ iff $V = \{\, [d', u'] \in \mathbb{IR} \mid d < d' \leq u' < u \,\}$.

Note that if $d = u$, then $[d, u]$ has an empty covering family.

The *rational interval posite*, denoted $S^{\mathbb{Q}}_{\mathbb{IR}}$, is the posite $S_{\mathbb{IR}_{\text{pre}}}$ as defined in Proposition 2.18. Explicitly:

- The underlying set is $\mathbb{IR}_{\text{pre}} = \{\, (d, u) \in \mathbb{Q} \times \mathbb{Q} \mid d < u \,\}$,
- The order is given by inclusion, $(d, u) \leq (d', u')$ iff $d' \leq d < u \leq u'$, and
- The coverage is given by $V \rhd (d, u)$ iff $V = \{\, (d', u') \in \mathbb{IR}_{\text{pre}} \mid d < d' < u' < u \,\}$.

Remark 2.26 Because we have $\mathbb{R} \subseteq \mathbb{IR}$, the set $\mathbb{R}$ of Dedekind reals inherits a subspace topology. There are posites $S_{\mathbb{R}}$ and $S^{\mathbb{Q}}_{\mathbb{R}}$ whose frame of $(0, 1)$-sheaves is the frame of opens in $\mathbb{R}$. These posites have the same underlying posets as $S_{\mathbb{IR}}$ and $S^{\mathbb{Q}}_{\mathbb{IR}}$ (respectively), but more covering families. We explain this connection for the reader's intuition; we will not need it for our work, though one will see reincarnations of it, e.g., in Corollary 5.69.

Recall from Definition 2.25 that the posite $S_{\mathbb{IR}}$ for $\mathbb{IR}$ has the coverage $V \rhd [d, u]$ iff $V = \{ [d', u'] \mid d < d' \leq u' < u \}$. We informally called these "continuity" coverings in Eq. (1.4). The posite $S_{\mathbb{R}}$ for $\mathbb{R}$ has an expanded coverage: $V \rhd [d, u]$ iff either:

- $V = \{ [d', u'] \mid d < d' \leq u' < u \}$ or
- $V = \{ [d, u'], [d', u] \}$ where $d \leq d' \leq u' \leq u$.

In other words, it includes the "composition" coverings of Eq. (1.3). The coverage for the posite $S^{\mathbb{Q}}_{\mathbb{R}}$ is similarly analogous to that of $S^{\mathbb{Q}}_{\mathbb{IR}}$.

We can now give several equivalent constructions for the same topos. Theorem 2.27 will be proven in Sect. 2.5

Theorem 2.27 *The following categories are equivalent:*

1. *The category of sheaves* $\mathsf{Shv}(\Omega(\mathbb{IR}))$ *on the frame of opens of* $\mathbb{IR}$*, with the canonical coverage* $\{V_i\}_{i\in I} \rhd U$ *iff* $\bigcup_{i\in I} V_i = U$.
2. *The category of sheaves* $\mathsf{Shv}(S_{\mathbb{IR}})$ *on the interval posite* $S_{\mathbb{IR}}$.
3. *The category of sheaves* $\mathsf{Shv}(S^{\mathbb{Q}}_{\mathbb{IR}})$ *on the rational interval posite* $S^{\mathbb{Q}}_{\mathbb{IR}}$.
4. *The category of* $\mathsf{Cont}(\mathbb{IR})$ *of* continuous functors $\mathbb{IR} \to \mathbf{Set}$, *i.e., functors taking directed suprema in* $\mathbb{IR}$ *to colimits in* **Set**.

2.4 $\mathbb{IR}$ and the Upper Half-Plane

Let $H := \{(x, y) \in \mathbb{R}^2 \mid y \geq 0\}$ denote the upper half-plane. It is common to consider the bijection $h\colon |\mathbb{IR}| \xrightarrow{\cong} H$ from the set of points in $\mathbb{IR}$ to H, sending each interval to the pair consisting of its midpoint and radius. For the sake of mnemonic, say $(m, r) := h([d, u])$ and call (m, r) the *half-plane representation* of $[d, u]$. Then, the bijection is given by:

$$m = \frac{u+d}{2}, \qquad r = \frac{u-d}{2} \qquad \text{and} \qquad d = m - r, \quad u = m + r$$

Clearly, the subset $\mathbb{R} \subseteq \mathbb{IR}$ corresponds to the horizontal axis of H. Given a subset $S \subseteq |\mathbb{IR}|$, we also write $h(S)$ to mean the image of S under h.

We can transport the specialization order and the way-below relation on $\mathbb{IR}$ across this bijection:

$$\begin{aligned} [d_1, u_1] \sqsubseteq [d_2, u_2] \quad &\text{iff} \quad d_1 \leq d_2 \leq u_2 \leq u_1 \quad \text{iff} \quad |m_1 - m_2| \leq r_1 - r_2 \\ [d_1, u_1] \ll [d_2, u_2] \quad &\text{iff} \quad d_1 < d_2 \leq u_2 < u_1 \quad \text{iff} \quad |m_1 - m_2| < r_1 - r_2 \end{aligned} \tag{2.1}$$

We denote the order on the half-plane just as we do for points in the interval domain, writing $(m_1, r_1) \sqsubseteq (m_2, r_2)$ and $(m_1, r_1) \ll (m_2, r_2)$ in the above cases. The *upper cone* of any $(m, r) \in H$ is denoted $\uparrow(m, r) := \{(m', r') \mid (m, r) \sqsubseteq (m', r')\}$. Similarly, the *way-up closure* of (m, r) is denoted $\twoheaduparrow(m, r) := \{(m', r') \mid (m, r) \ll (m', r')\}$. Both consist of points in the cone visually below (m, r), as shown here:

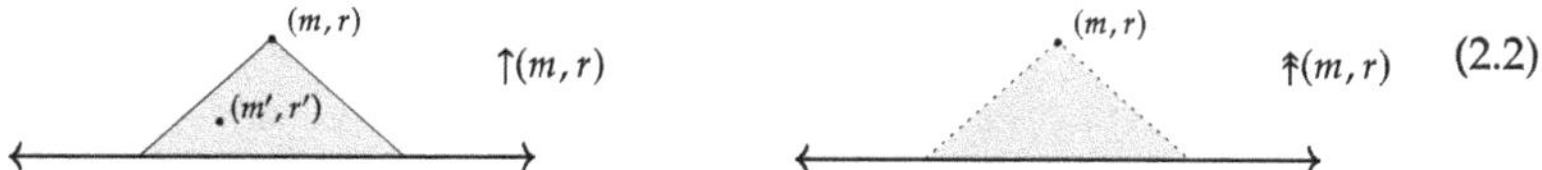

(2.2)

Warning 2.28 Despite their underlying sets being isomorphic, we consider the interval domain $\mathbb{IR}$ and the half-plane H to have distinct topologies: $\mathbb{IR}$ has the Scott topology and H has the Euclidean topology. For every Scott-open set $U \subseteq \mathbb{IR}$, its image $h(U)$ is open in the Euclidean topology—giving us Proposition 2.29—but the converse is not true. Thus, if $h([d, u]) = (m, r)$, then the sets $\twoheaduparrow[d, u]$ and $\twoheaduparrow(m, r)$ are isomorphic as sets, but they are not isomorphic as topological spaces.

Proposition 2.29 *Both h and h^{-1} are poset isomorphisms, and $h^{-1}\colon H \to \mathbb{IR}$ is continuous.*

Proof Since we defined the order on H in terms of that on $\mathbb{IR}$ and the bijection $|\mathbb{IR}| \cong H$, the first claim is true by definition. For the second, note that the Scott topology has a basis $\{\twoheaduparrow x\}_{x \in \mathbb{IR}}$ by Proposition 2.24. Clearly, the cone $\twoheaduparrow(m, r) = \{\,(m', r') \mid |m - m'| < r - r'\,\}$ is open in the upper half-plane. □

2.4.1 Open Sets in $\mathbb{IR}$ as Lipschitz Functions

Recall that for any two metric spaces (A, d_A) and (B, d_B) and number $M \geq 0$, an *M-Lipschitz* function is a function $f\colon A \to B$ such that for all $a_1, a_2 \in A$, the relation:

$$d_B(f(a_1), f(a_2)) \leq M * d_A(a_1, a_2)$$

holds. The same definition makes sense for extended metrics (where we allow $d(x_1, x_2) = \infty$). We will be interested in 1-Lipschitz functions, which we just call *Lipschitz*.

It is a standard fact that M-Lipschitz functions are absolutely continuous and hence differentiable almost everywhere (non-differentiable on a set with Lebesgue measure 0), and that their derivative is bounded between $-M$ and M.

Consider $\mathbb{R}$ as a metric space in the usual way, i.e., $d(x_1, x_2) := |x_1 - x_2|$, and consider the space $\mathbb{R}^{\infty}_{\geq 0} := \mathbb{R}_{\geq 0} \cup \{\infty\}$ with the metric:

$$d(y_1, y_2) := \begin{cases} |y_1 - y_2| & \text{if } y_1 \in \mathbb{R}_{\geq 0} \text{ and } y_2 \in \mathbb{R}_{\geq 0} \\ \infty & \text{if } y_1 \in \mathbb{R}_{\geq 0} \text{ and } y_2 = \infty \\ \infty & \text{if } y_1 = \infty \text{ and } y_2 \in \mathbb{R}_{\geq 0} \\ 0 & \text{if } y_1 = \infty \text{ and } y_2 = \infty \end{cases}$$

For any space X, the set of 1-Lipschitz functions $X \to \mathbb{R}^{\infty}_{\geq 0}$ has the structure of a poset: say that $f \leq g$ if $f(x) \leq g(x)$ for all x.

Our next goal is to show that the Scott-open sets of $\mathbb{IR}$ are in one-to-one correspondence with the Lipschitz functions $\mathbb{R} \to \mathbb{R}^{\infty}_{\geq 0}$; see Theorem 2.32.

Lemma 2.30 *For any Scott-open set $U \in \Omega_{\mathbb{IR}}$, the function $f_U : \mathbb{R} \to \mathbb{R}^{\infty}_{\geq 0}$, defined by:*

$$f_U(m) := \sup\{\, r \in \mathbb{R}^{\infty}_{\geq 0} \mid (m, r) \in U \,\}$$

is a 1-Lipschitz function and preserves order. Moreover, the following are equivalent:

- $U = \mathbb{IR}$,
- $f_U(m) = \infty$ *for all* $m \in \mathbb{R}$,
- $f_U(m_0) = \infty$ *for some* $m_0 \in \mathbb{R}$.

And also the following are equivalent for all $m \in \mathbb{R}$ and $r \in \mathbb{R}_{\geq 0}$:

- $f_U(m) = r$,
- $\Uparrow(m, r) \subseteq U$ *and* $(m, r) \notin U$.

Proof Clearly, $U \subseteq V$ implies $f_U(m) \leq f_V(m)$ for all m, so the assignment preserves order. We first prove the equivalences and then show that $f = f_U$ is a 1-Lipschitz function at the end.

For the first three, clearly the first implies the second, and the second implies the third. Suppose $f(m_0) = \infty$ for some $m_0 \in \mathbb{R}$. Then, $U = \mathbb{IR}$ because for any (m, r) we must have $(m_0, r + |m - m_0|) \in U$, which implies $(m, r) \in U$ since U is up-closed.

For the second pair, note that $\Uparrow(m, r) = \{(m', r') \mid |m - m'| < r - r'\}$. It is easy to see that $\Uparrow(m, r) \subseteq U$ iff $r \leq f(m)$. By definition, Scott-open sets are rounded, which implies $(m, r) \in U$ iff there is some r' with $r < r'$ and $(m, r') \in U$. From this, it is easy to see that $(m, r) \in U$ iff $r < f(m)$.

It remains to check that f is a 1-Lipschitz function. If $f(m) = \infty$ for any m, then f is constant, and hence a 1-Lipschitz function. So, suppose $f(m) \in \mathbb{R}_{\geq 0}$ for all m; we want to show that $|f(m') - f(m)| \leq |m' - m|$ for all $m, m' \in \mathbb{R}$. We may assume $m < m'$; we will show $f(m') - f(m) \leq m' - m$ and $f(m) - f(m') \leq m' - m$. For contradiction, suppose $m' - m + f(m) < f(m')$. Then, $(m', m' - m + f(m)) \in U$, which implies $(m, f(m)) \in U$, which is false as shown above. The other case is similar. □

Lemma 2.31 *For any 1-Lipschitz function $f\colon \mathbb{R} \to \mathbb{R}^{\infty}_{\geq 0}$, the set $U_f \subseteq \mathbb{IR}$ defined by:*

$$U_f := \{(m, r) \in \mathbb{IR} \mid r < f(m)\}$$

is Scott-open.

Proof To show that $U = U_f$ is Scott-open, we must show that $(m, r) \in U$ iff there exists $(m', r') \in U$ such that $|m' - m| < r' - r$. The forward direction is easy: if $r < f(m)$, then there exists r' such that $r < r' < f(m)$, and the conclusion follows by letting $m' := m$. For the backward direction, if (m', r') satisfies $r' < f(m')$ and $|m' - m| \leq r' - r$, then $f(m') - f(m) \leq |f(m') - f(m)| \leq |m' - m| \leq r' - r$, so $r < f(m)$ and hence $(m, r) \in U$. □

Theorem 2.32 *The assignments $U \mapsto f_U$ from Lemma 2.30 and $f \mapsto U_f$ from Lemma 2.31 are mutually inverse, giving an isomorphism between the poset $\Omega_{\mathbb{IR}}$ of open sets in the interval domain and the poset of 1-Lipschitz functions $\mathbb{R} \to \mathbb{R}^{\infty}_{\geq 0}$.*

Proof It is straightforward to check that these two constructions are mutually inverse. □

Subsets of the form $\Uparrow[d, u]$ form a basis of $\mathbb{IR}$, by Proposition 2.14, so it is useful to consider its poset $\Omega(\Uparrow[d, u])$ of open subsets.

Definition 2.33 (Dyck Path) For $d \leq u$, let $[d, u] \subseteq \mathbb{R}$ denote the closed interval in the usual topology. A *(continuous) Dyck path on* $[d, u]$ is a function $D\colon [d, u] \to \mathbb{R}_{\geq 0}$ such that:

- $D(d) = 0 = D(u)$, and
- $|D(x') - D(x)| \leq |x' - x|$ for all $x, x' \in [d, u]$.

The second condition is equivalent to saying that D is a 1-Lipschitz function. The order is given by $D \leq D'$ iff $D(x) \leq D'(x)$ for all $x \in [d, u]$.

Corollary 2.34 *For any $d \leq u$, there is an isomorphism of posets between $\Omega(\Uparrow[d, u])$ and the poset of Dyck paths on $[d, u]$.*

Proof By Theorem 2.32, $\{U \in \Omega_{\mathbb{IR}} \mid U \subseteq \Uparrow[d, u]\}$ is isomorphic to the set:

$$\{f\colon \mathbb{R} \to \mathbb{R}^{\infty}_{\geq 0} \mid f \leq f_{\Uparrow[d,u]} \text{ and } f \text{ is a 1-Lipschitz function}\}$$

where $f_{\Uparrow[d,u]}$ is the Lipschitz function associated to the open set $\Uparrow[d, u]$ as in Lemma 2.30. It is easy to see that $f_{\Uparrow[d,u]}$ is supported on the closed interval $[d, u]$, and in fact that it is the largest 1-Lipschitz function supported on this interval. Thus, the above set is indeed isomorphic to the set of length-ℓ Dyck paths. □

Here are the pictures of two Dyck paths on the interval $[d, u]$:

Remark 2.35 A *discrete Dyck path* is a path in the quadrant $\mathbb{N} \times \mathbb{N}$ from $(0, 0)$ to (n, n), for some $n \in \mathbb{N}$, such that every edge has length 1 and is oriented either upward or rightward. Discrete Dyck paths are discussed in the combinatorics literature because the number of discrete Dyck paths for varying n is the nth Catalan number. Just as continuous Dyck paths represent the subobject classifier for sheaves on $\mathbb{IR}$, it is similarly the case that discrete Dyck paths represent the subobject classifier for presheaves on the category $\mathsf{Tw}(\mathbb{Z}, \leq)$ of intervals in $\mathbb{Z}$. Discrete Dyck paths have also been discussed in the context of temporal logic; see [Fer16].

2.4.2 Real-Valued Functions on $\mathbb{IR}$

Recall that the underlying sets of the upper half-plane H and the interval domain $\mathbb{IR}$ are in bijection, but we endow H with the Euclidean topology and $\mathbb{IR}$ with the Scott topology; see Warning 2.28. In this section, we use the function $h^{-1} \colon H \to \mathbb{IR}$ from Proposition 2.29, which sends $(m, r) \mapsto [m - r, m + r]$, to better understand continuous functions from $\mathbb{IR}$ to various sorts of real number objects.

Let $\underline{\mathbb{R}}$ denote the space of *lower real numbers*. Its underlying set is $\mathbb{R} \cup \{\infty\}$ and a subset $U \subseteq \underline{\mathbb{R}}$ is open if either $U = \underline{\mathbb{R}}$ or there exists some $r \in \underline{\mathbb{R}}$ such that $U = \{r' \in \underline{\mathbb{R}} \mid r < r'\}$; in particular, $\{\infty\}$ is not open. In fact, $\underline{\mathbb{R}}$ forms a domain (see Example A.9); the above topology coincides with the Scott topology, the usual $\leq$ coincides with the domain order $\sqsubseteq$, and the usual $<$ coincides with the way-below relation $\ll$.

For topological spaces X, Y, let $\mathbf{Top}(X, Y)$ denote the set of continuous maps $X \to Y$.

Proposition 2.36 *Let H denote the upper half-plane, endowed with the $\sqsubseteq$ partial order from Eq.* (2.1). *The function $H \to \mathbb{IR}$ induces an injection $\mathbf{Top}(\mathbb{IR}, \underline{\mathbb{R}}) \to \mathbf{Top}(H, \underline{\mathbb{R}})$, whose image is precisely $\{ f \in \mathbf{Top}(H, \underline{\mathbb{R}}) \mid f \text{ is monotonic} \}$.*

Proof Since $H \to \mathbb{IR}$ is continuous and bijective on points, composing with it induces the injection above. Certainly anything in the image is monotonic, because Scott-continuous functions are. So, we need to show that if $f \colon H \to \underline{\mathbb{R}}$ is continuous and monotonic, then it is Scott-continuous. By Proposition 2.16, a function $f \colon \mathbb{IR} \to \underline{\mathbb{R}}$ is Scott-continuous iff $f(x) = \sup\{f(y) \mid y \ll x\}$ for all $x \in \mathbb{IR}$.

Suppose f is continuous and monotonic and take $x \in \mathbb{IR}$. If $f(x) < \infty$, then we want to show that for any $\epsilon > 0$ there exists $y \ll x$ with $f(x) - \epsilon < y$. By monotonicity, this is equivalent to showing that there exists $y \ll x$ with $|y - f(x)| < \epsilon$. But, the ϵ-ball around $f(x)$ is open, so its preimage U is an open neighborhood of x in H, and any open neighborhood of x in H intersects the open cone $\uparrow x$ (see Eq. (2.4)), and we are done. If instead $f(x) = \infty$, we want to show that for any $N > 0$ there exists $y \ll x$ with $N < y$; the proof is similar. □

Let $\bar{\mathbb{R}}$ denote the space of *upper real numbers*. Its underlying set is $\mathbb{R} \cup \{-\infty\}$ and a set $U \subseteq \bar{\mathbb{R}}$ is open if either $U = \bar{\mathbb{R}}$ or there exists some $r \in \underline{\mathbb{R}}$ such that $U = \{r' \in \underline{\mathbb{R}} \mid r' < r\}$; in particular, $\{-\infty\}$ is not open. In fact, $\bar{\mathbb{R}}$ forms a domain; (see Example A.9) the above topology coincides with the Scott topology, the usual $\geq$ coincides with the domain order $\sqsubseteq$, and the usual $>$ coincides with the way-below relation $\ll$.

The proof of the following proposition is analogous to that of Proposition 2.36. Note that $f \colon H \to \bar{\mathbb{R}}$ is monotonic iff $x \sqsubseteq x'$ implies $f(x) > f(x')$, in keeping with the order on $\bar{\mathbb{R}}$.

Proposition 2.37 *Let H denote the upper half-plane, endowed with the above $\sqsubseteq$ partial order. The function $H \to \mathbb{IR}$ induces an injection* $\mathbf{Top}(\mathbb{IR}, \bar{\mathbb{R}}) \to \mathbf{Top}(H, \bar{\mathbb{R}})$, *whose image is precisely* $\{ f \in \mathbf{Top}(H, \bar{\mathbb{R}}) \mid \forall x, x' \in H. x \leq x' \Rightarrow f(x) > f(x') \}$.

For any space X and function $f \colon X \to \mathbb{R} \cup \{\infty\}$, one says that f is *lower semi-continuous* (respectively, $f \colon X \to \mathbb{R} \cup \{-\infty\}$ is *upper semi-continuous*) if it is continuous as a map to the lower reals $\underline{\mathbb{R}} \cong \mathbb{R} \cup \{\infty\}$ (respectively, to the upper reals $\bar{\mathbb{R}} \cong \mathbb{R} \cup \{-\infty\}$). A more intuitive way to think about a lower semi-continuous function f is that for every point $x_0 \in X$ and number $\epsilon > 0$, there is a basic open neighborhood U of x_0 such that $-\epsilon < f(x) - f(x_0) < \infty$ for all $x \in U$, i.e., f may increase discontinuously, but it decreases continuously.

Let $\bar{\underline{\mathbb{R}}}$ denote the space of *extended intervals*, given by the product $\bar{\underline{\mathbb{R}}} = \underline{\mathbb{R}} \times \bar{\mathbb{R}}$. Then, $\mathbb{IR}$ and $\mathbb{R}$ can be identified with the following subsets of $\bar{\underline{\mathbb{R}}}$:

$$\mathbb{IR} \cong \{(x, y) \in \bar{\underline{\mathbb{R}}} \mid x \leq y\} \qquad \text{and} \qquad \mathbb{R} \cong \{(x, y) \in \bar{\underline{\mathbb{R}}} \mid x = y\}$$

We have proved the following.

Corollary 2.38 *Choose $[d, u] \in \mathbb{IR}$ and let $(m, r) \in H$ be the corresponding point in H. The Scott-open $\uparrow[d, u] \subseteq \mathbb{IR}$ corresponds to the subspace $\uparrow(m, r) \subseteq H$ (which we consider with the Euclidean topology). Then, we have isomorphisms:*

$$\mathbf{Top}(\uparrow[d, u], \underline{\mathbb{R}}) \cong \{ f \colon \uparrow(m, r) \to \mathbb{R} \cup \{\infty\} \mid$$
$$f \textit{ is lower semi-continuous and order-preserving} \}$$
$$\mathbf{Top}(\uparrow[d, u], \bar{\mathbb{R}}) \cong \{ f \colon \uparrow(m, r) \to \mathbb{R} \cup \{-\infty\} \mid$$
$$f \textit{ is upper semi-continuous and order-reversing} \}$$
$$\mathbf{Top}(\uparrow[d, u], \bar{\underline{\mathbb{R}}}) \cong \mathbf{Top}(\uparrow[d, u], \underline{\mathbb{R}}) \times \mathbf{Top}(\uparrow[d, u], \bar{\mathbb{R}})$$
$$\mathbf{Top}(\uparrow[d, u], \mathbb{IR}) \cong \{ (\underline{f}, \bar{f}) \in \mathbf{Top}(\uparrow[d, u], \bar{\underline{\mathbb{R}}}) \mid \forall x \in \uparrow(m, r).\, \underline{f}(x) \leq \bar{f}(x) \}$$
$$\mathbf{Top}(\uparrow[d, u], \mathbb{R}) \cong \{ (\underline{f}, \bar{f}) \in \mathbf{Top}(\uparrow[d, u], \bar{\underline{\mathbb{R}}}) \mid \forall x \in \uparrow(m, r).\, \underline{f}(x) = \bar{f}(x) \} \cong \mathbb{R}.$$

In the last case, we have $\mathbf{Top}(\uparrow[d,u],\mathbb{R}) \cong \mathbb{R}$ because $\underline{f} = \bar{f}$ is both order-preserving and order-reversing, and therefore constant since for any $x, y \in \uparrow(m,r)$, $\bar{f}(x) = \bar{f}(x \wedge y) = \bar{f}(y)$.

2.5 Grothendieck Posites

In this section, specifically in Theorem 2.54, we give the proof of Theorem 2.27. This material is not necessary for the rest of the book, so readers can feel free to skip to Chap. 3.

2.5.1 *Basic Theory of Grothendieck Posites and Dense Morphisms*

Definition 2.39 (Grothendieck Posite) Let $(S, \leq)$ be a poset. A *Grothendieck coverage* on S is a relation $\rhd \subseteq P(S) \times S$ such that:

- If $V \rhd u$, then $V \subseteq \downarrow u$.
- If $\downarrow V \rhd u$, then $V \rhd u$.
- If $V \rhd u$ and $u' \leq u$, then $(\downarrow V \cap \downarrow u') \rhd u'$.
- $\{u\} \rhd u$ for all $u \in S$.
- If $W \rhd u$ and $V \subseteq \downarrow u$ such that for all $w \in W$, $(V \cap \downarrow w) \rhd w$, then $V \rhd u$.

A *Grothendieck posite* is a poset equipped with a Grothendieck coverage. A Grothendieck coverage is in particular a coverage—in the sense of Definition 2.1—so a Grothendieck posite has an underlying posite. A (0, 1)-sheaf on a Grothendieck posite is simply a (0, 1)-sheaf on the underlying posite; see Definition 2.3.

Remark 2.40 It is easily shown that a Grothendieck posite additionally satisfies:

- If $V \rhd u$ and $V \subseteq V' \subseteq \downarrow u$, then $V' \rhd u$.
- If $V \rhd u$ and $V' \rhd u$, then $(\downarrow V \cap \downarrow V') \rhd u$.

Furthermore, if S is a poset, and we have a family of Grothendieck coverages $\{\rhd_i\}_{i\in I}$ indexed by an arbitrary set I, then one checks directly from Definition 2.39 that the intersection $\rhd$, defined by $V \rhd u$ iff $V \rhd_i u$ for all $i \in I$, is also a Grothendieck coverage.

Lemma 2.41 *Let S and T be posets, and let $D(S)$ denote the set of down-closed subsets of S, and similarly for $D(T)$. Any order-preserving map $F: S \to T$ induces an adjunction $F_* \dashv F^!$, where $F_*: D(T) \to D(S)$ and $F^!: D(S) \to D(T)$ are defined by:*

$$F_*(J) = F^{-1}(J) := \{u \in S \mid Fu \in J\} \qquad \text{and} \qquad F^!(I) := \{v \in T \mid F^{-1}(\downarrow v) \subseteq I\}. \tag{2.2}$$

Proof This is easily checked directly. Alternatively, note that if we consider S and T as categories, then $D(S)$ and $D(T)$ are the functor categories to the free-arrow category 2, F_* is precomposition by F, and $F^!$ is right Kan extension along F. □

Recall from Definition 2.3 the sheaf condition for down-closed subsets I of a posite: If $V \rhd u$ and $V \subseteq I$, then $u \in I$.

Lemma 2.42 *For any Grothendieck posite S, the inclusion $\mathsf{Shv}_0(S) \hookrightarrow D(S)$ has a left adjoint $a\colon D(S) \to \mathsf{Shv}_0(S)$, defined $a(I) := \{u \in S \mid (I \cap {\downarrow}u) \rhd u\}$.*

Proof First, we check that $a(I)$ is a (0, 1)-sheaf. For down-closure, suppose $u' \leq u \in a(I)$. Then, $(I \cap {\downarrow}u) \rhd u$, so $(I \cap {\downarrow}u') = (I \cap {\downarrow}u \cap {\downarrow}u') \rhd u'$ by Definition 2.39, and hence $u' \in a(I)$. For the sheaf condition, if $W \rhd u$ and $W \subseteq a(I)$, then for all $w \in W$ we have ${\downarrow}w \subseteq {\downarrow}u$ and $(I \cap {\downarrow}w) \rhd w$, so $(I \cap {\downarrow}u \cap {\downarrow}w) = (I \cap {\downarrow}w) \rhd w$. Hence, $(I \cap {\downarrow}u) \rhd u$ and $u \in a(I)$.

Finally, it is simple to check that $I \subseteq a(I)$ for any $I \in D(S)$, and that $a(J) = J$ for any $J \in \mathsf{Shv}_0(S)$, showing that a is left adjoint to the inclusion. □

Remark 2.43 It follows from Lemma 2.42 that the inclusion $\mathsf{Shv}_0(S) \hookrightarrow D(S)$ preserves all existing meets, and indeed it is simple to check that $\mathsf{Shv}_0(S)$ is closed under arbitrary intersections, so that $\mathsf{Shv}_0(S)$ is a complete lattice. The join of a family $X \subseteq \mathsf{Shv}_0(S)$ can be constructed $\sup X = \bigcap\{I \in \mathsf{Shv}_0(S) \mid \forall x \in X. x \subseteq I\}$. It also follows that this join can be constructed $\sup X = a(\bigcup X)$. Recall from Example 2.4 that the canonical coverage on $\mathsf{Shv}_0(S)$ is given by $X \rhd J \Leftrightarrow \sup X = J$.

Definition 2.44 Let $(S, \rhd)$ be a posite. We define a Grothendieck coverage $\widehat{\rhd}$ on S, called the *Grothendieck completion* of $\rhd$, by taking $\widehat{\rhd}$ to be the intersection of the set of Grothendieck coverages containing $\rhd$.

The next proposition shows that taking the Grothendieck completion of a posite doesn't change the (0, 1)-sheaves.

Proposition 2.45 *Let $(S, \rhd)$ be a posite. Let $\overline{\rhd}$ be any coverage on S containing $\rhd$ and contained in the Grothendieck completion $\widehat{\rhd}$, i.e., satisfying:*

- *$V \rhd u$ implies $V \mathrel{\overline{\rhd}} u$, and*
- *For any Grothendieck coverage $\widetilde{\rhd}$ containing $\rhd$, if $V \mathrel{\overline{\rhd}} u$ then $V \mathrel{\widetilde{\rhd}} u$.*

Let $I \subseteq S$ be any down-closed subset. Then, I satisfies the sheaf condition for $\rhd$ iff it satisfies the sheaf condition for $\overline{\rhd}$.

Hence, if $\hat{S}$ is the posite with the same underlying poset as S, but with coverage $\widehat{\rhd}$, then $\mathsf{Shv}_0(S) \cong \mathsf{Shv}_0(\hat{S})$. Similarly, $\mathsf{Shv}(S) \cong \mathsf{Shv}(\hat{S})$.

Proof Fix a down-closed $I \subseteq S$. Because $\overline{\rhd}$ contains $\rhd$, it is clear that if I satisfies the sheaf condition for $\overline{\rhd}$, then it does for $\rhd$ as well. For the converse, suppose I satisfies the sheaf condition for $\rhd$. We define a coverage $\widetilde{\rhd}$ by:

$$V \mathrel{\widetilde{\rhd}} u \Leftrightarrow (V \subseteq {\downarrow}u) \wedge \forall(v \leq u).\, ({\downarrow}V \cap {\downarrow}v) \subseteq I \Rightarrow v \in I.^4$$

Then, $\widetilde{\rhd}$ contains $\rhd$; indeed, suppose $V \rhd u$. Then, $V \subseteq {\downarrow}u$ and if $v \leq u$, then there exists a $V' \subseteq ({\downarrow}V \cap {\downarrow}v)$ such that $V' \rhd v$ by Definition 2.1. So, if $({\downarrow}V \cap {\downarrow}v) \subseteq I$, then $V' \subseteq I$ and so $v \in I$.

Moreover, $\widetilde{\rhd}$ is a Grothendieck coverage. We verify the last condition, leaving the other, simpler, checks to the reader. Suppose $W \mathrel{\widetilde{\rhd}} u$ and $V \subseteq {\downarrow}u$, such that for all $w \in W$, $(V \cap {\downarrow}w) \mathrel{\widetilde{\rhd}} w$; we want to show $V \mathrel{\widetilde{\rhd}} u$. Taking any $v \leq u$, we claim that $({\downarrow}V \cap {\downarrow}v) \subseteq I$ implies $({\downarrow}W \cap {\downarrow}v) \subseteq I$, which would then imply that $v \in I$ since $W \mathrel{\widetilde{\rhd}} u$. To see the claim, if $x \leq w \in W$ and $x \leq v$, then by assumption $(V \cap {\downarrow}w) \mathrel{\widetilde{\rhd}} w$, so $({\downarrow}V \cap {\downarrow}w \cap {\downarrow}x) \subseteq ({\downarrow}V \cap {\downarrow}v) \subseteq I$ implies $x \in I$. Hence, $\widetilde{\rhd}$ is a Grothendieck coverage containing $\rhd$.

We have assumed that I satisfies the sheaf condition for $\rhd$, and we want to show that it does so for $\overline{\rhd}$, i.e., that $V \mathrel{\overline{\rhd}} u$ and $V \subseteq I$ implies $u \in I$. But, $V \mathrel{\overline{\rhd}} u$ implies $V \mathrel{\widetilde{\rhd}}$, so if $V \subseteq I$ then $({\downarrow}V \cap {\downarrow}u) = {\downarrow}V \subseteq I$, hence $u \in I$.

The proof of the last claim is analogous; see also [Joh02, Prop. C2.1.9]. □

The following definition is adapted from [Shu12]:

Definition 2.46 Let $F \colon S \to T$ be an order-preserving map between Grothendieck posites. Call F a *dense morphism of posites* if:

- For all $u \in S$ and $V \subseteq {\downarrow}u$, $V \rhd u$ if and only if $FV \rhd Fu$,
- For all $u \in T$, $(FS \cap {\downarrow}u) \rhd u$,
- For all $u, v \in S$, if $Fu \leq Fv$ then $({\downarrow}u \cap {\downarrow}v) \rhd u$.

Proposition 2.47 *A dense morphism of Grothendieck posites $F \colon S \to T$ induces an isomorphism $\mathsf{Shv}_0(S) \cong \mathsf{Shv}_0(T)$.*

Proof From Lemma 2.41, we have a map $F_* \colon D(T) \to D(S)$ and its right adjoint $F^!$, and it easy to verify that F_* restricts to a map $F_* \colon \mathsf{Shv}_0(T) \to \mathsf{Shv}_0(S)$.

Given any sheaf $J \in \mathsf{Shv}_0(T)$, we first show $J = F^! F_* J$. The containment $J \subseteq F^! F_* J$ is the unit of the adjunction. In the other direction, suppose $v \in F^! F_* J$. Unwinding Eq. (2.2), for any $u \in S$, $Fu \leq v$ implies $Fu \in J$; hence, $(FS \cap {\downarrow}v) \subseteq J$. Then, $(FS \cap {\downarrow}v) \rhd v$ because F is dense, and we obtain $v \in J$ by the sheaf condition. Thus, indeed $J = F^! F_* J$.

Similarly, for any $I \in \mathsf{Shv}_0(S)$, we have $I = F_* F^! I$, where $F_* F^! I \subseteq I$ is simply the counit of the adjunction. To show the converse $I \subseteq F_* F^! I$, choose $u \in I$; we need to show $F^{-1}({\downarrow}Fu) \subseteq I$, so choose also $u' \in S$ with $Fu' \leq Fu$. Clearly, $({\downarrow}u' \cap {\downarrow}u) \subseteq I$, and by density $({\downarrow}u' \cap {\downarrow}u) \rhd u'$, hence $u' \in I$ by the sheaf condition on I.

All that remains is to verify that $F^!(I)$ is a (0, 1)-sheaf for any $I \in \mathsf{Shv}_0(S)$. We already know it is down-closed by Lemma 2.41, so suppose $V \subseteq F^!(I)$ and $V \rhd v$.

[4] We will neither need this nor prove this, but $\widetilde{\rhd}$ is the largest coverage for which I is a sheaf.

We want to show that $v \in F^!(I)$, so consider $u \in S$ with $Fu \leq v$; we want to show $u \in I$. Since $\rhd$ is a Grothendieck coverage, $V \rhd v$ and $Fu \leq v$ imply $(\downarrow V \cap \downarrow Fu) \rhd Fu$, which implies $(FS \cap \downarrow V \cap \downarrow Fu) \rhd Fu$ by density and Remark 2.40. But, $(FS \cap \downarrow V \cap \downarrow Fu) = F(F_*(\downarrow V) \cap F_*(\downarrow Fu))$, so $(F_*(\downarrow V) \cap F_*(\downarrow Fu)) \rhd u$. Since $(F_*(\downarrow V) \cap F_*(\downarrow Fu)) \subseteq F_*(\downarrow V) \subseteq F_* F^!(I) = I$, we have $u \in I$. Hence, $v \in F^!(I)$. □

Recall from Remark 2.43 the canonical coverage on $\mathsf{Shv}_0(S)$ and the associated sheaf functor $a \colon D(S) \to \mathsf{Shv}_0(S)$ from Lemma 2.42.

Proposition 2.48 *For any Grothendieck posite S, the map $i \colon S \to \mathsf{Shv}_0(S)$ given by:*

$$i(u) := a(\downarrow u) = \{v \in S \mid (\downarrow u \cap \downarrow v) \rhd v\}$$

defines a dense morphism of posites, regarding $\mathsf{Shv}_0(S)$ with its canonical coverage.

Proof First, note that for $I \in \mathsf{Shv}_0(S)$ and $v \in S$, we have $iv \subseteq I$ iff $v \in I$. To show that i is a dense morphism, there are three conditions to check. First, consider $u \in S$, $V \subseteq \downarrow u$, and we show that $V \rhd u$ iff $iV \rhd iu$, where $iV := \{iv \mid v \in V\}$. Since $V \subseteq \downarrow u \subseteq iu$, we have $iV \subseteq iu$, so by Remark 2.43:

$$\sup(iV) = \bigcap\{I \in \mathsf{Shv}_0(S) \mid \forall v \in V.\, iv \subseteq I\} = \bigcap\{I \in \mathsf{Shv}_0(S) \mid V \subseteq I\}$$

and $iV \rhd iu$ iff $\bigcap\{I \mid V \subseteq I\} = iu$. One containment is automatic, so $iV \rhd iu$ iff $iu \subseteq \bigcap\{I \mid V \subseteq I\}$, and this holds iff for each $I \in \mathsf{Shv}_0(S)$, $V \subseteq I$ implies $u \in I$. Thus by the sheaf condition, $V \rhd u$ implies $iV \rhd iu$. Conversely, suppose $V \subseteq I$ implies $u \in I$ for any $I \in \mathsf{Shv}_0(S)$. Then, $V \subseteq a(\downarrow V)$ implies $u \in a(\downarrow V)$, hence $\downarrow V = (\downarrow V \cap \downarrow u) \rhd u$, and therefore $V \rhd u$.

For the second condition, we want to show that for any $I \in \mathsf{Shv}_0(S)$, $(iS \cap \downarrow I) \rhd I$, i.e., $I = \sup\{iu \mid u \in I\}$, which is clear.

For the third and last condition, we want to show that for any $u, v \in S$, if $iu \subseteq iv$ then $(\downarrow u \cap \downarrow v) \rhd u$. But, $u \in iu \subseteq iv$ implies $(\downarrow u \cap \downarrow v) \rhd u$ by definition. □

Proposition 2.49 *For any posite S, consider the frame $\mathsf{Shv}_0(S)$ with the canonical coverage. Then, $\mathsf{Shv}(S) \cong \mathsf{Shv}(\mathsf{Shv}_0(S))$.*

Proof Let $\hat{S}$ be the Grothendieck completion of S as in Proposition 2.45, by which we obtain isomorphisms $\mathsf{Shv}_0(S) \cong \mathsf{Shv}_0(\hat{S})$ and $\mathsf{Shv}(S) \cong \mathsf{Shv}(\hat{S})$. Then, Propositions 2.47 and 2.48 together give the middle isomorphism in:

$$\mathsf{Shv}(S) \cong \mathsf{Shv}(\hat{S}) \cong \mathsf{Shv}(\mathsf{Shv}_0(\hat{S})) \cong \mathsf{Shv}(\mathsf{Shv}_0(S)).$$

□

2.5.2 *Equivalence of the Various Toposes*

We construct two Grothendieck coverages associated to a predomain B, then show that they are the Grothendieck completions of the posites S_B and S'_B from Propositions 2.18 and 2.20.

Proposition 2.50 *Let $(B, \prec)$ be a predomain and $\lessdot$ its upper specialization order. Then, there is a Grothendieck posite $\hat{S}_B$ with underlying poset $(B, \lessdot)^{\mathrm{op}}$, and with coverage defined by $V \vartriangleright b$ iff $\twoheaduparrow b \subseteq \uparrow V \subseteq \uparrow b$.*

Proof We only need to prove that $\vartriangleright$ is a Grothendieck coverage, i.e., the five conditions of Definition 2.39. The first four conditions are clear. For the final condition, suppose $W \vartriangleright b$, $V \subseteq \uparrow b$, and for all $w \in W$, $(V \cap \uparrow w) \vartriangleright w$. Thus, $\twoheaduparrow b \subseteq \uparrow W \subseteq \uparrow b$, and for all $w \in W$, $\twoheaduparrow w \subseteq \uparrow V$. Then, we want to show that $V \vartriangleright b$, i.e., $\twoheaduparrow b \subseteq \uparrow V$. Clearly, $\twoheaduparrow w \subseteq \uparrow V$ for all $w \in \uparrow W$, hence by the interpolative property of $\ll$ (see Proposition 2.14), we have $\twoheaduparrow b \subseteq \bigcup_{w \in \uparrow W} \twoheaduparrow w \subseteq \uparrow V$. □

Recall from Proposition 2.20 that for a predomain $(B, \prec)$, we say lower specialization implies upper specialization if $b \leqslant b'$ implies $b \lessdot b'$ for all $b, b' \in B$.

Proposition 2.51 *Suppose that for a predomain $(B, \prec)$, lower specialization implies upper specialization. Then, there is a Grothendieck posite $\hat{S}'_B$ with underlying poset $(B, \leqslant)^{\mathrm{op}}$, and with coverage defined by $V \vartriangleright b$ iff $\twoheaduparrow b \subseteq \uparrow V \subseteq \{b' \in B \mid b \leqslant b'\}$, and there is an isomorphism:*

$$\mathsf{Shv}_0(\hat{S}_B) \cong \mathsf{Shv}_0(\hat{S}'_B)$$

Proof The proof that $\vartriangleright$ defines a Grothendieck coverage is essentially the same as Proposition 2.18.

To see that $\mathsf{Shv}_0((B, \lessdot)^{\mathrm{op}}) \cong \mathsf{Shv}_0((B, \leqslant)^{\mathrm{op}})$, we show that the identity on elements map $(B, \leqslant)^{\mathrm{op}} \to (B, \lessdot)^{\mathrm{op}}$ is a dense morphism of sites. By assumption, $b \leqslant b'$ implies $b \lessdot b'$, so this map is order-preserving. Let us write $\bar{\uparrow} b = \{b' \in B \mid b \leqslant b'\}$ and $\uparrow b = \{b' \in B \mid b \lessdot b'\}$, so we have $\bar{\uparrow} b \subseteq \uparrow b$ for any $b \in B$, and we similarly write $\overline{\vartriangleright}$ for the coverage on $(B, \leqslant)^{\mathrm{op}}$ and $\vartriangleright$ for the coverage on $(B, \lessdot)^{\mathrm{op}}$. Then for any $V \subseteq \bar{\uparrow} b$, we clearly have $V \mathrel{\overline{\vartriangleright}} b$ iff $V \vartriangleright b$.

The remaining two conditions of Definition 2.46 are similarly trivial to verify. The proposition then follows from Proposition 2.47. □

Proposition 2.52 *For any predomain B, the Grothendieck posite $\hat{S}_B$ from Proposition 2.50 is the Grothendieck completion of the posite S_B defined in Proposition 2.18. Similarly, if $b \leqslant b'$ implies $b \lessdot b'$, then $\hat{S}'_B$ is the Grothendieck completion of S'_B.*

Proof Let us temporarily write $\widehat{\vartriangleright}$ for the coverage on $\hat{S}_B$, and $\vartriangleright$ for the coverage on S_B. Clearly, $V \vartriangleright b$ implies $V \mathrel{\widehat{\vartriangleright}} b$. Suppose $\widetilde{\vartriangleright}$ is any Grothendieck coverage containing $\vartriangleright$, i.e., such that $\twoheaduparrow b \mathrel{\widetilde{\vartriangleright}} b$ for any $b \in B$. Then, if $V \mathrel{\widehat{\vartriangleright}} b$, i.e., if $\twoheaduparrow b \subseteq$

$\uparrow V \subseteq \uparrow b$, then $\uparrow V \mathrel{\widetilde{\rhd}} b$ by Remark 2.40, hence $V \mathrel{\widetilde{\rhd}} b$. Thus, $\widetilde{\rhd}$ contains $\widehat{\rhd}$, so $\widehat{\rhd}$ is the smallest Grothendieck coverage containing $\rhd$.

The second case is similar. □

Corollary 2.53 *Let B be a predomain for which lower specialization implies upper specialization. Then,* $\mathsf{Shv}_0(S_B) \cong \mathsf{Shv}_0(S'_B)$.

Proof This follows from Propositions 2.45, 2.51, and 2.52. □

The following theorem was stated without proof in Theorem 2.27; we can now prove the equivalence of 1, 2, and 3 and sketch the proof that these are equivalent to 4.

Theorem 2.54 *The following categories are equivalent:*

1. *The category of sheaves* $\mathsf{Shv}(\Omega(\mathbb{IR}))$ *on the frame of opens of* $\mathbb{IR}$*, with the canonical coverage* $\{V_i\}_{i\in I} \rhd U$ *iff* $\bigcup_{i\in I} V_i = U$.
2. *The category of sheaves* $\mathsf{Shv}(S'_{\mathbb{IR}})$ *on the interval posite* $S_{\mathbb{IR}}$.
3. *The category of sheaves* $\mathsf{Shv}(S^{\mathbb{Q}}_{\mathbb{IR}})$ *on the rational interval posite* $S^{\mathbb{Q}}_{\mathbb{IR}}$.
4. *The category of* $\mathsf{Cont}(\mathbb{IR})$ *of* continuous functors $\mathbb{IR} \to$ **Set**, *i.e., functors taking directed suprema in* $\mathbb{IR}$ *to colimits in* **Set**.

Proof The proof of the equivalence $\mathsf{Shv}(\Omega(\mathbb{IR})) \cong \mathsf{Cont}(\mathbb{IR})$ is sketched in Remark B.27. For the rest, recall from Definition 2.25 that $S^{\mathbb{Q}}_{\mathbb{IR}} = S_{\mathbb{IR}_{\text{pre}}}$ and that $S_{\mathbb{IR}} = S'_{\mathbb{IR}}$, as defined in Proposition 2.20. It follows from Propositions 2.19 and A.12 and Corollary 2.53 that $\mathsf{Shv}_0(S'_{\mathbb{IR}}) \cong \mathsf{Shv}_0(S_{\mathbb{IR}}) \cong \Omega(\mathbb{IR}) \cong \mathsf{Shv}_0(S_{\mathbb{IR}_{\text{pre}}})$. Hence by Proposition 2.49:

$$\mathsf{Shv}(S'_{\mathbb{IR}}) \cong \mathsf{Shv}(\mathsf{Shv}_0(S'_{\mathbb{IR}})) \cong \mathsf{Shv}(\Omega(\mathbb{IR})) \cong \mathsf{Shv}(\mathsf{Shv}_0(S^{\mathbb{Q}}_{\mathbb{IR}})) \cong \mathsf{Shv}(S^{\mathbb{Q}}_{\mathbb{IR}}).$$

□

Chapter 3
Translation Invariance

As discussed in Sect. 1.2.2, the topos $\mathsf{Shv}(S_{\mathbb{IR}})$ of sheaves on the interval domain is slightly unsatisfactory as a model of behaviors. For example, to serve as a compositional model of dynamical systems, we do not want the set of possible behaviors in some behavior type to depend on any global time. In this chapter, we define a topos $\mathcal{B}$ of "translation-invariant behavior types" by defining a translation-invariant version of the interval domain and a corresponding site, denoted $\mathbb{IR}_{/\rhd}$ and $S_{\mathbb{IR}/\rhd}$, respectively, and letting $\mathcal{B} := \mathsf{Shv}(S_{\mathbb{IR}/\rhd})$. In an appendix, Appendix B, we prove that $\mathbb{IR}_{/\rhd}$ is a continuous category in the sense of Johnstone-Joyal [JJ82], giving us an analogue to Theorem 2.27; this is briefly discussed in Sect. 3.2.

3.1 Construction of the Translation-Invariant Topos $\mathcal{B}$

Recall the interval domain $(\mathbb{IR}, \sqsubseteq)$ from Definition 2.22. We will define $\mathcal{B}$ in terms of a site $\mathbb{IR}_{/\rhd}$, obtained as the quotient of $\mathbb{IR}$ by a free $\mathbb{R}$-action.

3.1.1 The Translation-Invariant Interval Domain, $\mathbb{IR}_{/\rhd}$

We have two primary definitions of the translation-invariant interval domain $\mathbb{IR}_{/\rhd}$, and it is easy to show their equivalence. On the one hand, there is a continuous action $\rhd$ of $\mathbb{R}$ on $\mathbb{IR}$,[1] given by $r \rhd [d, u] := [r + d, r + u]$, and we can consider the category of elements for this action. It is given by adjoining an isomorphism

[1]The $\rhd$ symbol used here (e.g., "$r \rhd [d, u]$") to denote a particular $\mathbb{R}$-action on $\mathbb{IR}$ is different from the $\triangleright$ symbol, used in Chap. 2 (e.g., "$V \triangleright u$") to denote a basic cover in a posite.

P. Schultz, D. I. Spivak, *Temporal Type Theory*, Progress in Computer Science and Applied Logic 29, https://doi.org/10.1007/978-3-030-00704-1_3

$(d, u, r) : [d, u] \to [d + r, u + r]$ to $\mathbb{IR}$ for each $r \in \mathbb{R}$, with $(d, u, r)^{-1} = (d + r, u + r, -r)$ and $(d + r, u + r, r') \circ (d, u, r) = (d, u, r + r')$.

On the other hand, and more conveniently, we can simply take $\mathbb{IR}_{/\rhd}$ to be the category of orbits of the $\rhd$ action. The equivalence class containing an interval $[d, u]$ can be represented by its length $\ell := u - d$, and the equivalence class of a specialization $[d, u] \sqsubseteq [d', u']$ can be represented by the nonnegative numbers $r := d' - d$ and $s := u - u'$, how much to "shave off" from the left and right of the longer interval. This leads to the following definition.

Definition 3.1 (The Translation-Invariant Interval Category $\mathbb{IR}_{/\rhd}$) Define the *translation-invariant interval category*, denoted $\mathbb{IR}_{/\rhd}$, to have

- objects $\{ \ell \in \mathbb{R} \mid \ell \geq 0 \}$,
- morphisms $\mathbb{IR}_{/\rhd}(\ell, \ell') = \{ \langle r, s \rangle \in \mathbb{R} \times \mathbb{R} \mid (r \geq 0) \wedge (s \geq 0) \wedge (\ell = r + \ell' + s) \}$,
- identities $\langle 0, 0 \rangle$, and
- composition $\langle r', s' \rangle \circ \langle r, s \rangle = \langle r' + r, s' + s \rangle$.

We refer to an object $\ell \in \mathbb{IR}_{/\rhd}$ as a *translation-invariant interval.*

Note that one can consider $\rhd$ as an $\mathbb{R}$-action on $\mathbb{IR}^{\text{op}}$ as well, and there is an isomorphism of categories $(\mathbb{IR}^{\text{op}})_{/\rhd} \cong (\mathbb{IR}_{/\rhd})^{\text{op}}$; in the future, we will elide the difference and denote them both simply by $\mathbb{IR}^{\text{op}}_{/\rhd}$.

Remark 3.2 The category $\mathbb{IR}_{/\rhd}$ is isomorphic to the twisted arrow category $\mathsf{Tw}(\mathsf{B}\mathbb{R}_{\geq 0})$, where $\mathbb{R}_{\geq 0}$ is the monoid of nonnegative real numbers and $\mathsf{B}\mathbb{R}_{\geq 0}$ denotes the corresponding category with one object. This category was fundamental in an earlier paper [SVS16] and was considered by Lawvere and others [Law86, BF00, Fio00] in the context of dynamical systems; see also [Joh99].

There is an evident quotient functor $p\colon \mathbb{IR} \to \mathbb{IR}_{/\rhd}$, which sends $[d, u] \in \mathbb{IR}$ to $p[d, u] := u - d \in \mathbb{IR}_{/\rhd}$, and sends $[d, u] \sqsubseteq [d', u']$ to $\langle r, s \rangle\colon \ell \to \ell'$, where $\ell := u - d$, $\ell' :- u' - d'$, $r := d' - d$, and $s := u - u'$. In Lemma 3.3 we will show that p is a discrete bifibration, i.e. both a discrete fibration and a discrete opfibration. This roughly means that if $[d, u]$ is an interval with extent $\ell := u - d$, then any extent containing ℓ (resp. contained in ℓ) will have a unique lift to an interval containing $[d, u]$ (resp. contained in $[d, u]$).

Lemma 3.3 *The functor $p\colon \mathbb{IR} \to \mathbb{IR}_{/\rhd}$ is a discrete bifibration, and it is surjective on objects. The same holds for $p^{\text{op}}\colon \mathbb{IR}^{\text{op}} \to \mathbb{IR}^{\text{op}}_{/\rhd}$.*

Proof Clearly, if p is a bifibration and surjective on objects, then so is p^{op}.

It is obvious that p is surjective on objects. Given $[d, u] \in \mathbb{IR}$ and $\langle r, s \rangle\colon (u - d) \to \ell$, a morphism $[d, u] \sqsubseteq [d', u']$ projects to $\langle r, s \rangle$ iff $r = d' - d$ and $s = u - u'$, hence $d' = d + r$ and $u' = u - s$ is the unique lift of $\langle r, s \rangle$. This shows that p is a discrete opfibration. The proof that p is a discrete fibration is similar. □

Recall the way-below relation on $\mathbb{IR}$ from Proposition 2.23: $[d, u] \ll [d', u']$ iff $d < d' \leq u' < d$ iff $r > 0$ and $s > 0$. We follow Johnstone and Joyal ([JJ82] and [Joh02, Definition C.4.2.12]) and give the following definition.

Definition 3.4 (Wavy Arrow) For any $\ell, \ell' \in \mathbb{IR}_{/\rhd}$, we say that a morphism $\langle r, s\rangle : \ell \to \ell'$ is a *wavy arrow* and denote it $\langle r, s\rangle : \ell \rightsquigarrow \ell'$, if $r > 0$ and $s > 0$.

We will use the same notation for morphisms in $\mathbb{IR}^{\text{op}}_{/\rhd}$, i.e. write $f : \ell' \rightsquigarrow \ell$, and call f a wavy arrow, iff f^{op} is a wavy arrow in $\mathbb{IR}_{/\rhd}$, as defined above.

It is tempting to think of $\ell' \rightsquigarrow \ell$ as a "strict subinterval," but one must remember that this strictness applies to both sides, r and s.

3.1.2 The Topos $\mathcal{B}$ of Behavior Types

We are now ready to define the site of translation-invariant intervals and the corresponding topos of sheaves.

Definition 3.5 (The Site $S_{\mathbb{IR}/\rhd}$ and the Topos $\mathcal{B}$ of Behavior Types) We define the site of *translation-invariant intervals*, denoted $S_{\mathbb{IR}/\rhd}$, as follows:

- The underlying category is $\mathbb{IR}^{\text{op}}_{/\rhd}$ as in Definition 3.1, and
- the coverage consists of one family, $\{\langle r, s\rangle : \ell' \to \ell \mid r, s > 0\}$, for each object ℓ.

That is, the collection of wavy arrows $\{\ell' \rightsquigarrow \ell\} \subseteq \mathbb{IR}^{\text{op}}_{/\rhd}(\ell', \ell)$ covers ℓ. We refer to $\mathcal{B} := \mathsf{Shv}(S_{\mathbb{IR}/\rhd})$ as the topos of *behavior types*.

Definition 3.5 given above is analogous to the posite version given in Definition 2.25. Indeed, in Theorem B.21 we will show that for any $\ell' \in \mathbb{IR}_{/\rhd}$, the functor $\mathbb{IR}^{\text{op}}_{/\rhd} \to \mathbf{Set}$ given by $\ell \mapsto \{f : \ell \rightsquigarrow \ell' \in \mathbb{IR}_{/\rhd}(\ell, \ell')\}$ is always flat, and hence an object of $\mathsf{Ind}\text{-}(\mathbb{IR}_{/\rhd})$. This assignment is functorial in ℓ' and is in fact left adjoint to colim: $\mathsf{Ind}\text{-}(\mathbb{IR}_{/\rhd}) \to \mathbb{IR}_{/\rhd}$. Thus it is analogous to the map $\downarrow : \mathbb{IR} \to \text{Id}(\mathbb{IR})$ which is left adjoint to sup: $\text{Id}(\mathbb{IR}) \to \mathbb{IR}$ (see Propositions 2.9 and 2.23).

Remark 3.6 The object $0 \in \mathbb{IR}_{/\rhd}$ has an empty covering family in $S_{\mathbb{IR}/\rhd}$. Thus any object X of the topos $\mathcal{B} = \mathsf{Shv}(S_{\mathbb{IR}/\rhd})$ has $X(0) \cong \{*\}$, and is hence completely determined by its values on objects $\ell > 0$. It is often more convenient to work with the subsite $S'_{\mathbb{IR}/\rhd} \subseteq S_{\mathbb{IR}/\rhd}$ whose underlying category is the full subcategory of $\mathbb{IR}_{/\rhd}$ spanned by the objects $\ell > 0$, and whose covering families match those in $S_{\mathbb{IR}/\rhd}$. In $S'_{\mathbb{IR}/\rhd}$, every covering family is filtered.

Proposition 3.7 *The quotient functor $p^{\text{op}} : \mathbb{IR}^{\text{op}} \to \mathbb{IR}^{\text{op}}_{/\rhd}$ (see Lemma 3.3) induces a geometric morphism $p_* : \mathsf{Shv}(S_{\mathbb{IR}}) \to \mathcal{B}$ with left exact left adjoint $p^* : \mathcal{B} \to \mathsf{Shv}(S_{\mathbb{IR}})$.*

Proof It suffices by [Joh02, C2.3.18] to check that p^{op} is cover-reflecting. This is immediate once one consults Definitions 2.25 and 3.5. □

A topos $\mathcal{E}$ is *locally connected* if the global sections geometric morphism $\Gamma : \mathcal{E} \to \mathbf{Set}$ is an essential geometric morphism, i.e. if the inverse image functor Cnst has a left adjoint Π_0; see [Joh02, Lemma C.3.3.6]. It is *connected* if Cnst

is fully faithful; this is equivalent to the condition that Π_0 preserves the terminal object. To prove that $\mathcal{E}$ is locally connected, it suffices to show that it is the topos of sheaves on a locally connected site, meaning that for each object U, all covering sieves of U are connected.

Proposition 3.8 *The topos $\mathcal{B}$ is locally connected and connected.*

Proof By Remark 3.6, there is a defining site $S'_{\mathbb{IR}/\rhd}$ for $\mathcal{B}$, in which all covering families are filtered; they are in particular connected, so $\mathcal{B}$ is locally connected.

The composite $\mathbf{Set} \xrightarrow{\mathsf{Cnst}} \mathsf{Shv}(S'_{\mathbb{IR}/\rhd}) \to \mathsf{Psh}(S'_{\mathbb{IR}/\rhd})$ is fully faithful, because the category underlying $S'_{\mathbb{IR}/\rhd}$ is connected. Since $\mathsf{Shv}(S'_{\mathbb{IR}/\rhd}) \to \mathsf{Psh}(S'_{\mathbb{IR}/\rhd})$ is fully faithful, being the direct image of a geometric inclusion, it follows that Cnst is fully faithful as well. □

Corollary 3.9 *If C, D are constant sheaves, then so is C^D.*

Proof We use the Yoneda lemma and Proposition 3.8. For any $X \in \mathcal{B}$, we have

$$\begin{aligned}[X, \mathsf{Cnst}(C)^{\mathsf{Cnst}(D)}] &\cong [X \times \mathsf{Cnst}(D), \mathsf{Cnst}(C)] \\ &\cong [\Pi_0(X \times \sqcup_{d\in D} 1), C] \\ &\cong [D \times \Pi_0(X), C] \cong [X, \mathsf{Cnst}(C^D)].\end{aligned}$$

□

3.2 $\mathbb{IR}_{/\rhd}$ as a Continuous Category

Continuous categories, defined by Johnstone and Joyal in [JJ82], are a generalization of domains (continuous posets). In fact, a poset is a domain iff it is continuous as a category. Under this generalization from posets to categories, ideals are replaced by Ind-objects, directed suprema are replaced by filtered colimits, the way-below relation is replaced by the set of wavy arrows (Definition 3.4), and Scott-continuous morphisms between domains are replaced by continuous functors: those which preserve filtered colimits.

Recall that a *point* of a topos $\mathcal{E}$ is a geometric morphism $p\colon \mathbf{Set} \to \mathcal{E}$, and a morphism of points $p \to q$ is a natural transformation of inverse images $p^* \to q^*$. Let $P_{\mathcal{E}}$ denote the category of points in $\mathcal{E}$. There is always a functor $s\colon \mathcal{E} \to \mathbf{Set}^{P_{\mathcal{E}}}$, sending a sheaf X and a point p to the set $p^*(X)$. When $\mathcal{E}$ is the topos of sheaves on a continuous category, s is fully faithful and its essential image is the category (topos) of continuous functors.

We discuss the above subject in greater detail in Appendix B. For example, in Theorem B.21 we prove that $\mathbb{IR}_{/\rhd}$ is a continuous category. We also prove an equivalence of categories

$$\mathsf{Shv}(S_{\mathbb{IR}/\rhd}) \cong \mathsf{Cont}(\mathbb{IR}_{/\rhd})$$

where $\mathsf{Cont}(\mathbb{IR}_{/\rhd})$ denotes the category of continuous—i.e., filtered-colimit preserving—functors $\mathbb{IR}_{/\rhd} \to \mathbf{Set}$. This gives two equivalent definitions of the topos $\mathcal{B}$, analogous to the equivalence of Theorem 2.27 3 and 4.

3.3 The Subobject Classifier

The subobject classifier of the topos $\mathsf{Shv}(S_{\mathbb{IR}})$ has a nice visual interpretation as the set of Lipschitz functions in the upper half-plane (see Sect. 2.4). We can transport this geometric picture to $\mathcal{B}$ because the semantics of the subobject classifier Ω in $\mathcal{B} = \mathsf{Shv}(S_{\mathbb{IR}/\rhd})$ is strongly related to those of the subobject classifier $\Omega_{\mathbb{IR}}$ for the topos $\mathsf{Shv}(S_{\mathbb{IR}})$, as we now explain.

For any object $\ell \in S_{\mathbb{IR}/\rhd}$ and $a \in \mathbb{R}$, there is a basic open set $\Uparrow[a, b]$ in $\mathbb{IR}$, where $b = a + \ell$; it is defined as

$$\Uparrow[a, b] := \{[x, x'] \mid a < x \leq x' < b\}.$$

Consider the frame $\Omega_{\mathbb{IR}}(\Uparrow[a, b])$ of open subsets $U \subseteq \Uparrow[a, b]$. It is independent of a in the sense that there is a canonical isomorphism $\Omega(\Uparrow[a, b]) \cong \Omega(\Uparrow[0, \ell])$. Recall from Proposition 2.14 that $\mathbb{IR}$ has a basis consisting of such open intervals $\Uparrow[a, b]$.

Proposition 3.10 *There is a canonical bijection* $\Omega(\ell) \xrightarrow{\cong} \Omega_{\mathbb{IR}}(\Uparrow[0, \ell])$.

Proof As in any sheaf topos, the set $\Omega(\ell)$ can be identified with the set of closed sieves on ℓ in $S_{\mathbb{IR}/\rhd}$. This set depends only on the slice category $(\ell \downarrow \mathbb{IR}_{/\rhd})$, and because the projection $p \colon \mathbb{IR} \to \mathbb{IR}_{/\rhd}$ is a discrete opfibration, as shown in Proposition 3.7, we indeed have $(\ell \downarrow \mathbb{IR}_{/\rhd}) \cong ([0, \ell] \downarrow \mathbb{IR})$. □

By the above proposition, any $\omega \in \Omega(\ell)$ corresponds to a Scott-open set $U_\omega \subseteq \Uparrow[0, \ell]$ in $\mathbb{IR}$. We can use Dyck paths (Corollary 2.34) to picture the subobject classifier Ω: for any ℓ, the poset $\Omega(\ell)$ is isomorphic to the set of Dyck paths on $[0, \ell]$. Given $\langle r, s\rangle \colon \ell' \to \ell$, the open set $U_{\omega'}$ corresponding to the restriction $\omega' := \omega|_{\langle r,s\rangle}$ is

$$U_{\omega'} = U_\omega \cap \Uparrow[r, \ell - s].$$

3.4 The Behavior Type Time

Let $p^* \dashv p_*$ be the geometric morphism from Proposition 3.7. We claim that it is *étale*, meaning that there is an object $T \in \mathcal{B}$ such that p_* is equivalent to the projection $\mathcal{B}/T \to \mathcal{B}$. It would follow that p^* has a further left adjoint $p_!$, and that $T = p_!(1)$.

3.4.1 The Geometric Morphism $\mathsf{Shv}(S_{\mathbb{IR}}) \to \mathcal{B}$ Is étale

To prove that p_* is étale, we show directly that its left adjoint p^* has a further left adjoint, and use this to show that p_* is isomorphic to the projection $\mathcal{B}/p_!(1) \to \mathcal{B}$.

The left adjoint $p_! \colon \mathsf{Shv}(S_{\mathbb{IR}}) \to \mathcal{B}$, if it exists, must be given by left Kan extension along $p^{\mathrm{op}} \colon \mathbb{IR} \to \mathbb{IR}_{/\rhd}$. The quotient functor p is a discrete opfibration, as shown in Lemma 3.3. It follows that the left Kan extension of $X \colon \mathbb{IR} \to \mathbf{Set}$ is simply fiberwise coproduct:

$$(p_!X)(\ell) = \coprod_{\{(d,u) \mid u-d=\ell\}} X([d,u]) \tag{3.1}$$

with restriction along $\langle r,s\rangle \colon \ell' \to \ell$ given by $((d,u),x)|_{\langle r,s\rangle} = ((d+r, u-s), x|_{(d+r,u-s)})$. It only remains to show that $(p_!X) \colon \mathbb{IR}_{/\rhd} \to \mathbf{Set}$ is indeed a sheaf, but this is clear given that X is a sheaf.

We define the sheaf $\mathtt{Time} := p_!(1)$. Using the above formula for $p_!$, we can identify it with the functor $\mathbb{IR}_{/\rhd} \to \mathbf{Set}$ whose length-ℓ sections are given by

$$\mathtt{Time}(\ell) = \{\, (d,u) \in \mathbb{R}^2 \mid u - d = \ell \,\}, \tag{3.2}$$

and whose restriction map for $\langle r,s\rangle \colon \ell \to \ell'$ is given by $\mathtt{Time}\langle r,s\rangle(d,u) = (d+r, u-s)$.

Proposition 3.11 *There is an equivalence of categories* $\mathsf{Shv}(S_{\mathbb{IR}}) \cong \mathcal{B}/\mathtt{Time}$, *such that the diagram*

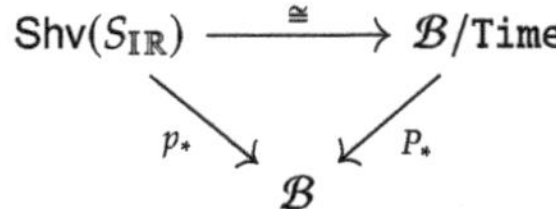

commutes, where (P^*, P_*) *is the base change geometric morphism along* $!_{\mathtt{Time}} \colon \mathtt{Time} \to 1$ *in* $\mathcal{B}$.

Proof The equivalence sends any sheaf $X \in \mathsf{Shv}(S_{\mathbb{IR}})$ to $p_!(!_X)$, where $!_X \colon X \to 1$ is the unique map to the terminal sheaf. Recall that $p_!(X)(\ell) = \coprod_{u-d=\ell} X(d,u)$; the function $p_!(!_X)_\ell$ sends $x \in X(d,u)$ to $(d,u) \in \mathtt{Time}(\ell)$. It is straightforward to check that this determines an equivalence.

The base change geometric morphism (P^*, P_*) is trivially étale, and in particular P^* has a further left adjoint $P_! \colon \mathcal{B}/\mathtt{Time} \to \mathcal{B}$, which is given by post-composition with P, i.e. $P_!$ is the standard projection out of the slice category. Then it suffices to show the diagram involving $p_!$ and $P_!$ commutes, but this is obvious. □

Corollary 3.12 *$\mathcal{B}$ is an étendue.*

Proof It is easy to see by Eq. (3.2) that the unique map $\mathtt{Time} \to 1$ is an epimorphism; so the result follows from Theorem 2.54 and Proposition 3.11. □

Remark 3.13 We think of the sheaf Time as the behavior type of a clock. Given a behavior of length ℓ, a $(d, u) \in \mathtt{Time}(\ell)$, we imagine this as the behavior of a clock which reads d at the beginning of the interval and reads u at the end. Clearly either one determines the other; see Remark 7.23. In temporal logic applications (see Sect. 8.6) it is sometimes helpful to think of the location where the clock reads zero as representing "now," even when that location is outside of the interval, i.e. when d and u are both positive or both negative.

3.4.2 $\mathcal{B}$ as a Quotient of $\mathsf{Shv}(S_{\mathbb{IR}})$

The topos $\mathcal{B}$ is the quotient of the localic topos $\mathsf{Shv}(S_{\mathbb{IR}})$. One way to see this is that $p^*\colon \mathcal{B} \to \mathsf{Shv}(S_{\mathbb{IR}})$ is faithful. Another way to see it is that $\mathcal{B}$ is equivalent to the category of coalgebras on the left-exact comonad $p^* p_*$ on $\mathcal{B}/\mathtt{Time}$.

With P as in Proposition 3.11, it is easy to see that for $X \in \mathcal{B}$, we have $P^*X = X \times \mathtt{Time}$, and for $(\tau\colon Y \to \mathtt{Time}) \in \mathcal{B}/\mathtt{Time}$, we have $P_!\tau = Y$ and $P_*\tau = \{f\colon \mathtt{Time} \to Y \mid \tau \circ f = \mathrm{id}_{\mathtt{Time}}\}$, i.e. $P_*\tau$ is the sheaf of sections of τ. The sections of $p_!X$ were given in Eq. (3.1), though we repeat it here for the reader's convenience, along with those of p^* and p_*. For $X \in \mathcal{B}$, we have

$$p^*X([d, u]) \cong X(u - d),$$

and for $Y \in \mathsf{Shv}(S_{\mathbb{IR}})$, we have

$$p_!Y(\ell) \cong \coprod_{\{(d,u) \mid u-d=\ell\}} Y([d, u]) \qquad \text{and} \qquad p_*Y(\ell) \cong \prod_{\{(d,u) \mid u-d=\ell\}} Y([d, u])$$

For any sheaf $Y \in \mathsf{Shv}(S_{\mathbb{IR}})$ and $r \in \mathbb{R}$, let $Y \rhd r$ be the sheaf $(Y \rhd r)([d, u]) := Y([d + r, u + r])$. The comonad $p^* p_*$ thus sends Y to $\prod_{r\in\mathbb{R}}(Y \rhd r)$; the counit is projection at $r = 0$, and the comultiplication is the obvious map $\prod_{r\in\mathbb{R}}(Y \rhd r) \to \prod_{r_1,r_2}(Y \rhd (r_1 + r_2))$. Thus an algebra for the comonad $p^* p_*$ is a sheaf $Y \in \mathsf{Shv}(S_{\mathbb{IR}})$ equipped with a homomorphism $h_r\colon Y \to (Y \rhd r)$ for each $r \in \mathbb{R}$, such that $h_0 = \mathrm{id}_Y$ and $h_{r_1+r_2} = h_{r_1} \circ h_{r_2}$. Since $\mathbb{R}$ is a group, this in particular implies that h_r is an isomorphism for all $r \in \mathbb{R}$.

Proposition 3.14 *A sheaf in $\mathcal{B}$ can be identified with a sheaf $Y \in \mathsf{Shv}(S_{\mathbb{IR}})$ equipped with a translation isomorphism $h_r\colon Y \xrightarrow{\cong} (Y \rhd r)$ for each $r \in \mathbb{R}$, such that $h_0 = \mathrm{id}_Y$ and $h_{r_1+r_2} = h_{r_1} \circ h_{r_2}$.*

Chapter 4
Logical Preliminaries

In this chapter, we transition from an external point of view to an internal one. In Chap. 2 we defined the interval domain $\mathbb{IR}$, and in Chap. 3 we defined a quotient $\mathcal{B} \cong \mathsf{Shv}(S_{\mathbb{IR}/\rhd})$ of its topos of sheaves. A main goal of this book is to define a temporal type theory—including one atomic predicate and ten axioms—that has semantics in $\mathcal{B}$; we do this in Chap. 5. In the present chapter, we attempt to provide the reader with a self-contained account of the sort of type theory and logic we will be using, as well as some important concepts definable therein.

We begin in Sect. 4.1 with an informal introduction to our type theory and logic, as well as its relation to toposes. We then discuss modalities in Sect. 4.2, which are the internal view of what in topos theory literature are often called Lawvere-Tierney topologies or local operators. Logically, modalities are internal monads on the type `Prop` of propositions, whereas semantically they correspond to subtoposes. Finally in Sect. 4.3, we discuss numeric types, e.g. the Dedekind real numbers and related types, relative to an arbitrary modality. In this section we also discuss inequalities between, and the arithmetic of, numeric types.

4.1 Informal Introduction to Type Theory

In this section we will informally introduce the type theory we use in the remainder of the book. We justify the lack of formality in three ways. First, most of what we do is relatively standard material: higher-order logic on top of the simply-typed lambda calculus with sum types and a type of natural numbers, plus subtypes and quotient types. However, we also make very limited use of dependent types, so providing full details of our type theory would be a major endeavor. This is the second reason: a full account would simply take too much space to be worthwhile. And the third reason is that we want to be users of logic—not logicians—here.

P. Schultz, D. I. Spivak, *Temporal Type Theory*, Progress in Computer Science and Applied Logic 29, https://doi.org/10.1007/978-3-030-00704-1_4

Thus we will present enough information to get readers started; for those who want a full account of categorical logic and type theory, we recommend [Jac99].

What will matter most is that our type theory is constructive and has an object `Prop` of propositions—and hence has a higher-order logic. It is well-known that this logic has semantics in any topos [MM92]. Sometimes we use dependent types, subtypes, and quotient types; it is less well-known but true that these also have semantics in any topos [Mai05]. While we were not able to find a reference, we believe that the Calculus of Constructions [CH88]—used in the automated proof assistants Lean and Coq—also has sound semantics in any topos, so the system we present below could be formalized in either proof assistant.[1]

4.1.1 Notions from the Simply-Typed Lambda Calculus

In this section (Sect. 4.1.1) we give an informal account of the simply-typed lambda calculus. The material is standard; see, e.g., [LS88, Jac99]. The key words we will discuss are: *types*, *variables*, *contexts*, *terms*, *substitution*, and *conversion rules*. As we get into Sect. 4.1.2 and begin to talk about logic, we will add *propositions*, *premises*, and *truth judgments*.

It is important to understand the relationship between a type theory and its semantics in the appropriate kind of category, say a topos $\mathcal{E}$, without conflating the two. Each has value independent of the other; for example, computers do not work on semantics, they work on syntax. Having a syntax that "compiles" to our chosen semantics can be quite freeing, because proofs in the logic are far easier to check than those in the semantics they compile to. Moreover, there is not always a perfect correspondence between what is important in type theory and what is important in semantics. For example, variables and contexts in a type theory have little semantic meaning. Still, the connection between the type theory and its $\mathcal{E}$-semantics is important, so we summarize the touch points now, and will continue to do so throughout the section.

Types in our type theory correspond to objects in $\mathcal{E}$. Variables and contexts—which are lists of variables and types—mainly serve a bookkeeping role. They gain semantic meaning when they are attached to terms. That is, a well-formed term always has a context and a type; it looks like this

$$a : A, b_1 : B, b_2 : B \vdash f(a, g(b_1, b_2), a) : C \tag{4.1}$$

[1] The dependent type theory in [Mai05] is sound and complete for 1-toposes, relying heavily on extensivity of types. This is fine for 1-toposes, but it makes computability impossible and hence is not useful for a proof assistant like Coq. On the other hand, the Calculus of Constructions [CH88] and related formalisms used in Coq and Lean are formulated using a hierarchy of universes. We believe this aspect of the Calculus of Constructions is sound for toposes, and we believe this follows from [Str05], but could not find an explicit reference.

Here, A, B, and C are types; a, b_1, and b_2 are variables; a, b_1, b_2, $g(b_1, b_2)$, and $f(a, g(b_1, b_2), a)$ are terms. These terms appear in the context $a : A, b_1 : B, b_2 : B$ and, for example the term $f(a, g(b_1, b_2), a)$ has type C. A term corresponds to a morphism in $\mathcal{E}$; its domain is the product of types in its context, and its codomain is its type. So in the above example, $f(a, g(b_1, b_2), a)$ presents a morphism $A \times B \times B \to C$.

There is a grammar for types and a grammar for terms, meaning that new types and terms can be produced from old using what are called *type constructors* and *term constructors*. There are several rules that dictate when the resulting types and terms are well-formed and when two terms are declared equal; the latter are the conversion rules. Issues like which variables are free and how substitution works for replacing a variable by a term (one must be careful that variables are not accidentally bound in the process) can be carefully worked out. This is important for computer applications, and may be interesting, but it will not concern us much here because mathematicians have generally developed an engineer's sense of how to work with types and terms without needing to know much about what goes on under the hood.

We now expand on the above summary of the simply-typed lambda calculus and its semantics.

Types We begin with types. One may declare atomic types $\mathtt{T}_1, \mathtt{T}_2, \ldots$ (though we will not need to do so in this book). The type $\mathtt{Prop}$ of propositions, the type $\mathbb{N}$ of natural numbers, the empty type $\mathtt{empty}$, and the unit type 1 are given as a base types, without needing to declare them. From these, one may form new types by taking finite products, finite sums, and arrow types; these are the type constructors. So suppose we declare $\mathtt{T}$ to be our only atomic type. Then we also have, for example, the three following types:

$$\mathtt{T} \times \mathtt{T} \times \mathtt{T}, \quad 1, \quad \mathtt{T} \times \mathtt{Prop} \to (\mathbb{N} \to (\mathtt{T} + 1)).$$

To give types semantics in $\mathcal{E}$, one must choose an object for each atomic type ($\mathtt{T}_i$). The rest is automatic: the type $\mathtt{Prop}$ corresponds to the subobject classifier in **E**, the type $\mathbb{N}$ corresponds to the natural numbers object in $\mathcal{E}$, the types $\mathtt{empty}$ and 1 correspond to the initial and final objects of $\mathcal{E}$, the product types correspond to products in $\mathcal{E}$, and arrow types correspond to exponential objects in $\mathcal{E}$.

Before moving on, we make two notes. First, there is a common convention to reduce parenthetical clutter: a sequence of arrow-types is parsed by "right associativity." That is, $\tau_1 \to (\tau_2 \to \tau_3)$ can be written simply as $\tau_1 \to \tau_2 \to \tau_3$. Second, we mentioned that one can declare some atomic types. We will later see that one can declare atomic terms and atomic predicates as well. All of these declarations make up what is called a *signature*. We will present a specific signature for temporal type theory in Chap. 5.

Variables and Contexts One assumes that an infinite set of symbols V has been fixed in advance for use as variables. Then saying "a is a variable" just means $a \in V$. We also assume that V is disjoint from all other sets of symbols we use, to avoid ambiguity.

On its own, a variable $a \in V$ has no type. For expressions involving a to be meaningful, a type for a must first be declared. As mentioned above, a context is a list of distinct variables, each with a type declaration. For example, if τ_1 and τ_2 are types and x, y, and z are variables, then $x : \tau_1,\ y : \tau_2,\ z : \tau_2$ is a context. Thus every expression will be considered relative to some (explicit or implicit) context, which defines the type of each variable used in the expression. To make contexts explicit, we use the standard notation $x : \tau_1,\ y : \tau_2 \vdash e$ to state that the expression e is being considered in the context $x : \tau_1,\ y : \tau_2$. While contexts are technically lists, type theories such as ours have rules that ensure that order does not matter.

A context, such as $x_1 : \tau_1, \ldots, x_n : \tau_n$, is often denoted Γ; if Γ' is another context, we may write Γ, Γ' to denote the union of these contexts.

Terms Terms are certain expressions which represent "elements" of types. A term t may contain variables as in Eq. (4.1), in which case t represents a parameterized family of elements. We require that every term t be explicitly paired with a context declaring the types of all free variables occurring in t, and we also require that every term itself has a unique type. We write $\Gamma \vdash t : \tau$ to record that the term t has type τ in context Γ.

As part of a signature, one may declare a finite number of atomic terms $c : \tau$, each having a specified type, in this case c has type τ. Each declared atomic term is valid in the empty context, e.g. $\vdash c : \tau$ would be judged "valid." A variable of a given type is also a term, but it is only valid in a context containing it. For example, $a : \mathbb{N}, b : \mathtt{T} \vdash a : \mathbb{N}$ is judged valid, assuming a and b are variables.

For the base types $\mathtt{Prop}$, $\mathbb{N}$, $\mathtt{empty}$, and 1, as well as for each type constructor, there are associated term constructors. We will hold off discussing $\mathtt{Prop}$ until Sect. 4.1.2, because it is the most complex, in order to focus on the main idea here.

The type $\mathbb{N}$ has three term constructors:

- $\vdash 0 : \mathbb{N}$,
- $a : \mathbb{N} \vdash s(a) : \mathbb{N}$, and
- $e : \tau,\ f : \tau \to \tau,\ g : \mathbb{N} \vdash \mathtt{rec}(e, f, g) : \tau$,

where the last exists for any type τ. Similarly,

- product types have term constructors called *tupling* and *projecting*;
- sum types have term constructors called *co-tupling* and *co-projecting*; and
- arrow types have term constructors called *lambda abstraction* and *application*.

All of these are standard,[2] but we discuss the last of these to give more of the flavor.

We begin with the term constructor for evaluation, because it is straightforward:

$$f : \tau_1 \to \tau_2,\ a : \tau_1 \vdash \mathtt{ap}(f, a) : \tau_2.$$

[2]Note that what we call sum types are called "coproduct types" in [Jac99, Section 2.3].

We often denote $\mathtt{ap}(f, a)$ simply by $f(a)$ or fa.

The lambda abstraction term constructor is a bit more subtle. The corresponding mathematical idiom is something like this: "Suppose a is an integer. Using a we can define a rational b as follows .... This defines a function $\mathbb{Z} \to \mathbb{Q}$, given by $a \mapsto b$." For any context Γ, types τ_1, τ_2, and term $\Gamma,\ a : \tau_1 \vdash b : \tau_2$, where obviously b is allowed to use variables from the context, there is a term $\Gamma \vdash \lambda(a : \tau_1).\, b : (\tau_1 \to \tau_2)$. In case the notation is not clear, this says $\lambda(a : \tau_1).\, b$ has arrow type $\tau_1 \to \tau_2$.

In the expression $\lambda(a : \tau_1).\, b$, the variable a has been *bound*, meaning it is no longer free. This has two consequences. The first is that the variable a can be replaced by any other variable without changing the term; this is called α-conversion and we mention it briefly when we discuss conversion rules; see Eq. (4.2). The second is that one can no longer substitute an arbitrary expression for a; we discuss substitution next.

Substitution As always, a variable of type τ represents an indeterminate value of type τ. Thus, if $x : \tau \vdash e : \tau'$ is a term—so e may contain the variable x—then we are free to replace x throughout e with any concrete term of type τ.

In this sense, terms can be substituted for free variables. Suppose $a : X$ is a variable of some type X and $\Gamma \vdash e : X$ is a term of the same type. Then $[a := e]$ denotes a sort of search-and-replace function, sending terms to terms. That is, for any term of the form $\Gamma', a : X \vdash f : \tau$, we can search for free occurrences of a in f and replace them with e, and the result is the term $\Gamma', \Gamma \vdash [a := e]f : \tau$.

As the $[a := e]$ notation can get unwieldy, we introduce a convenient shorthand. If $\Gamma, x : \tau \vdash e : \tau'$ is a term, then we can write $\Gamma, x : \tau \vdash e(x) : \tau'$ to draw attention to the fact that the variable x occurs in e (even though other variables from Γ might also occur in e). Then instead of $[x := t]e$, we can more simply write $e(t)$.

Conversion Rules The term constructors are meant to express universal properties about our various type constructors, and we want to force these to be true in the $\mathcal{E}$-semantics. So far, we have constructed the terms—which correspond to morphisms in $\mathcal{E}$—but we have not specified anything that says the appropriate diagrams commute. This is the job of conversion rules.

A conversion rule is a rule of the form

$$\Gamma \vdash e_1 \equiv e_2 : \tau,$$

where $\Gamma \vdash e_1 : \tau$ and $\Gamma \vdash e_2 : \tau$ are terms. There are conversion rules for each type constructor, but again they are standard as well as straightforward, so we only write out a few, namely those for natural numbers and arrow types. We also leave out the context and the typing to clarify the idea:

$$\begin{aligned}
\texttt{rec}(e, f, 0) &\equiv e && \text{base case}\\
\texttt{rec}(e, f, s(g)) &\equiv f(\texttt{rec}(e, f, g)) && \text{recursive step}\\
(\lambda(a : \tau_1).\, f(a))(e) &\equiv f(e) && \beta\text{-reduction}\\
\lambda(a : \tau_1).\, f(a) &\equiv f && \eta\text{-conversion}\\
\lambda(a : \tau_1).\, f(a) &\equiv \lambda(b : \tau_1).\, f(b) && \alpha\text{-conversion}
\end{aligned} \tag{4.2}$$

If we want to be a bit more pedantic and add back in the context and type, the first above is actually

$$e : \tau,\ \ f : \tau \to \tau,\ \ g : \mathbb{N} \vdash \texttt{rec}(e, f, 0) \equiv e : \tau,$$

etc. Hopefully the necessary context in each of the cases from Eq. (4.2) is clear enough.

Example 4.1 (Arithmetic of Natural Numbers) Suppose we want to construct a term $m : \mathbb{N}, n : \mathbb{N} \vdash \texttt{plus}(m, n) : \mathbb{N}$ for addition of natural numbers. Define it to be

$$m : \mathbb{N}, n : \mathbb{N} \vdash \texttt{rec}(m, \lambda(x : \mathbb{N}).\, s(x), n) : \mathbb{N}, \tag{4.3}$$

where s is the successor. We use the usual infix notation $m + n$ as shorthand for this term. In fact, we could have written Eq. (4.3) before discussing conversion rules, but now that we have conversion rule we check that our definition actually does what we want: $m + 0 \equiv m$ and $m + s(n) \equiv s(m + n)$. Multiplication of natural numbers is given by

$$m : \mathbb{N}, n : \mathbb{N} \vdash \texttt{rec}(0, \lambda(p : \mathbb{N}).\, m + p, n).$$

We can also define a function $+\colon (\mathbb{N} \times \mathbb{N}) \to \mathbb{N}$ by lambda abstraction: $+(m, n) := \lambda((m, n) : \mathbb{N} \times \mathbb{N}).\, \texttt{plus}(m, n)$, and similarly for multiplication.

Propositions, Premises, and Truth Judgments From the perspective of specifying a type theory and logic, one has—in addition to types, terms, etc.—things called *propositions* P, which have a status similar to types and terms. While a type is something that intuitively has "elements," and the type theory provides rules for constructing terms of that type, a proposition is something that intuitively has a "truth value," and the type theory provides rules for constructing proofs of that proposition.

We are presenting a higher-order logic, which means there is a special type `Prop`, and propositions are identified with terms of type `Prop`. We will describe how to build new propositions from old in Sect. 4.1.2.

Before getting to that, we will quickly explain premises and truth judgments. A *premise* is like a context: it is a finite list of propositions. That is, if $P_1, \ldots, P_n$ are valid propositions, then $\Theta = \{P_1, \ldots, P_n\}$ is a premise. Finally a *truth judgment* consists of a context Γ, a premise Θ, and a proposition Q; it looks like this:

$$\Gamma \mid \Theta \vdash Q. \tag{4.4}$$

Equation (4.4) basically says that, in the context Γ (where all the variables used in $P_1, \ldots, P_n$ and Q are defined), the conjunction of P_1 through P_n is enough to conclude Q. But we will explain this more in the next section, where we discuss how one decides which truth judgments are valid.

4.1.2 Higher Order Logic

Logic is a set of rules about propositions and predicates, by which we can decide validity. We discussed the type `Prop` of propositions above; a *predicate* is just a term of type `Prop`, say $\Gamma \vdash P : \texttt{Prop}$. We often refer to P simply as a proposition, since it is a term of type `Prop`, rather than as a predicate. In other words, the fact that there is a context—possibly empty—is assumed, as usual.

Just like a type theory begins with a set of atomic types and a set of atomic terms, it also begins with a set of atomic predicates, again of the form $\Gamma \vdash P : \texttt{Prop}$. Together the atomic types, terms, and predicates make up a signature.

Obtaining New Propositions from Old

One important way to form a proposition is by equality. Namely, given a type τ and terms $a : \tau$ and $b : \tau$, there is a proposition $(a = b) : \texttt{Prop}$. There are also the logical connectives: given propositions $\Gamma \vdash P : \texttt{Prop}$ and $\Gamma \vdash Q : \texttt{Prop}$, we have

$$\top : \texttt{Prop} \qquad \bot : \texttt{Prop} \qquad P \wedge Q : \texttt{Prop} \qquad P \vee Q : \texttt{Prop} \qquad P \Rightarrow Q : \texttt{Prop}$$

all in the context Γ.[3] These are pronounced *true, false, P and Q, P or Q*, and *P implies Q*, respectively. Given a predicate $\Gamma, x : \tau \vdash P : \texttt{Prop}$, one also has the quantifiers

$$\Gamma \vdash \forall(x : \tau).\, P : \texttt{Prop} \qquad \Gamma \vdash \exists(x : \tau).\, P : \texttt{Prop}$$

pronounced *for all x of type τ, P* and *there exists x of type τ such that P*. Note that, just like lambda abstraction binds a variable in a term, the quantifiers bind a variable in a proposition.

Before moving on, we note a common convention to reduce parenthetical clutter: a sequence of implications is parsed by "right associativity." That is, $P \Rightarrow (Q \Rightarrow R)$ can be written simply as $P \Rightarrow Q \Rightarrow R$. Once we have rules for deduction, we will be able to deduce that $P \Rightarrow Q \Rightarrow R$ is equivalent to $(P \wedge Q) \Rightarrow R$.

[3]The propositions $\neg P$ and $P \Leftrightarrow Q$ are just shorthands for $P \Rightarrow \bot$ and $(P \Rightarrow Q) \wedge (Q \Rightarrow P)$.

Valid Truth Judgments

To do logic, one begins with a set of axioms, and proceeds to judge which propositions are true. The technical form of this is the truth judgment Eq. (4.4). In this section we give the rules for determining which truth judgments are valid in a given premise context $\Gamma \mid \Theta$, given a set $\mathcal{A}$ of axioms. Validity of a truth judgment is inductively *derived* using certain allowable steps, called *proof rules*. We divide the proof rules into four main subgroups: premises and axioms, equalities, the connectives, and the quantifiers.

Premises and Axioms An axiom is a truth judgment, i.e. a statement of the form $\Gamma \mid \Theta \vdash P$. Axioms are automatically valid as truth judgments. Similarly, $\top$ is always valid. If $P : \texttt{Prop}$ is in the list of propositions defining a premise Θ, then $\Gamma \mid \Theta \vdash P$ is valid. Technically, these three rules could be written as follows:

$$\frac{\Gamma \vdash \Theta\ \mathit{Prem}}{\Gamma \mid \Theta \vdash \top} \qquad {\scriptstyle P\in\Theta}\ \frac{\Gamma \vdash \Theta\ \mathit{Prem}}{\Gamma \mid \Theta \vdash P} \qquad {\scriptstyle (\Gamma\mid\Theta\vdash P)\in\mathcal{A}}\ \frac{}{\Gamma \mid \Theta \vdash P} \tag{4.5}$$

The first says that if Θ is a valid premise in context Γ (i.e., if all the symbols used in the propositions in Θ are defined in Γ), then $\Gamma \mid \Theta \vdash \top$ is valid. The second is similar, but adds a side condition that P is one of the propositions in Θ, in which case $\Gamma \mid \Theta \vdash P$ is valid. The third says that axioms are automatically valid.

As expected, one can prove by induction that if one increases the premise by adding a new proposition, the set of valid truths one can derive does not decrease. This is called *weakening*. Thus if $\Gamma \vdash Q : \texttt{Prop}$ is a term and $\Gamma \mid \Theta \vdash Q$ is a valid truth judgment, then so is $\Gamma \mid \Theta, P \vdash Q$ for any proposition P in the same context, $\Gamma \vdash P : \texttt{Prop}$.

Equalities Recall the notion of conversion rules from Sect. 4.1.1, page 51. These are of the form $\Gamma \vdash e_1 \equiv e_2 : \tau$. Every such conversion rule gives rise to a valid truth judgment $\Gamma \vdash e_1 = e_2$. Written more technically,

$$\frac{\Gamma \vdash \Theta\ \mathit{Prem} \qquad \Gamma \vdash e_1 \equiv e_2 : \tau}{\Gamma \mid \Theta \vdash e_1 = e_2}$$

New valid truth judgments of the form $\Gamma \vdash e_1 = e_2$ can be generated from old by reflexivity, symmetry, and transitivity, as well as by substitution. For example, we also have

$$\frac{\Gamma \mid \Theta \vdash e_1 = e_2 \qquad \Gamma \mid \Theta \vdash e_2 = e_3}{\Gamma \mid \Theta \vdash e_1 = e_3}$$

The Connectives Most of the valid truth judgments coming from connectives and quantifiers are fairly obvious, but there are differences between classical logic and constructive logic, so we should be a bit careful. This difference only makes an appearance in disjunction $\vee$ and existential quantification $\exists$. Instead of starting there, we begin with a familiar case, namely conjunction. The rules for conjunction are

$$\frac{\Gamma \mid \Theta \vdash \phi_1 \quad \Gamma \mid \Theta \vdash \phi_2}{\Gamma \mid \Theta \vdash \phi_1 \wedge \phi_2} \qquad \frac{\Gamma \mid \Theta \vdash \phi_1 \wedge \phi_2}{\Gamma \mid \Theta \vdash \phi_1} \qquad \frac{\Gamma \mid \Theta \vdash \phi_1 \wedge \phi_2}{\Gamma \mid \Theta \vdash \phi_2}$$

These correspond to mathematical idioms such as "We have ϕ_1 and ϕ_2, so in particular we have ϕ_1."

The rules for implication are:

$$\frac{\Gamma \mid \Theta, \phi_1 \vdash \phi_2}{\Gamma \mid \Theta \vdash \phi_1 \Rightarrow \phi_2} \qquad \frac{\Gamma \mid \Theta \vdash \phi_1 \Rightarrow \phi_2 \quad \Gamma \mid \Theta \vdash \phi_1}{\Gamma \mid \Theta \vdash \phi_2}$$

The first corresponds to the mathematical idiom "To prove $\phi_1 \Rightarrow \phi_2$, we assume ϕ_1 and attempt to show ϕ_2," while the second corresponds to the idiom "We know that ϕ_1 implies ϕ_2, so since ϕ_1 holds, so does ϕ_2."

The constant $\bot$, representing falsehood, may seem at first sight rather useless, but it plays an important role in constructive logic. Its only rule is the following, which says intuitively "if we can prove false, then we can prove anything":

$$\frac{\Gamma \vdash \phi : \texttt{Prop} \quad \Gamma \mid \Theta \vdash \bot}{\Gamma \mid \Theta \vdash \phi}$$

In constructive logic, negation is not a primitive connective. Rather, $\neg P$ is simply shorthand for $P \Rightarrow \bot$. Thus $\neg P$ means precisely "if P were true, we could prove false." So trivially, we have that $P \Rightarrow \bot$ implies $\neg P$, but it is emphatically *not* the case that $\neg P \Rightarrow \bot$ implies P. That is, in constructive logic there is no "proof-by-contradiction": just because assuming P is false leads to a contradiction, that does not provide a proof that P is true. Instead, all we can say is that $\neg P \Rightarrow \bot$ implies $\neg\neg P$. As an exercise, the reader might try proving that $\neg\neg\neg P$ implies $\neg P$.

There are three rules for disjunction. The first corresponds to the idiom "We know that either ϕ_1 or ϕ_2 holds. Either way we can prove ϕ_3, so ϕ_3 holds."

$$\frac{\Gamma \mid \Theta \vdash \phi_1 \vee \phi_2 \quad \Gamma \mid \Theta, \phi_1 \vdash \phi_3 \quad \Gamma \mid \Theta, \phi_2 \vdash \phi_3}{\Gamma \mid \Theta \vdash \phi_3}$$

The next two are more obvious, but note that—other than via axioms—there is no way to derive $\phi \vee \neg\phi$. The only way to derive a disjunction is to derive one of the disjuncts. Similarly, given $\phi_1 \Rightarrow \phi_2$, it does not necessarily follow that $\neg\phi_1 \vee \phi_2$.

$$\frac{\Gamma \vdash \phi_2 : \texttt{Prop} \quad \Gamma \mid \Theta \vdash \phi_1}{\Gamma \mid \Theta \vdash \phi_1 \vee \phi_2} \qquad \frac{\Gamma \vdash \phi_1 \texttt{Prop} \quad \Gamma \mid \Theta \vdash \phi_2}{\Gamma \mid \Theta \vdash \phi_1 \vee \phi_2}$$

The Quantifiers Like lambda-abstraction, the quantifiers change the context by binding variables. Here are the proof rules for the universal quantifier:

$$x \notin \mathrm{fv}(\Theta)\ \frac{\Gamma, x : \tau \mid \Theta \vdash \phi}{\Gamma \mid \Theta \vdash \forall(x : \tau).\, \phi} \qquad \frac{\Gamma \mid \Theta \vdash \forall(x : \tau).\, \phi(x) \qquad \Gamma \vdash e : \tau}{\Gamma \mid \Theta \vdash \phi(e)}$$

The first rule corresponds to the mathematical idiom, "To prove $\forall(x : \tau).\, \phi$, it suffices to take an arbitrary element x of type τ and prove ϕ." The word "arbitrary" signifies that x is not already being used in the assumptions Θ, which is formalized by saying x is not in the set $\mathrm{fv}(\Theta)$ of free variables occurring in Θ. The second rule corresponds to the idiom, "Since ϕ holds for all x of type τ, it holds in particular for e."

Finally, we have the existential quantifier. The first proof rule corresponds to the idiom, "To prove that $\exists(x : \tau).\, \phi$ holds, it suffices to find a witness e such that $\phi(e)$ holds."

$$\frac{\Gamma, x : \tau \vdash \phi(x) : \texttt{Prop} \qquad \Gamma \vdash e : \tau \qquad \Gamma \mid \Theta \vdash \phi(e)}{\Gamma \mid \Theta \vdash \exists(x : \tau).\, \phi(x)}$$

The second proof rule corresponds to the idiom, "We know that there exists an x for which ϕ_1 holds, and we know that for an arbitrary x, if ϕ_1 holds then so does ϕ_2, so it follows that ϕ_2 also holds."

$$x \notin \mathrm{fv}(\Theta, \phi_2)\ \frac{\Gamma \mid \Theta \vdash \exists(x : \tau).\, \phi_1 \qquad \Gamma, x : \tau \mid \Theta, \phi_1 \vdash \phi_2}{\Gamma \mid \Theta \vdash \phi_2}$$

Remark 4.2 Higher order logic is extremely expressive, and the logic we have presented has considerable redundancy. In fact, with the higher order proposition type `Prop`, only implication and universal quantification are needed. That is, we have the following equivalences:

$$\begin{aligned}
\bot &\Leftrightarrow \forall(\alpha : \texttt{Prop}).\, \alpha \\
\top &\Leftrightarrow \bot \Rightarrow \bot \\
\phi \vee \psi &\Leftrightarrow \forall(\alpha : \texttt{Prop}).\, (\phi \Rightarrow \alpha) \Rightarrow (\psi \Rightarrow \alpha) \Rightarrow \alpha \\
\phi \wedge \psi &\Leftrightarrow \forall(\alpha : \texttt{Prop}).\, (\phi \Rightarrow \psi \Rightarrow \alpha) \Rightarrow \alpha \\
\exists(x : \tau).\, \phi &\Leftrightarrow \forall(\alpha : \texttt{Prop}).\, (\forall(x : \tau).\, \phi \Rightarrow \alpha) \Rightarrow \alpha.
\end{aligned}$$

Even equality can be defined in this way. Given terms $e_1, e_2 : \tau$,

$$e_1 = e_2 \quad \Leftrightarrow \quad \forall(P : \tau \to \texttt{Prop}).\, P(e_1) \Rightarrow P(e_2).$$

We mention this to give the reader an idea of some of the unexpected consequences of the type `Prop`. However, the rules stated in Sect. 4.1.2, before this remark, are easier to work with and—unlike the above higher-order reformulations—trivially generalize to the non-higher-order setting.

The Extensionality Axioms

Propositional extensionality says that equivalent propositions are automatically equal: $\forall(P, Q : \texttt{Prop}).(P \Leftrightarrow Q) \Rightarrow (P = Q)$. This is an axiom that we will assume throughout the book. It is sound in any Grothendieck topos because propositions correspond externally to closed sieves (see, e.g., [MM92, Prop III.3.3], which form a partial order, not just a preorder). Moreover, propositional extensionality is assumed in proof assistants like Coq and Lean.

Function extensionality, like propositional extensionality, is sound in any topos and commonly assumed in proof assistants. It says that two functions $X \to Y$ are equal if and only if their values are equal on all $x : X$.

Axiom 0 (Extensionality Axioms)

Propositional extensionality

$$P : \texttt{Prop}, Q : \texttt{Prop} \mid P \Leftrightarrow Q \vdash P = Q$$

Function extensionality Let X and Y be types. Then

$$f : X \to Y, g : X \to Y \mid \forall(x : X).\, fx = gx \vdash f = g$$

Using Eq. (4.5), the extensionality axioms can be used in derivations of valid truth judgments. The specific type signature we give in Chap. 5 will contain ten more axioms, all of which fit into the logic at the same point as this one. Again, the reason we do not include propositional extensionality with the other ten is that it is sound in any topos, and is not specific to ours.

Often we will write axioms, propositions, definitions, etc. more informally, e.g. replacing the formal syntax of the statements in Axiom 0 with something like the following.

Propositional extensionality Suppose that P and Q are propositions, and assume that $P \Leftrightarrow Q$ holds. Then $P = Q$.

Function extensionality Suppose that X and Y are types and that $f, g : X \to Y$ are functions. If $\forall(x : X).\, fx = gx$ holds, then $f = g$.

4.1.3 Subtypes and Quotient Types

Here we add two new type constructors: subtypes and quotient types. We could not have covered these earlier, when discussing the other type constructors, because these two depend on the logical layer.

Subtypes Let τ be a type, and suppose given a predicate in the one-variable context, $x : \tau \vdash \phi : \texttt{Prop}$. The corresponding subtype is denoted $\{x : \tau \mid \phi\}$.[4]

There are two term constructors, two conversion rules, and a truth judgment for subtypes. The first term constructor says that if $\Gamma \vdash e : \tau$ is a term and $\Gamma \mid \varnothing \vdash \phi(e)$ is a valid truth judgment, then there is a new term $\Gamma \vdash \mathsf{i}(e) : \{x : \tau \mid \phi\}$. The second term constructor says that if $\Gamma \vdash e' : \{x : \tau \mid \phi\}$ then there is a new term $\Gamma \vdash \mathsf{o}(e') : \tau$. Then conversion rules say $\mathsf{o}(\mathsf{i}(e)) \equiv e$ and $\mathsf{i}(\mathsf{o}(e')) \equiv e'$. However, in practice it is more convenient to simply drop the i's and o's.

The truth judgment for subtypes is that if $\Gamma, x : \tau \mid \Theta(x), \phi(x) \vdash \psi(x)$ is a valid truth judgment, then so is $y : \{z : \tau \mid \phi(z)\} \mid \Theta(\mathsf{o}(y)) \vdash \psi(\mathsf{o}(y))$. With the informal notation dropping i's and o's, this can be simplified to $y : \{z : \tau \mid \phi(z)\} \mid \Theta(y) \vdash \psi(y)$.

Remark 4.3 We saw in Remark 4.2 that all of the first-order logic connectives can be defined in terms of just $\Rightarrow$ and $\forall$, by making clever use of the higher-order proposition type `Prop`. With the addition of subtypes, this surprising expressivity of higher-order logic extends to the type theory. For example, the rules for sum types are redundant. The sum of two types A and B can be *defined* as

$$\begin{aligned} A + B \quad := \quad & \{(\phi, \psi) : (A \to \texttt{Prop}) \times (B \to \texttt{Prop}) \mid \\ & (\text{is\_sing}(\phi) \wedge \text{is\_empty}(\psi)) \vee (\text{is\_empty}(\phi) \wedge \text{is\_sing}(\psi))\} \end{aligned} \tag{4.6}$$

where

$$\begin{aligned} \text{is\_sing}(\phi) &:= \exists(a : A).\, \phi(a) \wedge \forall(a' : A).\, \phi(a') \Rightarrow a = a' \\ \text{is\_empty}(\phi) &:= \forall(a : A).\, \neg\phi(a). \end{aligned}$$

This can be made precise in two ways. First, if we have explicit sum types in the type theory, then it is possible to construct an isomorphism between the "real" sum $A + B$ and the type defined in (4.6). Second, even without explicit sum types, it is possible to prove that the type defined in (4.6) satisfies all of the rules that define a sum type.

Again, it is more convenient to just use the explicit sum type rules. However, there are properties of sum types which hold in a higher-order logic—and which can be proven from (4.6)—which are not provable in non-higher-order logic. We will see an example of such a property in Corollary 5.8, where we show that the sum of two types with decidable equality has decidable equality.

[4]It is natural to want to extend this idea to contexts with more than one variable, but that takes us into dependent types; see Sect. 4.1.4.

Quotient Types Let τ be a type, and suppose given a predicate in the two-variable context, also known as a binary relation, $x : \tau, y : \tau \vdash R(x, y) : \texttt{Prop}$. The corresponding quotient type is denoted τ/R.[5]

There are two term constructors, two conversion rules, and a truth judgment for quotient types. The first term constructor says that if $\Gamma \vdash e : \tau$ is a term, then there is a new term $\Gamma \vdash [e]_R : \tau/R$. We read $[e]_R$ as "the equivalence class of e." The second says that if $\Gamma, z : \tau \vdash e'(z) : \tau'$ is a term and $\Gamma, x : \tau, y : \tau \mid R(x, y) \vdash e'(x) = e'(y)$ is a valid truth judgment, then there is a new term $\Gamma, a : \tau/R \vdash \textsf{pick } x \textsf{ from } a \textsf{ in } e'(x) : \tau'$. The intuition is that a represents an equivalence class, so we can pick some representative x from a and form the term $e'(x)$, with the guarantee that it doesn't matter which representative we choose. The first conversion rule says $\textsf{pick } x \textsf{ from } [e]_R \textsf{ in } e'(x) \equiv e'(e)$, and the second says $\textsf{pick } x \textsf{ from } Q \textsf{ in } e'([x]_R) \equiv e'(Q)$.

The truth judgment for quotient types is that if $\Gamma \vdash e : \tau$ and $\Gamma \vdash e' : \tau$ are terms, then $\Gamma \mid R(e, e') \vdash [e]_R = [e']_R$ is a valid truth judgment.

Example 4.4 Recall the definition of addition and multiplication of natural numbers from Example 4.1. We can also define the inequality $\leq$ for natural numbers, $m : \mathbb{N}, n : \mathbb{N} \vdash \texttt{leq}(m, n)$, to be the term $\texttt{leq}(m, n) := \exists(p : \mathbb{N}). n + p = m$. Write $m \leq n$ if $\texttt{leq}(m, n)$.

With quotient types in hand, we can define the types $\mathbb{Z}$ and $\mathbb{Q}$. For the former, let R be the following equivalence relation on $\mathbb{N} \times \mathbb{N}$,

$$(p, m) : \mathbb{N} \times \mathbb{N}, (p', m') : \mathbb{N} \times \mathbb{N} \vdash p + m' = p' + m : \texttt{Prop}$$

Then $\mathbb{Z}$ is defined to be the type $(\mathbb{N} \times \mathbb{N})/R$. One can construct addition, subtraction, and multiplication for $\mathbb{Z}$ using the term constructors for quotient types. For example, addition in $\mathbb{Z}$ is given by the term

$$z : \mathbb{Z}, z' : \mathbb{Z} \vdash \textsf{pick } (p, m) \textsf{ from } z \textsf{ in } (\textsf{pick } (p', m') \textsf{ from } z' \textsf{ in } [p{+}p', m{+}m']_R : \mathbb{Z}.$$

To construct this term—as mentioned above—one must check that the terms $p + p'$ and $m + m'$ are well-defined, i.e. independent of the choice of (p, m) and (p', m').

Similarly, for $\mathbb{Q}$, let $\mathbb{N}_+$ denote the subset type $\mathbb{N}_+ := \{n : \mathbb{N} \mid n \geq 1\}$. Then define S be the following equivalence relation on $\mathbb{Z} \times \mathbb{N}_+$,

$$(n, d) : \mathbb{Z} \times \mathbb{N}_+, (n', d') : \mathbb{Z} \times \mathbb{N}_+ \vdash d' * n = d * n' : \texttt{Prop}$$

and let $\mathbb{Q} = (\mathbb{Z} \times \mathbb{N}_+)/S$. Again, one can construct addition, subtraction, and multiplication for $\mathbb{Q}$, as well as the partial reciprocal function, using the term constructors for quotient types.

[5] As with subtypes, it is natural to want to extend the above idea to contexts with more than two variables of the same type, but doing so takes us into dependent types; see Sect. 4.1.4.

4.1.4 Dependent Types

Consider the inequality $x : \mathbb{N}, y : \mathbb{N} \vdash x \leq y$. Then it is natural to form the subtype $\mathbb{N}_{\leq y} := \{x : \mathbb{N} \mid x \leq y\}$. However, this type *depends* on a variable y, which the type theory we have sketched so far is unable to handle. Intuitively, it is best to think of the type $\{x : \mathbb{N} \mid x \leq y\}$ as a *family* of types, parameterized by $y : \mathbb{N}$. In particular, for each concrete term such as $4 : \mathbb{N}$ (where 4 is shorthand for $ssss0$; see page 50), there is a type $\mathbb{N}_{\leq 4} := \{x : \mathbb{N} \mid x \leq 4\}$ obtained by substitution $[y := 4]$.

While such type families are an intuitively natural and useful concept, formalizing dependent type theory (where types are allowed to depend on terms) is significantly more subtle than the simple type theory we have so far described. As an illustration of the extra difficulty, consider the type families $\{x : \mathbb{N} \mid x \leq y\}$ and $\{x : \mathbb{N} \mid x \leq (y + 0)\}$. We know that $y \equiv y + 0$ by the definition of addition—i.e. we consider y and $y + 0$ to be *the same* term—and as such we should consider $\mathbb{N}_{\leq y}$ and $\mathbb{N}_{\leq (y+0)}$ to be *the same* type, even though they are syntactically different. Thus in a dependent type theory, one must extend the conversion relation to types as well as terms.

In the theory presented in the rest of this book, there are a few occasions where we need to consider subtypes and quotient types which are technically only well-formed in a dependent type theory, such as $\mathbb{N}_{\leq y}$. We trust that it will be intuitively clear how to work with these types, and we refer the reader interested in the details of dependent type theory to sources such as [Jac99, Voe+13].

Recall the basic idea of the semantics of type theory in a topos $\mathcal{E}$: each type is assigned an object of $\mathcal{E}$, each term is assigned a morphism, and propositions $a : A \vdash \phi : \texttt{Prop}$ correspond to subobjects of A, i.e. monomorphisms $\{a : A \mid \phi\} \hookrightarrow A$. The semantics of a dependent type $a : A \vdash B$ is an arbitrary morphism $B \to A$. So one can think of dependent types as generalizing subtypes: a proposition that is dependent on a context determines a monomorphism into the context, whereas a type that is dependent on a context determines an arbitrary morphism into the context. For example, the dependent type $y : \mathbb{N} \vdash \mathbb{N}_{\leq y}$—which we think of as a family of types parameterized by y—is represented by the composite

$$\{x : \mathbb{N}, y : \mathbb{N} \mid x \leq y\} \hookrightarrow \mathbb{N} \times \mathbb{N} \xrightarrow{\text{pr}_2} \mathbb{N}.$$

A concrete member of the family, such as $\mathbb{N}_{\leq 4}$, is represented by a *fiber* of this map, e.g. $\mathbb{N}_{\leq 4}$ is represented by the pullback

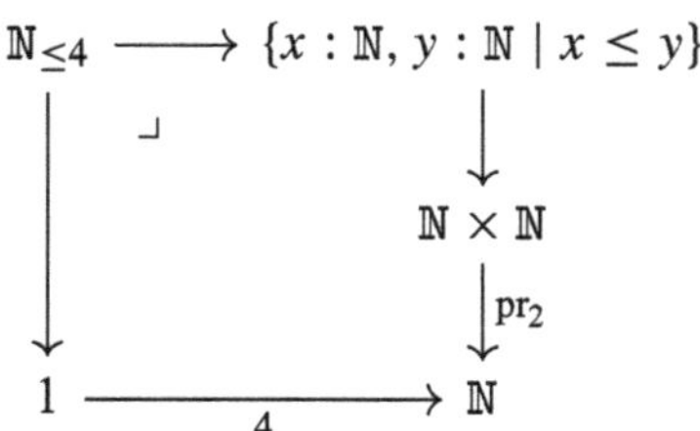

In general, a helpful intuition is that non-dependent types (with no free variables) are represented by objects of $\mathcal{E}$, while dependent types in the context $(x_1 : A_1, \ldots, x_n : A_n)$ are represented by objects of the slice category $\mathcal{E}/(A_1 \times \cdots \times A_n)$.

4.2 Modalities

Now that we have informally laid out the basics of the type theory we use in this book, we proceed to discuss some constructions one can do with it. In this section we discuss how modalities allow one to consider subtoposes within the type theory of the larger topos.

A modality is a term of type $\mathtt{Prop} \to \mathtt{Prop}$ satisfying a few axioms. Topos-theoretically, a modality is precisely what [MM92] calls a *Lawvere-Tierney topology* and what [Joh02] calls a *local operator*; we use the term "modality" to emphasize the logical viewpoint.

We define modalities in Sect. 4.2.1, discuss various related notions—such as closed propositions and sheaves—in Sect. 4.2.2, and finally discuss the relationship between modalities and subtoposes in Sect. 4.2.3.

4.2.1 *Definition of Modality*

In any topos, a modality is an internal monad on $\mathtt{Prop}$, considered as an internal poset. The definition can be given type-theoretically as follows:

Definition 4.5 A *modality* is a map $j : \mathtt{Prop} \to \mathtt{Prop}$ satisfying the following three conditions for any $P, Q : \mathtt{Prop}$,

- $P \Rightarrow jP$,
- $jjP \Rightarrow jP$, and
- $(P \Rightarrow Q) \Rightarrow jP \Rightarrow jQ$.

The third condition can be replaced by various equivalent statements, as we now show.

Lemma 4.6 *Suppose* $j : \mathtt{Prop} \to \mathtt{Prop}$ *satisfies* $P \Rightarrow jP$ *and* $jjP \Rightarrow jP$ *for any* $P : \mathtt{Prop}$. *Then the following are equivalent:*

1. $(P \Rightarrow Q) \Rightarrow jP \Rightarrow jQ$,
2. $jP \Rightarrow (P \Rightarrow jQ) \Rightarrow jQ$,
3. $j(P \Rightarrow Q) \Rightarrow jP \Rightarrow jQ$.
4. $(jP \wedge jQ) \Leftrightarrow j(P \wedge Q)$.

Proof 2 $\Rightarrow$ 1 and 3 $\Rightarrow$ 1 are easy. 1 $\Rightarrow$ 2 follows from $jjQ \Rightarrow jQ$. We next show 1 $\Rightarrow$ 3. But 3 is equivalent to $jP \Rightarrow j(P \Rightarrow Q) \Rightarrow jQ$, so first assume

jP. Assuming 1, we have $(P \Rightarrow Q) \Rightarrow jQ$. So if we apply 1 again on this last implication and assume $j(P \Rightarrow Q)$ then we have jjQ hence jQ.

Next we show 3 ⇒ 4. It is easy to show $j(P \wedge Q) \Rightarrow jP$ and hence $j(P \wedge Q) \Rightarrow (jP \wedge jQ)$. We need to show the converse $(jP \wedge jQ) \Rightarrow j(P \wedge Q)$, which is equivalent to $jP \Rightarrow jQ \Rightarrow j(P \wedge Q)$. Clearly $P \Rightarrow (Q \Rightarrow P \wedge Q)$, so applying j and using 3 with the assumption jP and again with jQ, we obtain $j(P \wedge Q)$ as desired.

Finally for 4 ⇒ 1, note that $(P \Rightarrow Q)$ is equivalent to $(P \wedge Q) \Leftrightarrow P$ and hence to $(P \wedge Q) = P$ by propositional extensionality, Axiom 0. Assuming $(P \wedge Q) = P$ and jP we have $j(P \wedge Q)$ and hence jQ by 4. □

Example 4.7 Modalities can be ordered by reverse implication, $j \leq j'$ iff $j'P \Rightarrow jP$ for all $P : \texttt{Prop}$. The top element in this ordering is the identity modality $P \mapsto P$. The bottom element is the constant modality $P \mapsto \top$. Somewhere in between is the double-negation modality $P \mapsto \neg\neg P$. We will discuss the relationship between modalities and subtoposes in Sect. 4.2.3.

4.2.2 *j*-Closed Propositions, *j*-Separated Types, and *j*-Sheaves

Given a modality j and a type X, we can internally express the semantic notion that the sheaf associated to X is j-separated or a j-sheaf.

Definition 4.8 A proposition $P : \texttt{Prop}$ is called *j-closed* if it satisfies $jP \Rightarrow P$. A predicate $P : X \to \texttt{Prop}$ is called *j-closed* if $\forall (x : X).\, jPx \Rightarrow Px$.

For any modality j, the reflection sending a proposition to a j-closed proposition is given by $P \mapsto jP$.

The following remark is straightforward to verify but *very useful in practice*. We hope the reader takes careful note of it, because we will use it often. Reading a proof and wondering "where did the j's go?" is a sign that the reader should revisit this remark. The goal is to understand how logic in the j-subtopos is constructed in the larger topos.

Remark 4.9 (j-Logic) Let j be a modality. If $P : \texttt{Prop}$ is written as a complex expression, jP can often be simplified recursively. For example, we have the following equivalences and implications:

$$j\top \Leftrightarrow \top \qquad j(\phi \wedge \psi) \Leftrightarrow j\phi \wedge j\psi \qquad j(\psi \Rightarrow \phi) \Rightarrow (\psi \Rightarrow j\phi)$$
$$j\forall (x : X).\, \phi(x) \Rightarrow \forall (x : X).\, j\phi(x) \qquad \exists (x : X).\, j\phi(x) \Rightarrow j\exists (x : X).\, \phi(x) \tag{4.7}$$

Now assume ϕ is j-closed. We can replace two of the implications by equivalences:

$$j(\psi \Rightarrow \phi) \Leftrightarrow (\psi \Rightarrow \phi) \qquad j\forall (x : X).\, \phi(x) \Leftrightarrow \forall (x : X).\, \phi(x)$$

and, quite usefully, when proving ϕ we can drop j from the front of all hypotheses:

$$\big((\psi_1 \wedge \cdots \wedge \psi_n) \Rightarrow \phi\big) \Leftrightarrow \big((j\psi_1 \wedge \cdots \wedge j\psi_n) \Rightarrow \phi\big).$$

Here is an example of a proof that uses the above remark. We go through it slowly.

Proposition 4.10 *Suppose* $P : \texttt{Prop}$ *is decidable, i.e.* $P \vee \neg P$ *holds. For any* $Q : \texttt{Prop}$ *we have* $j(P \vee Q) \Leftrightarrow P \vee jQ$.

Proof For one direction we suppose $j(P \vee Q)$ and prove $P \vee jQ$. If P holds we are done, so suppose $\neg P$. We will prove jQ, so by Remark 4.9 we can drop j from the hypothesis, at which point we have $P \vee Q$ and thus Q.

The other direction is easier and does not depend on P being decidable. In case P, we have $P \vee Q$ so $j(P \vee Q)$. In case jQ we can drop the j, and Q implies $P \vee Q$ and hence $j(P \vee Q)$. □

Definition 4.11 (j-Separated, j-Sheaf) A type X is *j-separated* iff it satisfies $\forall (x, x' : X).\, j(x = x') \Rightarrow (x = x')$. The *$j$-separification of* X, denoted $\mathsf{sep}_j X$ is the quotient of X by the internal equivalence relation $j(x = x')$. The *j-sheafification of* X, denoted $\mathsf{sh}_j(X)$, is the subtype of predicates $\phi : X \to \texttt{Prop}$ satisfying

j-closed predicate: $\forall (x : X).\, j\phi(x) \Rightarrow \phi(x)$,
j-local singleton: $j\exists (x : X).\, \forall (x' : X).\, \phi(x') \Leftrightarrow j(x = x')$.

If X is separated, one may drop the inner j from the j-local singleton condition. If $\{\cdot\} : X \to (X \to \texttt{Prop})$ is the usual singleton function, then the j-local singleton condition is equivalent to $j\exists (x : X).\, j(\phi = \{x\})$.

There is a map $\eta_j : X \to \mathsf{sh}_j(X)$; it sends $x : X$ to the predicate $\eta_j(x) : X \to \texttt{Prop}$ given on $x' : X$ by

$$\eta_j(x)(x') := j(x = x'). \tag{4.8}$$

It is easy to check that $\eta_j(x)$ is j-closed and j-locally singleton, and X is j-separated iff η_j is an internal injection. X is called a *j-sheaf* if η_j is also an internal surjection, i.e. satisfies $\forall (\phi : \mathsf{sh}_j(X)).\, \exists (x : X).\, \phi = \eta_j(x)$.

Proposition 4.12 *A type* X *is a* j*-sheaf iff it is* j*-separated and, for any* j*-closed predicate* $\phi : X \to \texttt{Prop}$ *the singleton condition on* ϕ *is* j*-closed in the sense that*

$$j\big(\exists (x : X).\, \forall (x' : X).\, \phi(x') \Leftrightarrow (x = x')\big) \Rightarrow \exists (x : X).\, \forall (x' : X).\, \phi(x') \Leftrightarrow (x = x'). \tag{4.9}$$

Proof By Definition 4.11, we may assume X is j-separated, and we need to show that η_j is surjective iff Eq. (4.9) holds for every j-closed predicate ϕ. First suppose η_j is surjective and that ϕ is a j-closed predicate that satisfies the hypothesis of implication (4.9). Then by definition we have $\phi : \mathsf{sh}_j(X)$, so $\exists (x : X).\, \phi = \eta_j(x)$, which is the conclusion of (4.9). For the other direction, if every j-closed predicate $\phi : X \to \texttt{Prop}$ satisfies (4.9), then η_j is surjective. □

Proposition 4.13 *For any modality j, the type* $\texttt{Prop}_j := \{P : \texttt{Prop} \mid jP \Rightarrow P\}$ *is a j-sheaf.*

Proof It is easy to check that $\texttt{Prop}_j$ is j-separated. So choose a j-closed predicate $\phi : \texttt{Prop}_j \to \texttt{Prop}$ and assume the hypothesis of (4.9),

$$j\big(\exists(x : \texttt{Prop}_j).\, \forall(x' : \texttt{Prop}_j).\, \phi(x') \Leftrightarrow (x = x')\big). \tag{4.10}$$

To obtain the existential statement required for the conclusion of (4.9), begin by defining $\overline{x} := \forall(x' : \texttt{Prop}_j).\, \phi(x') \Rightarrow x'$, which is j-closed by Remark 4.9. It is our candidate for the required existential; i.e. using Proposition 4.12 we will be done if we can show $\forall(x' : \texttt{Prop}_j).\, \phi(x') \Leftrightarrow (\overline{x} = x')$. One checks using Remark 4.9 that this statement is j-closed, so we can drop the j from our hypothesis (4.10). Thus we have some $x : \texttt{Prop}_j$ satisfying $\forall(x' : \texttt{Prop}_j).\, \phi(x') \Leftrightarrow (x = x')$. Note in particular that $\phi(x)$ holds.

First we show $\phi(\overline{x})$. By definition of $\overline{x}$ we have $\overline{x} \Rightarrow x$. But we also have $x \Rightarrow \overline{x}$ because for any $x' : \texttt{Prop}_j$ such that $\phi(x')$, we have $x = x'$. Thus $x = \overline{x}$, so $\phi(\overline{x})$, as desired.

Choosing $x' : \texttt{Prop}_j$, it remains to show that $\phi(x') \Rightarrow (\overline{x} = x')$. But given $\phi(x')$ and the already-established $\phi(\overline{x})$, we have $\overline{x} = x = x'$ by hypothesis, so we are done. □

Proposition 4.14 *Given a j-sheaf X and a j-closed predicate $\psi : X \to \texttt{Prop}_j$, the subtype $\{x : X \mid \psi x\}$ is a j-sheaf.*

Proof Any subtype of a separated type is separated, so it suffices to prove (4.9) for any $\phi : \{x : X \mid \psi x\} \to \texttt{Prop}_j$. The hypothesis says $j\exists(x : X).\, \psi x \wedge \forall(x' : X).\, \psi x' \Rightarrow (\phi x \Leftrightarrow (x = x'))$. It follows easily that $j\exists(x : X).\, \forall(x' : X).\, (\psi x' \wedge \phi x) \Leftrightarrow (x = x')$, and we may drop the j because X is a j-sheaf. The conclusion of (4.9) follows directly. □

Proposition 4.15 *The product of j-sheaves is a j-sheaf.*

Proof We use Proposition 4.12. It is easy to see that the product of separated sheaves is separated. Supposing $j\exists((x_1, x_2) : X_1 \times X_2).\, \forall((x_1', x_2') : X_1 \times X_2).\, \phi(x_1', x_2') \Leftrightarrow (x_1, x_2) = (x_1', x_2')$, one proves

$$\begin{aligned} &j\exists(\phi_1 : X_1 \to \texttt{Prop})(\phi_2 : X_2 \to \texttt{Prop}).\, \forall(x_1' : X_1)(x_2' : X_2).\, \phi(x_1', x_2') \\ &\quad \Leftrightarrow \phi_1(x_1') \wedge \phi_2(x_2') \end{aligned}$$

by taking (x_1, x_2) from the hypothesis and letting $\phi_i(x_i') := (x_i = x_i')$ for $i = 1, 2$. We now obtain $j\exists(x_i : X_i)\forall(x_i' : X_i).\, \phi_1(x_i') \Leftrightarrow (x_i = x_i')$, and the result follows from Proposition 4.12. □

Proposition 4.16 *Let X a type. Suppose that j is a modality satisfying*

$$\forall(P : X \to \texttt{Prop}).\, \big(j\exists(x : X).\, Px\big) \Rightarrow \big(\exists(x : X).\, jPx\big).$$

Then the map $\eta_j : X \to \mathsf{sh}_j(X)$ is surjective.

Proof Suppose $\phi : X \to \texttt{Prop}_j$ satisfies $j\exists(x : X).\, \forall(x' : X).\, \phi(x') \Leftrightarrow (x = x')$. Then by hypothesis, there is some $x : X$ satisfying $j\forall(x' : X).\, \phi(x') \Leftrightarrow (x = x')$, and this implies $\forall(x' : X).\, j(\phi(x') \Leftrightarrow (x = x'))$, which in turn implies $\forall(x' : X).\, \phi(x') \Leftrightarrow j(x = x')$. □

Proposition 4.17 *Suppose X has decidable equality. If j is a dense modality (i.e., $j\bot = \bot$), then X is j-separated.*

Proof If X has decidable equality, then it is easy to check $j(x_1 = x_2) \Rightarrow (x_1 = x_2)$. □

Proposition 4.18 *Suppose X decidable equality. Let j be a modality and let $X \to \mathrm{sep}_j(X)$ denote its j-separification. Then the natural map below is an isomorphism:*

$$(\mathrm{sep}_j(X) \to \texttt{Prop}_j) \xrightarrow{\cong} (X \to \texttt{Prop}_j).$$

Proof The map $f : (X \to \texttt{Prop}_j) \to (\mathrm{sep}_j(X) \to \texttt{Prop}_j)$, applied to any $P : X \to \texttt{Prop}_j$ is given as follows. Any $y : \mathrm{sep}_j(X)$ is represented by some $x : X$, so we may let $f(P)(y) = P(x)$ and it remains to prove that this respects the equivalence relation: $j(x = x') \Rightarrow (P(x) = P(x'))$. $x = x'$ implies $Px \Leftrightarrow Px'$, so it suffices to see that $Px \Leftrightarrow Px'$ is j-closed; see Eq. (4.7). □

4.2.3 Modalities and Subtoposes

In this section we discuss how the logical notion of modality corresponds to the semantic notion of subtopos. Much of the following is taken from [Joh02, A.4.4 and A.4.5]; again, what we call modalities are called *local operators* there.

There are several equivalent definitions of subtoposes of $\mathcal{E}$. One is a subcategory $\mathcal{E}' \subseteq \mathcal{E}$ with a left-exact left-adjoint. Another is a modality j on $\mathcal{E}$, by which one can define the subcategory $\mathcal{E}_j$ of j-sheaves and the left-exact left adjoint sh_j defined in Definition 4.11.

As mentioned in Example 4.7, there is a partial order on modalities: $j \leq j'$ iff $\forall(P : \texttt{Prop}).\, j'P \Rightarrow jP$. With this partial order, modalities on a topos $\mathcal{E}$ form a lattice (in fact a co-Heyting algebra), and there is a poset isomorphism between that of modalities and that of subtoposes of $\mathcal{E}$ under inclusion.

To every proposition $U : \texttt{Prop}$ we can associate three modalities: the open modality $o(U)$, the closed modality $c(U)$, and the quasi-closed modality $q(U)$. A topos of sheaves for one of these is, respectively, called an *open subtopos*, a *closed subtopos*, and a *quasi-closed subtopos*. If $\mathcal{E} = \mathsf{Shv}(L)$ for some locale L, then U can be identified with an open subspace $U \subseteq L$, and its topos of sheaves is equivalent to $\mathsf{Shv}(L)_{o(U)}$. Similarly, $\mathsf{Shv}(L)_{c(U)}$ is equivalent to the topos of sheaves on the closed subspace complementary to U, and $\mathsf{Shv}(L)_{q(U)}$ is equivalent to the topos of sheaves on the localic intersection of all dense subspaces of the complement of U. Logically, these modalities are defined as follows:

- $o(U)P := U \Rightarrow P$,
- $c(U)P := U \vee P$,
- $q(U)P := (P \Rightarrow U) \Rightarrow U$.

We say that a modality j is *open* if there exists U : Prop such that $j = o(U)$, and similarly that j is *closed* or *quasi-closed* if there exists U : Prop such that $j = c(U)$ or $j = q(U)$.

Remark 4.19 In a boolean topos, the closed modality $U \vee P$ is the same as the open modality $\neg U \Rightarrow P$, but not in an arbitrary topos.

Example 4.20 Consider the case $U = \bot$, corresponding to the empty subtopos. The associated open modality $o(U)$ sends P to $\top$, i.e. it is the terminal modality. The associated closed modality $c(U)$ is the identity, i.e. it is the initial modality.

The most interesting is the associated quasi-closed modality $q(U)$, which in this case is double-negation, sending P to $\neg\neg P$. It corresponds to a boolean subtopos which is dense in $\mathcal{E}$. Semantically, the double-negation modality can be thought of as "almost always." Every quasi-closed modality $q(U)$ is a double-negation modality, relative to the closed modality $j = c(U)$, in the sense that $q(U)P = (P \Rightarrow j\bot) \Rightarrow j\bot$.

As mentioned above, j_1 corresponds to a subtopos of j_2 iff $j_2(P) \Rightarrow j_1(P)$ for all propositions P. For example, it is easy to check that for any proposition U, we have $\forall(P : \texttt{Prop}).\, c(U)(P) \Rightarrow q(U)(P)$. In the lattice of modalities, the join of j_1 and j_2 is given by $(j_1(P) \wedge j_2(P))$ of modalities. If terms c : C of a constant sheaf index modalities j_c, then $\forall(c : \mathtt{C}).\, j_c$ is a modality, and it corresponds to their C-indexed join.

If j_1 and j_2 are modalities, then it is not necessarily the case that $j_1 j_2$ will be a modality. It is easy to prove directly from definitions that $j_1 j_2$ is a modality if either j_1 is open or j_2 is closed. If $j_1 j_2$ is a modality, then it is the meet of j_1 and j_2. Modalities j_1 and j_2 are called *disjoint* if their meet j is the bottom element, i.e. if $j\bot$ holds.

4.3 Dedekind j-Numeric Types

We continue to work within the type theory discussed in Sect. 4.1. Using the notion of modality discussed in Sect. 4.2, we define various sorts of numeric objects that exist and are internally definable for any topos $\mathcal{E}$ and modality j.

4.3.1 Background

From our work in Sect. 4.1, and in particular Example 4.4, we have seen that the types $\mathbb{N}$, $\mathbb{Z}$, and $\mathbb{Q}$, together with their additive and multiplicative structures (see Example 4.1) exist in our type theory. It turns out that, semantically, the sheaves corresponding to $\mathbb{N}$, $\mathbb{Z}$, and $\mathbb{Q}$ are (locally) constant in any topos, corresponding to the images of the sets $\mathbb{N}$, $\mathbb{Z}$, $\mathbb{Q}$ under the inverse image part of the unique geometric morphism $\mathcal{E} \leftrightarrows \mathbf{Set}$.

One manifestation of this constancy is that the $=$ and $<$ relations on $\mathbb{N}$, $\mathbb{Z}$, or $\mathbb{Q}$ are decidable in any topos. Another is that one can get away with blurring the distinction between internal rational numbers—i.e. terms of the type $\mathbb{Q}$ in the internal language—and external rational numbers; similarly for $\mathbb{N}$ and $\mathbb{Z}$.

With $\mathbb{Q}$ in hand, the standard construction of the type $\mathbb{R}$ of real numbers from the rationals by Dedekind cuts can also be carried out in the type theory with good results. However, unlike $\mathbb{N}$, $\mathbb{Z}$, and $\mathbb{Q}$, one should not expect the type $\mathbb{R}$ of real numbers to be semantically constant in an arbitrary topos, and similarly one should expect neither the $=$ nor the $<$ relation to be decidable.[6] In particular one cannot generally regard a section of $\mathbb{R}$ as an ordinary real number. Perhaps the best intuition in an arbitrary topos is that terms of the type $\mathbb{R}$ correspond semantically to continuous real-valued functions on the topos (as a generalized topological space).

In Dedekind's construction of the real numbers [Ded72], a real number $x : \mathbb{R}$ is defined in terms of two subsets of rational numbers: those that are less than x and those that are greater than x. In our context, we replace the external notion, "two subsets of $\mathbb{Q}$" with the internal notion, "two subtypes of $\mathbb{Q}$". They are classified by predicates $\delta : \mathbb{Q} \to \texttt{Prop}$ and $\upsilon : \mathbb{Q} \to \texttt{Prop}$, which roughly correspond to the predicates "is less than x" and "is greater than x".[7]

These satisfy several axioms: for example, for any $q_1 < x$, there exists q_2 such that $q_1 < q_2$ and $q_2 < x$. This is translated into the logical statement

$$\forall(q_1 : \mathbb{Q}).\, \delta q_1 \Rightarrow \exists(q_2 : \mathbb{Q}).\, (q_1 < q_2) \wedge \delta q_2. \tag{4.11}$$

This is one of the usual Dedekind axioms—often called *roundedness*—which forces δ to act like $<$ rather than $\leq$. As mentioned above, the Dedekind axioms make sense in any topos $\mathcal{E}$, giving the standard construction of the type $\mathbb{R}$ of real numbers in $\mathcal{E}$; see, e.g., [MM92, VI.8].

However, we make two slightly non-standard moves. The first is that we will be interested in several numeric types—not just the real numbers—which are defined

[6] It will turn out that the type $\mathbb{R}$ of real numbers *is* constant in our main topos of interest, $\mathcal{B}$, but not in many of the subtoposes we consider. For example, the type $\mathbb{R}_\pi$ of real numbers in $\mathcal{B}_\pi$ does not have decidable equality, roughly because continuous functions that are unequal globally may become equal locally.

[7] Throughout this book, we use δ (delta) and d for "down" and υ (upsilon) and u for "up." Thus δ classifies a "down-set" of $\mathbb{Q}$, and υ classifies an "up-set."

by relaxing some of the Dedekind axioms. Doing so in the topos **Set** would yield a definition of the sets $\underline{\mathbb{R}}$, $\bar{\mathbb{R}}$, $\bar{\underline{\mathbb{R}}}$, $\mathbb{IR}$, and $\mathbb{R}$, discussed in Sect. 2.4.2; in this book, we will want to consider the analogous objects in other toposes.

The other non-standard move is that we want to work relative to an arbitrary modality j, as defined in Definition 4.5. To do so, we take each of the standard Dedekind axioms and apply j "at every position"; the result will be an axiom for the associated numeric type in the j-subtopos. For example, the roundedness axiom (4.11) becomes

$$j\forall(q_1 : \mathbb{Q}).\, j\big(j\delta q_1 \Rightarrow j\exists(q_2 : \mathbb{Q}).\, j(q_1 < q_2) \wedge j\delta q_2\big).$$

But if δ is a j-closed proposition, then we can use j-logic (see Remark 4.9) to find an equivalent statement that is much simpler:

$$\forall(q_1 : \mathbb{Q}).\, \delta q_1 \Rightarrow j\exists(q_2 : \mathbb{Q}).\, (q_1 < q_2) \wedge \delta q_2.$$

We make similar replacements of all the axioms from [MM92, VI.8].

The result of our two non-standard moves is that we obtain what we call the Dedekind j-numeric types, which we now formally define.

4.3.2 *Definition of the Dedekind j-Numeric Types*

In the definition below, one can simply remove the j's (i.e., use the identity modality $jP = P$) to obtain the usual notion of Dedekind real numbers, and related numeric types.

Definition 4.21 (Dedekind j-Numeric Types) Let $\mathcal{E}$ be a topos, let $\mathbb{Q}$ be the type of rational numbers, and let j be a modality on $\mathcal{E}$. Consider the following conditions on predicates $\delta : \mathbb{Q} \to \texttt{Prop}$ and $\upsilon : \mathbb{Q} \to \texttt{Prop}$,

0a. $\forall(q : \mathbb{Q}).\, j(\delta q) \Rightarrow \delta q$ 0b. $\forall(q : \mathbb{Q}).\, j(\upsilon q) \Rightarrow \upsilon q$
1a. $\forall q_1, q_2.\, (q_1 < q_2) \Rightarrow \delta q_2 \Rightarrow \delta q_1$ 1b. $\forall q_1, q_2.\, (q_1 < q_2) \Rightarrow \upsilon q_1 \Rightarrow \upsilon q_2$
2a. $\forall q_1.\, \delta q_1 \Rightarrow j\exists q_2.\, (q_1 < q_2) \wedge \delta q_2$ 2b. $\forall q_2.\, \upsilon q_2 \Rightarrow j\exists q_1.\, (q_1 < q_2) \wedge \upsilon q_1$
3a. $j\exists q.\, \delta q$ 3b. $j\exists q.\, \upsilon q$
4. $\forall q.\, (\delta q \wedge \upsilon q) \Rightarrow j\bot$
5. $\forall q_1, q_2.\, (q_1 < q_2) \Rightarrow j(\delta q_1 \vee \upsilon q_2)$

We refer to these conditions as j-closed (0), down/up-closed (1), j-rounded (2), j-bounded (3), j-disjoint (4), and j-located (5). It is easy to check that the j-disjointness condition is equivalent to $\forall(q_1, q_2 : \mathbb{Q}).\, \delta q_1 \Rightarrow \upsilon q_2 \Rightarrow j(q_1 < q_2)$.

When $j = \mathrm{id}$ is the trivial modality, we refer to the conditions simply as *rounded*, *bounded*, etc. In particular, axioms 0a and 0b can be dropped entirely when $j = \mathrm{id}$.

Table 4.1 Ten Dedekind j-numeric types

Name of type in $\mathcal{E}$	Notation	Definition
j-Local unbounded lower reals	$\underline{\mathbb{R}}_j^\infty$	$\{\delta \mid 0a, 1a, 2a\}$
j-Local unbounded upper reals	$\bar{\mathbb{R}}_j^\infty$	$\{\upsilon \mid 0b, 1b, 2b\}$
j-Local unbounded improper intervals	$\underline{\bar{\mathbb{R}}}_j^\infty$	$\{(\delta, \upsilon) \mid 0, 1, 2\}$
j-Local unbounded (proper) intervals	$\mathbb{IR}_j^\infty$	$\{(\delta, \upsilon) \mid 0, 1, 2, 4\}$
j-Local unbounded reals	$\mathbb{R}_j^\infty$	$\{(\delta, \upsilon) \mid 0, 1, 2, 4, 5\}$
j-Local lower reals	$\underline{\mathbb{R}}_j$	$\{\delta \mid 0a, 1a, 2a, 3a\}$
j-Local upper reals	$\bar{\mathbb{R}}_j$	$\{\upsilon \mid 0b, 1b, 2b, 3b\}$
j-Local improper intervals	$\underline{\bar{\mathbb{R}}}_j$	$\{(\delta, \upsilon) \mid 0, 1, 2, 3\}$
j-Local (proper) intervals	$\mathbb{IR}_j$	$\{(\delta, \upsilon) \mid 0, 1, 2, 3, 4\}$
j-Local real numbers	$\mathbb{R}_j$	$\{(\delta, \upsilon) \mid 0, 1, 2, 3, 4, 5\}$

We define ten real-number-like types, including the usual type of Dedekind real numbers $\mathbb{R}_j$ for the subtopos $\mathcal{E}_j$, by using various subsets of these axioms. We call these *Dedekind j-numeric types*; see Table 4.1. We refer to those having only one cut ($\underline{\mathbb{R}}_j$, $\bar{\mathbb{R}}_j$, $\underline{\mathbb{R}}_j^\infty$, and $\bar{\mathbb{R}}_j^\infty$) as *one-sided* and to those having both cuts ($\mathbb{R}_j$, $\mathbb{IR}_j$, $\underline{\bar{\mathbb{R}}}_j$, $\mathbb{R}_j^\infty$, $\mathbb{IR}_j^\infty$, and $\underline{\bar{\mathbb{R}}}_j^\infty$) as *two-sided*. We refer to those without the locatedness axiom ($\underline{\mathbb{R}}_j$, $\bar{\mathbb{R}}_j$, $\underline{\mathbb{R}}_j^\infty$, $\bar{\mathbb{R}}_j^\infty$, $\mathbb{IR}_j$, $\underline{\bar{\mathbb{R}}}_j$, $\mathbb{IR}_j^\infty$, and $\underline{\bar{\mathbb{R}}}_j^\infty$) as the *$j$-numeric domains*, a name that will be justified in Proposition 4.30.

Given unbounded reals $r = (\delta, \upsilon) : \mathbb{R}_j^\infty$ and $r' = (\delta', \upsilon') : \mathbb{R}_j^\infty$, we often use the more familiar notation for inequalities between them or involving a rational $q : \mathbb{Q}$:

$$q < r := \delta q \qquad r < q := \upsilon q \qquad r <_j r' := j\exists(q : \mathbb{Q}).\,(r < q) \wedge (q < r')$$
$$r \leq_j q := \delta q \Rightarrow j\bot \quad q \leq_j r := \upsilon q \Rightarrow j\bot \quad r' \leq_j r := (r <_j r') \Rightarrow j\bot$$

The same notation makes sense when $r, r' : \mathbb{R}_j$ are (bounded) reals, but one should be a bit careful with other j-numeric types. We will discuss this more in Sect. 4.3.5.

Remark 4.22 We associate to any $q : \mathbb{Q}$ a pair of cuts (δ_q, υ_q), defined as follows on $q' : \mathbb{Q}$

$$\delta_q q' \Leftrightarrow j(q < q') \qquad \text{and} \qquad \upsilon_q q' \Leftrightarrow j(q' < q). \tag{4.12}$$

It is easy to check that (δ_q, υ_q) is a j-local real number, so we have a map $\mathbb{Q} \to \mathbb{R}_j$. In particular, for $r = (\delta_r, \upsilon_r) : \mathbb{R}_j$, we write $q = r$ to mean $\forall(q' : \mathbb{Q}).\,(\delta_q q' \Leftrightarrow \delta_r q') \wedge (\upsilon_q q' \Leftrightarrow \upsilon_r q')$.

There are also obvious isomorphisms $\underline{\bar{\mathbb{R}}}_j \cong \underline{\mathbb{R}}_j \times \bar{\mathbb{R}}_j$ and $\underline{\bar{\mathbb{R}}}_j^\infty \cong \underline{\mathbb{R}}_j^\infty \times \bar{\mathbb{R}}_j^\infty$. Here is a diagram of relationships among the Dedekind j-numeric types:

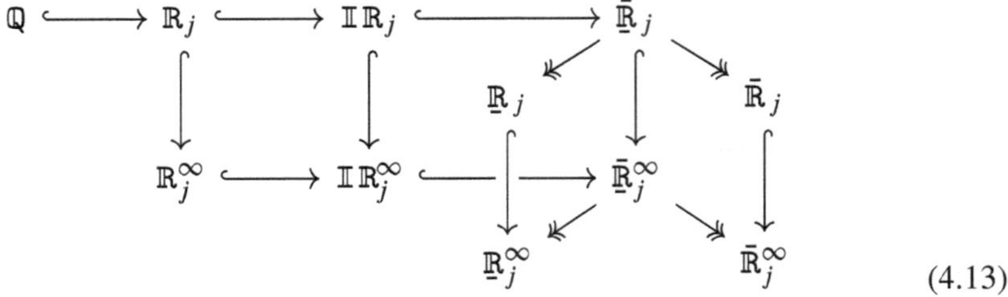

(4.13)

Remark 4.23 Each of the Dedekind j-numeric types, $\mathbb{R}_j$, $\mathbb{IR}_j$, $\bar{\underline{\mathbb{R}}}_j$, $\underline{\mathbb{R}}_j$, $\bar{\mathbb{R}}_j$, $\mathbb{R}_j^\infty$, $\mathbb{IR}_j^\infty$, $\bar{\underline{\mathbb{R}}}_j^\infty$, $\underline{\mathbb{R}}_j^\infty$, and $\bar{\mathbb{R}}_j^\infty$ is a j-sheaf. Proving this requires function extensionality; see Axiom 0.

4.3.3 Some Preliminary Numerical Facts

Most of the material in this section is taken from [BT09]. The results here will mainly be used in the next section, to show that addition and multiplication preserve locatedness; see Theorem 4.50. Throughout the section, j is an arbitrary modality.

Proposition 4.24 ($\mathbb{Q}$ Is Archimedean)

$$\forall (q_1, q_2 : \mathbb{Q}).\, (q_1 > 0 \wedge q_2 > 0) \Rightarrow \exists (k : \mathbb{N}).\, q_2 < k * q_1.$$

Proof We may assume $q_1 = \frac{a+1}{m+1}$ and $q_2 = \frac{b+1}{n+1}$ for $a, b, m, n \in \mathbb{N}$. Then we can set $k := (m+1) * (b+1)$. □

Proposition 4.25 *For any $n : \mathbb{N}$, rationals $q_0 < q_1 < \cdots < q_{n+2}$, and real $r : \mathbb{R}_j$, we have*

$$q_0 < r < q_{n+2} \Rightarrow j\exists (k : \mathbb{N}).\, (0 \leq k \leq n) \wedge (q_k < r < q_{k+2}).$$

Proof We proceed by induction on n. The claim is trivial for $n = 0$, so suppose it is true for arbitrary n. Suppose further that there are rationals $q_0 < q_1 < \cdots < q_{n+3}$ and $q_0 < r < q_{n+3}$. By locatedness, $j(q_{n+1} < r \vee r < q_{n+2})$; we can drop the j by Remark 4.9,[8] to obtain $q_{n+1} < r \vee r < q_{n+2}$. In the first case we have $q_{n+1} < r < q_{n+3}$ and in the second we use the inductive hypothesis. □

The next two propositions say that any real number can be bounded between rationals that are arbitrarily close together, in the additive and the multiplicative sense.

[8] Throughout this section, we will use Remark 4.9 without mentioning it.

Proposition 4.26 *Any* $r : \mathbb{R}_j$ *is* j-arithmetically located *in the sense that*

$$\forall(p : \mathbb{Q}).\, p > 0 \Rightarrow j\exists(d, u : \mathbb{Q}).\, (d < r < u) \wedge (0 < u - d < p).$$

Proof By boundedness, $j\exists(d', u' : \mathbb{Q}).\, d' < r < u'$, and by the Archimedean property (Proposition 4.24), $\exists(k : \mathbb{N}).\, u' - d' < k * \frac{p}{3}$. For each $i : \mathbb{N}$, let $q_i := d' + i * \frac{p}{3}$, so $d' = q_0 < q_1 < \cdots < q_{k+2}$ and $u' < q_k < q_{k+2}$. By Proposition 4.25, there is some i with $0 \leq i \leq k$ with $q_i < r < q_{i+2}$. The conclusion is satisfied using $d := q_i$ and $u := q_{i+2}$, because $u - d = \frac{2p}{3} < p$. □

Proposition 4.27 *Any* $r : \mathbb{R}_j$ *with* $0 < r$ *is* j-multiplicatively located *in the sense that*

$$\forall(p, p' : \mathbb{Q}).\, 0 < p < p' \Rightarrow j\exists(d, u : \mathbb{Q}).\, (0 < d < r < u) \wedge \left(\frac{u}{d} < \frac{p'}{p}\right).$$

Proof By roundedness $j\exists(q : \mathbb{Q}).\, 0 < q < r$. Choose $q' : \mathbb{Q}$ such that $0 < q' < q(p' - p)$. By arithmetic locatedness, $j\exists(d, u : \mathbb{Q}).\, (d < r < u) \wedge (0 < u - d < \frac{q'}{p'})$. Then $q < u$ by disjointness, so $p'(u - d) < q' < q(p' - p) < u(p' - p)$, hence $up < p'd$ and $\frac{u}{d} < \frac{p'}{p}$. □

The standard definition of addition of unbounded improper intervals, relativized to the j-modality, is given by $(\delta_1, \upsilon_1) + (\delta_2, \upsilon_2) = (\delta', \upsilon')$, where

$$\begin{aligned} \delta' q &\Leftrightarrow j\exists(q_1, q_2 : \mathbb{Q}).\, \delta_1 q_1 \wedge \delta_2 q_2 \wedge (q < q_1 + q_2) \\ \upsilon' q &\Leftrightarrow j\exists(q_1, q_2 : \mathbb{Q}).\, \upsilon_1 q_1 \wedge \upsilon_2 q_2 \wedge (q_1 + q_2 < q). \end{aligned} \tag{4.14}$$

Theorem 4.28 *The type* $\bar{\underline{\mathbb{R}}}_j^\infty$ *has the structure of a commutative monoid, with the addition operation* $+$ *from Eq.* (4.14)*. Each of the types in Eq.* (4.13) *is closed under this operation, and each of the maps shown there is a monoid homomorphism.*

Proof We begin by working only with lower cuts; upper cuts are similar. It is easy to see that $+$ is commutative and associative. For example, the lower cut for both sides of $r_1 + (r_2 + r_3) = (r_1 + r_2) + r_3$ will be equivalent to

$$j\exists(q_1, q_2, q_3 : \mathbb{Q}).\, \delta_1 q_1 \wedge \delta_2 q_2 \wedge \delta_3 q_3 \wedge (q < q_1 + q_2 + q_3).$$

The additive unit is $0 : \mathbb{Q}$; indeed, it suffices to check the following equivalence for any lower cut δ and rational q:

$$\delta q \Leftrightarrow j\exists(q_1, q_2 : \mathbb{Q}).\, q_1 < 0 \wedge \delta q_2 \wedge (q < q_1 + q_2).$$

For the backwards direction, the existential is equivalent to $\exists(q_2 : \mathbb{Q}).\, \delta q_2 \wedge (q < q_2)$, so we obtain δq by down-closure (1a) and j-closure (0a). For the forwards direction, suppose δq. Then by roundedness (2a) twice, $\exists(q', q'' : \mathbb{Q}).\,(q < q' < q'') \wedge \delta q''$, and the conclusion is satisfied by letting $q_1 :- q' - q''$ and $q_2 := q''$.

Thus $\bar{\underline{\mathbb{R}}}_j^\infty$ has the structure of a commutative monoid. It is straightforward to check that if δ_1 and δ_2 are bounded (3a) then so is their sum, so the operation restricts to one on $\bar{\underline{\mathbb{R}}}_j$, which then also has the structure of a commutative monoid.

For j-disjointness (4) and j-locatedness (5), we consider both lower and upper cuts. If r_1 and r_2 are j-disjoint, then we want to see that $(\delta', \upsilon') = r_1 + r_2$ is too. Suppose $\delta' q \wedge \upsilon' q$; we want to show $j\bot$. From

$$j\exists(q_1, q_2, q_1', q_2' : \mathbb{Q}).\, \delta_1 q_1 \wedge \delta_2 q_2 \wedge (q < q_1 + q_2) \wedge \upsilon_1 q_1' \wedge \upsilon_2 q_2' \wedge (q_1' + q_2' < q)$$

one obtains $q_1' + q_2' < q_1 + q_2$ and by Trichotomy of rationals, $(q_1' < q_1) \vee (q_2' < q_2)$. Either way, we obtain $j\bot$ from down/up closure and disjointness.

Suppose $r_1, r_2 : \mathbb{R}_j$; we need to show $r_1 + r_2$ is j-located, so take $q < q'$ and let $p := \frac{q'-q}{2}$. By Proposition 4.26, we can find rationals $q_1 < r_1 < q_1'$ and $q_2 < r_2 < q_2'$ such that $(q_1' - q_1) < p$ and $(q_2' - q_2) < p$. It is easy to check that $q_1 + q_2 < r_1 + r_2 < q_1' + q_2'$. Either $q < q_1 + q_2$ or $q \geq q_1 + q_2$; in the first case, we get $q < r_1 + r_2$ and in the second case we can get $r_1 + r_2 < q'$, as desired. □

Corollary 4.29 *The monoid structure on $\mathbb{R}_j$ is in fact that of a group.*

Proof Take any $(\delta_1, \upsilon_1) : \mathbb{R}_j$ and define cuts δ_2, υ_2 on $q : \mathbb{Q}$ by

$$\delta_2 q \Leftrightarrow \upsilon_1(-q) \qquad \text{and} \qquad \upsilon_2 q \Leftrightarrow \delta_1(-q).$$

It is easy to check that (δ_2, υ_2) is a j-local real. Let $(\delta', \upsilon') := (\delta_1, \upsilon) + (\delta_2, \upsilon_2)$ be their sum, as in Eq. (4.14). We want to show $\delta' q \Leftrightarrow j(q < 0)$ and $\upsilon' q \Leftrightarrow j(0 < q)$. Consider the latter, the former being similar; we must check

$$j(0 < q) \Leftrightarrow j\exists(q_1, q_2 : \mathbb{Q}).\, \upsilon_1 q_1 \wedge \delta_1(-q_2) \wedge (q_1 + q_2 < q).$$

The forward direction is Proposition 4.26; the backwards direction comes down to the fact that (δ_1, υ_1) is j-disjoint. □

4.3.4 Numeric Domains

In Definition 4.21 we referred to eight of the j-numeric types as *j-numeric domains*. In this section, specifically Proposition 4.30, we show that they actually are domains internal to $\mathcal{E}$.

In fact, these eight domains correspond to the eight predomains we defined in Example A.45. Though the subject of predomains is a bit technical, its value to us it that predomains provide a very useful tool for defining arithmetic operations between numeric types.

Again, the eight predomains from Example A.45 can be defined internally to any topos. Write $\underline{\mathbb{R}}_{\text{pre}}$, $\bar{\mathbb{R}}_{\text{pre}}$, $\bar{\underline{\mathbb{R}}}_{\text{pre}}$, $\mathbb{IR}_{j,\text{pre}}$, $\underline{\mathbb{R}}^{\infty}_{\text{pre}}$, $\bar{\mathbb{R}}^{\infty}_{\text{pre}}$, $\bar{\underline{\mathbb{R}}}^{\infty}_{\text{pre}}$, and $\mathbb{IR}^{\infty}_{j,\text{pre}}$ to denote the corresponding internal constructions in any topos $\mathcal{E}$ and modality j. For example, $\underline{\mathbb{R}}_{\text{pre}}$ is just the rationals with the usual $<$ ordering: $\underline{\mathbb{R}}_{\text{pre}} = (\mathbb{Q}, <)$. Because our type theory has sum types (see Sect. 4.1.1) we can also make sense of the unbounded predomains, such as $\underline{\mathbb{R}}^{\infty}_{\text{pre}} = (\mathbb{Q} + \{\infty\}, <)$.

Recall the j-local rounded ideal construction from Definition A.44.

Proposition 4.30 *Let $\mathcal{E}$ be a topos, and let j be a modality on $\mathcal{E}$. Then we have isomorphisms*

$$\underline{\mathbb{R}}_j \cong \mathrm{RId}_j(\underline{\mathbb{R}}_{\text{pre}}) \quad \bar{\mathbb{R}}_j \cong \mathrm{RId}_j(\bar{\mathbb{R}}_{\text{pre}}) \quad \bar{\underline{\mathbb{R}}}_j \cong \mathrm{RId}_j(\bar{\underline{\mathbb{R}}}_{\text{pre}}) \quad \mathbb{IR}_j \cong \mathrm{RId}_j(\mathbb{IR}_{j,\text{pre}})$$

$$\underline{\mathbb{R}}^{\infty}_j \cong \mathrm{RId}_j(\underline{\mathbb{R}}^{\infty}_{\text{pre}}) \quad \bar{\mathbb{R}}^{\infty}_j \cong \mathrm{RId}_j(\bar{\mathbb{R}}^{\infty}_{\text{pre}}) \quad \bar{\underline{\mathbb{R}}}^{\infty}_j \cong \mathrm{RId}_j(\bar{\underline{\mathbb{R}}}^{\infty}_{\text{pre}}) \quad \mathbb{IR}^{\infty}_j \cong \mathrm{RId}_j(\mathbb{IR}^{\infty}_{j,\text{pre}})$$

Thus each j-numeric domain in the sense of Definition 4.21 is in fact an internal domain—and in particular an internal topological space—in the subtopos $\mathcal{E}_j$.

Proof Consider the four conditions in the definition of j-rounded ideal found in Definition A.44; below we will refer to them 0', 1', 2', 3'. They match exactly with the j-closed, up/down-closed, j-roundedness, and j-boundedness conditions of Definition 4.21. The first two isomorphisms are hence direct. Proposition A.8 is easily modified to the context of a modality j, showing that

$$\mathrm{RId}_j(\bar{\underline{\mathbb{R}}}_{\text{pre}}) = \mathrm{RId}_j(\underline{\mathbb{R}}_{\text{pre}} \times \bar{\mathbb{R}}_{\text{pre}}) \cong \mathrm{RId}_j(\underline{\mathbb{R}}_{\text{pre}}) \times \mathrm{RId}_j(\bar{\mathbb{R}}_{\text{pre}}) \cong \underline{\mathbb{R}} \times \bar{\mathbb{R}} \cong \bar{\underline{\mathbb{R}}}, \tag{4.15}$$

giving the third isomorphism. It is not quite as automatic, but still easy to see that we have the fourth isomorphism $\mathbb{IR}_j \cong \mathrm{RId}_j(\mathbb{IR}_{j,\text{pre}})$, because a rounded ideal $I \subseteq \bar{\underline{\mathbb{R}}}_{\text{pre}}$ is contained in $\mathbb{IR}_{j,\text{pre}}$ iff $(q, q') \in I \Rightarrow j(q < q')$ iff $\delta q \Rightarrow \upsilon q' \Rightarrow j(q < q')$ iff I is j-disjoint.

The bottom row is more interesting; we consider only the case for unbounded lower reals, the others being similar. We want to obtain the isomorphism

$$\{\delta' : \mathbb{Q} \sqcup \{-\infty\} \to \texttt{Prop} \mid 0', 1', 2', 3'\} \cong \{\delta : \mathbb{Q} \to \texttt{Prop} \mid 0\text{a}, 1\text{a}, 2\text{a}\}$$

because the left-hand side is $\mathrm{RId}_j(\underline{\mathbb{R}}^{\infty}_{\text{pre}})$ and the right-hand side is $\underline{\mathbb{R}}^{\infty}$.

Given δ' on the left, define $\delta q := \delta' q$; it is easy to check that it satisfies 0a, 1a, 2a. Conversely, given δ on the right, define $\delta' q'$ to be $\delta q'$ if $q' \in \mathbb{Q}$ and $\top$ if $q' = -\infty$. Recall from Example A.3 that we defined the order relation so that $-\infty < -\infty$. With that in mind, the conditions 0', 1', 2', 3' are easy.

The second statement follows directly from Proposition 4.30 and Corollary A.46. □

Domains have two useful relations, $\sqsubseteq$ and $\ll$; see Sect. 2.2. Thus by Proposition 4.30, these symbols make sense for each of the eight j-numeric domains.

Proposition 4.31 *Suppose that* $\delta, \delta' : \underline{\mathbb{R}}_j$ *or* $\delta, \delta' : \underline{\mathbb{R}}_j^\infty$. *And suppose that* $\upsilon, \upsilon' : \bar{\mathbb{R}}_j$ *or* $\upsilon, \upsilon' : \bar{\mathbb{R}}_j^\infty$. *Then*

$$\delta' \sqsubseteq \delta \Leftrightarrow \forall(q : \mathbb{Q}).\, \delta' q \Rightarrow \delta q \qquad \delta' \ll \delta \Leftrightarrow j\exists(q : \mathbb{Q}).\, \delta q \wedge (\delta' q \Rightarrow j\bot)$$

$$\upsilon' \sqsubseteq \upsilon \Leftrightarrow \forall(q : \mathbb{Q}).\, \upsilon' q \Rightarrow \upsilon q \qquad \upsilon' \ll \upsilon \Leftrightarrow j\exists(q : \mathbb{Q}).\, \upsilon q \wedge (\upsilon' q' \Rightarrow j\bot)$$

For (δ, υ) *and* (δ', υ') *in* $\bar{\underline{\mathbb{R}}}_j$, $\bar{\underline{\mathbb{R}}}_j^\infty$, $\mathbb{IR}_j$, *or* $\mathbb{IR}_j^\infty$, *we have*

$$\big((\delta, \upsilon) \sqsubseteq (\delta', \upsilon')\big) \Leftrightarrow \big((\delta \sqsubseteq \delta') \wedge (\upsilon \sqsubseteq \upsilon')\big) \quad \text{and} \quad \big((\delta, \upsilon) \ll (\delta', \upsilon')\big)$$
$$\Leftrightarrow \big((\delta \ll \delta') \wedge (\upsilon \ll \upsilon')\big)$$

Proof Consider the δ case; the rest are similar. The statement for $\sqsubseteq$ is basically just the definition of these domains, and comes directly from Corollary A.46. The statement about $\ll$ will also follow from Corollary A.46. Indeed, it says

$$\delta' \ll \delta \Leftrightarrow j\exists(q : \mathbb{Q}).\, \delta q \wedge \forall(q' : \mathbb{Q}).\, \delta' q' \Rightarrow j(q' < q),$$

and the right-hand side is equivalent to $j\exists(q : \mathbb{Q}).\, \delta q \wedge (\delta' q \Rightarrow j\bot)$ by trichotomy for rationals, $(q < q') \vee (q = q') \vee (q > q')$. □

Remark 4.32 Rather than define Dedekind j-numeric types as predicates on $\mathbb{Q}$, as we did in Definition 4.21, one could also define them as predicates on $\mathsf{sh}_j\mathbb{Q}$, the j-sheafification of rationals. However, it turns out that doing so makes no difference: semantically, it defines the same topos of sheaves.

Indeed, it is enough by Proposition 4.18 to check semantically that j-separification agrees with j-sheafification, $\text{sep}_j\boldsymbol{Q} \cong \mathsf{sh}_j\boldsymbol{Q}$, for the constant type $\boldsymbol{Q}$ on any set Q. So suppose given a j-covering family $\{V_i \to V\}_{i \in I}$ and sections $q_i \in \boldsymbol{Q}(V_i)$ such that $j(q_i = q_{i'})$ for $i, i' \in I$. For each i, we may assume each $q_i \in Q$, so it has an extension $\overline{q_i} \in \boldsymbol{Q}(V)$. We want to show $j(\overline{q_i} = \overline{q_{i'}})$. We have $(\overline{q_i} = \overline{q_{i'}}) \vee \neg(\overline{q_i} = \overline{q_{i'}})$, so we may assume $\neg(\overline{q_i} = \overline{q_{i'}})$ from which we obtain $\neg(q_i = q_{i'})$ and hence $j\bot$, which proves the result.

Before concluding this section, we record a lemma that will become useful in Sect. 7.3.3, when we discuss differentiability.

Lemma 4.33 *Let* $x, y : \bar{\underline{\mathbb{R}}}_j$ *be* j*-improper intervals such that* $x \sqsubseteq y$. *If* x *is* j*-located and* y *is* j*-disjoint, then* $x = y$.

Proof Write $x = (\delta_x, \upsilon_x)$ and $y = (\delta_y, \upsilon_y)$. For all $q : \mathbb{Q}$ we have $\delta_x q \Rightarrow \delta_y q$ and $\upsilon_x q \Rightarrow \upsilon_y q$, so it suffices to show $\delta_y q \Rightarrow \delta_x q$ for any $q : \mathbb{Q}$ (the υ case is similar). Since δ_x is j-closed, we may drop j from the front of all hypotheses by Remark 4.9.

Assuming $\delta_y q$, we have $\exists q'.\, q < q' \wedge \delta_y q'$ by j-roundedness. Since x is j-located we have $\delta_x q \vee \upsilon_x q'$. In the first case we are done; in the second case, we have $\upsilon_y q'$ which is a contradiction because y is j-disjoint. □

4.3.5 Inequalities

In this section we will discuss inequalities for Dedekind j-numeric objects, and in Sect. 4.3.7 we discuss their arithmetic. We learned some of this from [Kau80] and [Sai+14], which in particular discuss the order-theoretic and algebraic properties of the extended interval domain $\bar{\underline{\mathbb{R}}}$.

The proofs in the remainder of this chapter, Sects. 4.3.5–4.3.7, will depend heavily on the work on predomains discussed in Appendix A, however the results can usually be stated without it. We write $\mathbb{Q}$ to denote the type of rational numbers, together with the usual $<$ ordering, and we write $\mathbb{Q}^{\mathrm{op}}$ for the opposite ordering. Both of these are predomains, and we will often consider the predomain $\mathbb{Q} \times \mathbb{Q}^{\mathrm{op}}$.

We defined inequalities for reals in Definition 4.21. We can easily generalize the definition to j-local improper intervals. Given two j-local unbounded improper intervals $x_1, x_2 : \bar{\underline{\mathbb{R}}}_j^\infty$, we define $x_1 <_j x_2$ by

$$x_1 <_j x_2 \Leftrightarrow j\exists(q : \mathbb{Q}).\,(x_1 < q) \wedge (q < x_2), \tag{4.16}$$

where as usual $x_1 < q$ and $q < x_2$, respectively, mean $\upsilon_1 q$ and $\delta_2 q$, for $x_i = (\delta_i, \upsilon_i)$.[9] The same definition also works for the other two-sided j-numeric types $\mathbb{IR}_j^\infty$, $\mathbb{R}_j^\infty$, $\bar{\underline{\mathbb{R}}}_j$, $\mathbb{IR}_j$, and $\mathbb{R}_j$.

Proposition 4.34 *For any modality j, the set $\{(x_1, x_2) \mid x_1 <_j x_2\} \subseteq \bar{\underline{\mathbb{R}}}_j^\infty \times \bar{\underline{\mathbb{R}}}_j^\infty$ is an open subset. The same is true when $\bar{\underline{\mathbb{R}}}_j^\infty$ is replaced by any of the other two-sided j-numeric types.*

Proof The other two-sided j-numeric types are subspaces of $\bar{\underline{\mathbb{R}}}_j^\infty$, so it suffices to check that one. Begin by defining an open of the predomain $\bar{\underline{\mathbb{R}}}_{\mathrm{pre}}^\infty \times \bar{\underline{\mathbb{R}}}_{\mathrm{pre}}^\infty$ by

$$U_< := \{(d_1, u_1), (d_2, u_2) \mid u_1 < d_2\} \in \Omega(\bar{\underline{\mathbb{R}}}_{\mathrm{pre}}^\infty \times \bar{\underline{\mathbb{R}}}_{\mathrm{pre}}^\infty).$$

This is easily seen to be an open in the sense of Definition A.4: it is up-closed (decreasing u_1 and increasing d_2 stays in the set) and rounded (if $u_1 < d_2$, then we can easily find d_1', u_2', d_2', u_2' with $d_1' < d_1$, $u_1 < u_1'$, $d_2' < d_2$, $u_2 < u_2'$, and still $u_1' < d_2'$).

By Theorem A.25, this determines a Scott open subset of $\mathrm{RId}_j(\bar{\underline{\mathbb{R}}}_{\mathrm{pre}}^\infty \times \bar{\underline{\mathbb{R}}}_{\mathrm{pre}}^\infty)$ given by $\mathcal{U}_{U_<} = \{I \mid \exists(d_1, u_1, d_2, u_2) \in I.\, u_1 < d_2\}$. Under the isomorphism $\bar{\underline{\mathbb{R}}}_j^\infty \times \bar{\underline{\mathbb{R}}}_j^\infty \cong \mathrm{RId}_j(\bar{\underline{\mathbb{R}}}_{\mathrm{pre}}^\infty \times \bar{\underline{\mathbb{R}}}_{\mathrm{pre}}^\infty)$ from Propositions 4.30 and A.8, this is equivalent to

---

[9] In Definition 4.21 we gave the warning that one should be a bit careful with the $<$ relation on other numeric types. As an example of what can violate standard intuition, let $x = (\delta, \upsilon) : \bar{\underline{\mathbb{R}}}$ be given by $\delta q \Leftrightarrow q < 1$ and $\upsilon q \Leftrightarrow -1 < q$. Then $x < x$ holds.

For the one-sided numeric types, e.g. $\underline{\mathbb{R}}$, there is a good notion of $\leq$, namely $\forall(q : \mathbb{Q}).\,\delta q \Rightarrow \delta' q$. To obtain $<$ one might try $(\delta \leq \delta') \wedge (\delta \neq \delta')$, but this is semantically too strong in a general topos.

$$\left\{((\delta_1, \upsilon_1), (\delta_2, \upsilon_2)) \;\middle|\; j\exists(d_1, u_1, d_2, u_2).\, \delta_1 d_1 \wedge \upsilon_1 u_1 \wedge \delta_2 d_2 \wedge \upsilon_2 u_2 \wedge u_1 < d_2\right\},$$

or more simply $\{(\delta_1, \upsilon_1), (\delta_2, \upsilon_2) \mid j\exists q.\, \upsilon_1 q \wedge \delta_2 q\}$, for which Eq. (4.16) is shorthand. In the language of Definition A.44, $(x_1 <_j x_2)$ iff $(x_1, x_2) \models_j U_<$. □

We now move on to discussing the non-strict inequality, $\leq_j$.

Proposition 4.35 *The following are equivalent for all* $a, b : \mathbb{R}_j$*:*

1. $(b <_j a) \Rightarrow j\bot$
2. $\forall(q : \mathbb{Q}).\, q < a \Rightarrow q < b$
3. $\forall(q : \mathbb{Q}).\, b < q \Rightarrow a < q$

Proof This is proven for an arbitrary topos in [Joh02, Lemma D.4.7.6], so it easily applies to the j-subtopos. However it is also straightforward, so we show the proof for $1 \Rightarrow 2$; the other implications are similarly easy.

Write $a = (\delta_a, \upsilon_a)$ and $b = (\delta_b, \upsilon_b)$. To prove $1 \Rightarrow 2$, assume $(b <_j a) \Rightarrow j\bot$, and choose q such that $\delta_a q$. It suffices to prove $j\delta_b q$ because $\delta_b q$ is j-closed; thus we may drop j from the front of all hypotheses. By j-roundedness, we have $j\exists q'.\, q < q' \wedge \delta_a q'$, so (dropping j) we choose such a q' and, by j-locatedness, we have $\upsilon_b q' \vee \delta_b q$. In one case we are done; in the other, we have $\delta_a q' \wedge \upsilon_b q'$, meaning $b <_j a$, and we obtain $j\bot$ by assumption. □

Suppose $a, b : \mathbb{R}_j$ are j-local reals. We write

$$a \leq_j b \tag{4.17}$$

iff any of the equivalent conditions of Proposition 4.35, e.g. $(b <_j a) \Rightarrow j\bot$, is satisfied.

It is easy to check that the relation $\leq_j$ on $\mathbb{R}_j$ is reflexive, transitive, and anti-symmetric, $(a \leq_j b) \wedge (b \leq_j a) \Rightarrow a = b$. It is also easy to check that $(a <_j b) \Rightarrow (a \leq_j b)$, as well as the following facts:

$$(a <_j b) \wedge (b \leq_j c) \Rightarrow (a <_j c) \qquad \text{and} \qquad (a \leq_j b) \wedge (b <_j c) \Rightarrow (a <_j c).$$

As mentioned above, Proposition 4.35 is about j-local reals, *not* j-local improper intervals.

Remark 4.36 The above story also works for unbounded reals and one-sided numeric types. For $a, b : \mathbb{R}_j^\infty$ one can define $a \leq_j b$ to be any of the following equivalent conditions:

1. $(b <_j a) \Rightarrow j\bot$
2. $\forall(q : \mathbb{Q} \cup \{-\infty\}).\, q <_j a \Rightarrow q <_j b$
3. $\forall(q : \mathbb{Q} \cup \{\infty\}).\, b <_j q \Rightarrow a <_j q$

For $a, b : \underline{\mathbb{R}}$ or $a, b : \underline{\mathbb{R}}^\infty$, only condition 2 is defined, and so we define $a \leq_j b$ as condition 2; this agrees with the domain order on $\underline{\mathbb{R}}$ and $\underline{\mathbb{R}}^\infty$. For $a, b : \bar{\mathbb{R}}$ or

$a, b : \bar{\mathbb{R}}^\infty$, only condition 3 is defined, and so we define $b \geq_j a$ as condition 3; this agrees with the domain order on $\bar{\mathbb{R}}$ and $\bar{\mathbb{R}}^\infty$.

4.3.6 j-Constant Numeric Types

The present section (Sect. 4.3.6) is non-standard material, regarding what we call *constant numeric types*. The technical underpinnings of these ideas is given in Appendix A.3.2.

Definition 4.37 **(j-Constant Numeric Types)** Let j be a modality. Say that $\phi : \mathbb{Q} \to \mathtt{Prop}$ is *j-decidable* if it satisfies

$$\forall (q : \mathbb{Q}).\, \phi q \vee (\phi q \Rightarrow j\bot).$$

Let $c\underline{\mathbb{R}}_j$ denote the subtype of $\underline{\mathbb{R}}_j$ consisting of those δ that are j-decidable, and similarly define $c\bar{\mathbb{R}}_j$, $c\mathbb{I}\mathbb{R}_j$, $c\underline{\bar{\mathbb{R}}}_j$, $c\mathbb{R}_j$ $c\underline{\mathbb{R}}_j^\infty$, $c\bar{\mathbb{R}}_j^\infty$, $c\mathbb{I}\mathbb{R}_j^\infty$, $c\underline{\bar{\mathbb{R}}}_j^\infty$, and $c\mathbb{R}_j^\infty$. We refer to these as the *j-constant numeric types*.

Note that $c\underline{\bar{\mathbb{R}}}_j = c\underline{\mathbb{R}}_j \times c\bar{\mathbb{R}}_j$ and $c\underline{\bar{\mathbb{R}}}_j^\infty = c\underline{\mathbb{R}}_j^\infty \times c\bar{\mathbb{R}}_j^\infty$.

Remark 4.38 For each of the eight j-numeric domains (see Proposition 4.31), the constant types form a basis in the sense of Definition A.20; indeed, this follows from Example A.57 and Proposition A.58. This fact implies, for example, that for any $x, y : \underline{\bar{\mathbb{R}}}_j$,

$$(x \ll y) \Leftrightarrow \exists (c : c\underline{\bar{\mathbb{R}}}_j).\, x \ll c \ll y \text{ and } x \sqsubseteq y \Leftrightarrow \forall (c : c\underline{\bar{\mathbb{R}}}_j).\, (c \ll x) \Rightarrow (c \sqsubseteq y)$$

and similarly when $\underline{\bar{\mathbb{R}}}$ is replaced with any of the other j-numeric domains.

Proposition 4.39 *For any j there are internal bijections*

$$c\underline{\mathbb{R}}_j^\infty \cong c\mathbb{R}_j^\infty \cong c\bar{\mathbb{R}}_j^\infty.$$

Proof For the first claim, consider the functions $c\mathbb{R}_j^\infty \to c\underline{\mathbb{R}}_j^\infty$ and $c\mathbb{R}_j^\infty \to c\bar{\mathbb{R}}_j^\infty$ given by sending (δ, υ) to δ and υ, respectively. In the first case, the inverse sends δ to (δ, υ) where for any $q : \mathbb{Q}$,

$$\upsilon q \Leftrightarrow j\exists (q' : \mathbb{Q}).\, (q' < q) \wedge (\delta q \Rightarrow j\bot) \tag{4.18}$$

It is easy to check that υ is j-closed, up-closed, and j-rounded, and that the pair (δ, υ) is j-disjoint and j-located. To see that υ is j-decidable, we proceed as in Proposition 5.6. Take $q : \mathbb{Q}$ and consider the constant type $\mathbb{Q}_{<q} = \{q' : \mathbb{Q} \mid q' < q\}$. We have $\forall (q' : \mathbb{Q}_{<q}).\, (\delta q' \Rightarrow j\bot) \vee \delta q'$, so $\forall (q' : \mathbb{Q}_{<q}).\, \exists (q'' : \mathbb{Q}_{<q}).\, (\delta q'' \Rightarrow j\bot) \vee \delta q'$, and thus $\upsilon q \vee \forall (q' : \mathbb{Q}_{<q}).\, \delta q'$ by Axiom 2. It is easy to show that

$\forall(q' : \mathbb{Q}_{<q}).\, \delta q'$ implies $\upsilon q \Rightarrow j\bot$, so υ is indeed j-constant. Finally, to see that the two functions are mutually inverse, one takes $(\delta, \upsilon) : \mathrm{c}\mathbb{R}_j^\infty$ and shows using j-locatedness that Eq. (4.18) holds for any $q : \mathbb{Q}$. □

Corollary 4.40 *There are internal bijections*

$$\mathrm{c}\underline{\bar{\mathbb{R}}}_j^\infty \cong \mathrm{c}\mathbb{R}_j^\infty \times \mathrm{c}\mathbb{R}_j^\infty \qquad \textit{and} \qquad \mathrm{c}\mathbb{I}\mathbb{R}_j^\infty \cong \{(d,u) \in \mathrm{c}\mathbb{R}_j^\infty \times \mathrm{c}\mathbb{R}_j^\infty \mid d \le_j u\}$$

Proof We obtain the left-hand isomorphism from Proposition 4.39 and the isomorphisms $\mathrm{c}\underline{\bar{\mathbb{R}}}_j^\infty \cong \mathrm{c}\underline{\mathbb{R}}_j^\infty \times \mathrm{c}\bar{\mathbb{R}}_j^\infty$ and $\mathrm{c}\underline{\bar{\mathbb{R}}}_j \cong \mathrm{c}\underline{\mathbb{R}}_j \times \mathrm{c}\bar{\mathbb{R}}_j$. Using the left-hand isomorphism, we will also have the right-hand isomorphism if we can show that for any $d, u : \mathbb{R}^\infty$ we have $\big(\forall q.\,((q < d \wedge u < q) \Rightarrow j\bot)\big) \Leftrightarrow d \le_j u$. The left-hand side is constructively equivalent to $(\exists q.\, q < d \wedge u < q) \Rightarrow j\bot$ and the right-hand side is by definition $(u <_j d) \Rightarrow j\bot$, so it suffices to show $u <_j d \Leftrightarrow j\exists(q : \mathbb{Q}).\, q < d \wedge u < q$, but that is the definition (4.21). □

Notation 4.41 For $d, u : \mathrm{c}\mathbb{R}_j^\infty$, we may write $[d, u] : \mathrm{c}\underline{\bar{\mathbb{R}}}_j^\infty$ to denote the constant interval $(\delta, \upsilon) : \mathrm{c}\underline{\bar{\mathbb{R}}}_j^\infty$ given by

$$\delta q \Leftrightarrow j(q < d) \qquad \text{and} \qquad \upsilon q \Leftrightarrow j(u < q)$$

Thus we have $[-,-] : \mathrm{c}\mathbb{R}_j^\infty \times \mathrm{c}\mathbb{R}_j^\infty \to \mathrm{c}\underline{\bar{\mathbb{R}}}_j^\infty$.

Recall from Example A.3 that $\underline{\bar{\mathbb{R}}}_{\text{pre}}^\infty = \mathbb{Q} \times \mathbb{Q}$ with $(q_1, q_2) \prec (q_1', q_2') \Leftrightarrow (q_1 < q_1') \wedge (q_2' < q_2)$. Composing $[-,-]$ with the square of the map $i : \mathbb{Q} \to \mathbb{R}_j^\infty$, one obtains $[i, i] : \underline{\bar{\mathbb{R}}}_{\text{pre}}^\infty \to \underline{\bar{\mathbb{R}}}_j^\infty$ which sends (q, q') to $[q, q']$; it is the usual inclusion $\twoheaddownarrow$ of the predomain into its associated domain (see Lemma A.10).

Let $f : X \to Y$ be a function. One might say it is j-injective if it satisfies $\forall(x, x' : X).\, f(x) = f(x') \Rightarrow j(x = x')$, but for Proposition 4.42 we want a stronger notion. So say that f is a *strongly j-injective* if it satisfies

$$\forall(x, x' : X).\, f(x) = f(x') \Rightarrow (x = x') \vee j\bot. \tag{4.19}$$

Proposition 4.42 *There are functions*

$$\mathrm{c}\mathbb{R}_j \sqcup \{\infty\} \to \mathrm{c}\underline{\mathbb{R}}_j \qquad \textit{and} \qquad \mathbb{R}_j \sqcup \{-\infty\} \to \mathrm{c}\bar{\mathbb{R}}_j.$$

which are internally surjective and strongly j-injective.

Proof The map $\mathrm{c}\mathbb{R}_j \sqcup \{\infty\} \to \mathrm{c}\underline{\mathbb{R}}_j$ sends $(\delta, \upsilon) : \mathrm{c}\mathbb{R}_j$ to δ, and sends ∞ to the unique predicate δ satisfying $\forall q.\, \delta q$. We will show it is an internal surjection and a j-injection, the claim for $\mathrm{c}\bar{\mathbb{R}}_j$ being similar. Take $\delta : \mathrm{c}\underline{\mathbb{R}}_j$ and define υ as in Eq. (4.18), and the pair (δ, υ) is in $\mathrm{c}\mathbb{R}_j$ iff υ is j-bounded. We already have shown that $\forall q.\, \upsilon q \vee (\upsilon q \Rightarrow j\bot)$ and again follows from Proposition 5.6 that $(\exists q.\, \upsilon q) \vee \forall q.\, (\upsilon q \Rightarrow j\bot)$. One can prove that the second case is equivalent to $\forall q.\, \delta q$, and so our function is indeed surjective.

To see that the map is strongly j-injective, take $x, x' : \mathrm{c}\mathbb{R}_j \sqcup \{\infty\}$. We need only consider two of the four cases, namely when $x, x' : \mathrm{c}\mathbb{R}_j$, and when $x : \mathrm{c}\mathbb{R}_j$ and $x' = \infty$. For the former we need to show that if both $x = (\delta, \upsilon)$ and $x' = (\delta, \upsilon')$ are in $\mathrm{c}\mathbb{R}_j$ then $(\upsilon = \upsilon') \vee j\bot$. So take $q : \mathbb{Q}$; we want to show $\upsilon q \Leftrightarrow \upsilon' q$. Because υ and υ' are constant, it suffices to consider the case $\upsilon q \wedge (\upsilon' q \Rightarrow j\bot)$. We obtain $j\bot$ by j-roundedness of υ, j-disjointness of (δ, υ), and j-locatedness of (δ, υ').

Finally, suppose $x = (\delta, \upsilon) : \mathrm{c}\mathbb{R}_j$ and $x' = \infty$. It suffices to show that if δq holds for all q then $j\bot$. We get this because υ is j-bounded and (δ, υ) is j-disjoint. □

Proposition 4.43 *For any j, the subtype of j-constant elements of any j-numeric domain is again a domain, namely the domain of constant rounded ideals on the associated predomain:*

$$\mathrm{c}\underline{\mathbb{R}}_j \cong \mathrm{cRId}_j(\underline{\mathbb{R}}_{\mathrm{pre}}) \quad \mathrm{c}\bar{\mathbb{R}}_j \cong \mathrm{cRId}_j(\bar{\mathbb{R}}_{\mathrm{pre}}) \quad \mathrm{c}\bar{\underline{\mathbb{R}}}_j \cong \mathrm{cRId}_j(\bar{\underline{\mathbb{R}}}_{\mathrm{pre}}) \quad \mathrm{c}\mathbb{IR}_j \cong \mathrm{cRId}_j(\mathbb{IR}_{j,\mathrm{pre}})$$

$$\mathrm{c}\underline{\mathbb{R}}_j^\infty \cong \mathrm{cRId}_j(\underline{\mathbb{R}}_{\mathrm{pre}}^\infty) \quad \mathrm{c}\bar{\mathbb{R}}_j^\infty \cong \mathrm{cRId}_j(\bar{\mathbb{R}}_{\mathrm{pre}}^\infty) \quad \mathrm{c}\bar{\underline{\mathbb{R}}}_j^\infty \cong \mathrm{cRId}_j(\bar{\underline{\mathbb{R}}}_{\mathrm{pre}}^\infty) \quad \mathrm{c}\mathbb{IR}_j^\infty \cong \mathrm{cRId}_j(\mathbb{IR}_{j,\mathrm{pre}}^\infty)$$

Proof This is just a matter of seeing that Definitions 4.37 and A.56 are equivalent. □

4.3.7 Arithmetic

The usual arithmetic operations on the usual set $\bar{\underline{\mathbb{R}}}$ of extended intervals are well-known (see [Kau80] and [Gol11], who calls them "modal intervals"). For example, addition is straightforward, $(a_1, b_1) + (a_2, b_2) = (a_1 + a_2, b_1 + b_2)$, whereas multiplication involves multiple cases depending on signs. It turns out that disjoint intervals and located disjoint intervals—i.e., real numbers—are each closed under these operations.

Here we discuss the internal j-local arithmetic for an arbitrary modality j. Our main task is to understand when 2-variable functions in $\bar{\underline{\mathbb{R}}}_{\mathrm{pre}} := (\mathbb{Q} \times \mathbb{Q}^{\mathrm{op}})$ give rise to continuous 2-variable functions on $\bar{\underline{\mathbb{R}}}$. For elements $b_1 = (q_1, q_1')$ and $b_2 = (q_2, q_2')$ in $\mathbb{Q} \times \mathbb{Q}^{\mathrm{op}}$ we write $b_1 \prec b_2$ to mean $q_1 < q_2$ and $q_2' < q_1'$. We write $b_1 \leqslant b_2$ to mean $q_1 \leq q_2$ and $q_2' \leq q_1'$. This notation is aligned with that in Appendix A.1.

Proposition 4.44 is a bit technical, but it paves the way for the j-arithmetic, from Theorem 4.46 through the end of the chapter.

Proposition 4.44 *Let $f : \bar{\underline{\mathbb{R}}}_{\mathrm{pre}} \times \bar{\underline{\mathbb{R}}}_{\mathrm{pre}} \to \bar{\underline{\mathbb{R}}}_{\mathrm{pre}}$ be a function satisfying the following four conditions:*

$$\begin{aligned}
&(b_1' \prec b_1) \Rightarrow f(b_1', b_2) \leqslant f(b_1, b_2)(c \prec f(b_1, b_2)) \Rightarrow \exists b_1'.\,(b_1' \prec b_1) \wedge (c \prec f(b_1', b_2)) \\
&(b_2' \prec b_2) \Rightarrow f(b_1, b_2') \leqslant f(b_1, b_2)(c \prec f(b_1, b_2)) \Rightarrow \exists b_2'.\,(b_2' \prec b_2) \wedge (c \prec f(b_1, b_2'))
\end{aligned} \tag{4.20}$$

Then f determines an approximable mapping $f^*\colon \bar{\underline{\mathbb{R}}}_{\text{pre}} \times \bar{\underline{\mathbb{R}}}_{\text{pre}} \to \bar{\underline{\mathbb{R}}}_{\text{pre}}$, given by $f^*(b_1, b_2, b') \Leftrightarrow f(b_1, b_2) < b'$. In turn, f^* determines a continuous morphism of domains

$$F_j := \mathrm{RId}_j(f^*)\colon \bar{\underline{\mathbb{R}}}_j \times \bar{\underline{\mathbb{R}}}_j \to \bar{\underline{\mathbb{R}}}_j$$

for any modality j, and this morphism preserves constants.

Exactly the same statements are true for the other j-numeric domains, i.e. when the predomain $\bar{\underline{\mathbb{R}}}_{\text{pre}}$ is replaced by $\underline{\mathbb{R}}_{\text{pre}}$, $\bar{\mathbb{R}}_{\text{pre}}$, $\mathbb{IR}_{j,\text{pre}}$, $\underline{\mathbb{R}}^{\infty}_{\text{pre}}$, $\bar{\mathbb{R}}^{\infty}_{\text{pre}}$, $\bar{\underline{\mathbb{R}}}^{\infty}_{\text{pre}}$, or $\mathbb{IR}^{\infty}_{j,\text{pre}}$, and the domain $\bar{\underline{\mathbb{R}}}_j$ is replaced by the corresponding domain of j-rounded ideals, $\underline{\mathbb{R}}_j$, $\bar{\mathbb{R}}_j$, $\mathbb{IR}_j$, $\underline{\mathbb{R}}^{\infty}_j$, $\bar{\mathbb{R}}^{\infty}_j$, $\bar{\underline{\mathbb{R}}}^{\infty}_j$, or $\mathbb{IR}^{\infty}_j$.

Proof Both $\mathbb{Q}$ and $\mathbb{Q}^{\text{op}}$ are linear and unbounded, so the predomain $\mathbb{Q} \times \mathbb{Q}^{\text{op}}$ is rounded and has binary joins given by (max, min). The four conditions are exactly those required by Proposition A.41, so have an approximable mapping $f^*\colon \bar{\underline{\mathbb{R}}}_{\text{pre}} \times \bar{\underline{\mathbb{R}}}_{\text{pre}} \to \bar{\underline{\mathbb{R}}}_{\text{pre}}$. It is decidable (in the sense of Definition A.56) because the $<$-relation on $\mathbb{Q}$ is decidable.

By Propositions A.49 and A.8, we obtain a morphism of domains

$$\mathrm{RId}_j(f^*)\colon \mathrm{RId}_j(\bar{\underline{\mathbb{R}}}_{\text{pre}}) \times \mathrm{RId}_j(\bar{\underline{\mathbb{R}}}_{\text{pre}}) \to \mathrm{RId}_j(\bar{\underline{\mathbb{R}}}_{\text{pre}}),$$

or simply $\bar{\underline{\mathbb{R}}}_j \times \bar{\underline{\mathbb{R}}}_j \to \bar{\underline{\mathbb{R}}}_j$, since by definition $\mathrm{RId}_j(\bar{\underline{\mathbb{R}}}_{\text{pre}}) = \bar{\underline{\mathbb{R}}}_j$. This map preserves constants by Proposition A.60. □

For any $f\colon \bar{\underline{\mathbb{R}}}_{\text{pre}} \times \bar{\underline{\mathbb{R}}}_{\text{pre}} \to \bar{\underline{\mathbb{R}}}_{\text{pre}}$, we write $\underline{f}$ and $\bar{f}$ for its first and second projections via the isomorphism $\bar{\underline{\mathbb{R}}}_{\text{pre}} \to \underline{\mathbb{R}}_{\text{pre}} \times \bar{\mathbb{R}}_{\text{pre}}$. Then the continuous map F_j from Proposition 4.44 is given by the formula $F_j\big((\delta_1, \upsilon_1), (\delta_2, \upsilon_2)\big) := (\delta', \upsilon')$ where

$$\begin{aligned}
\delta' d' &\Leftrightarrow j\exists(d_1, d_2, u_1, u_2 : \mathbb{Q}).\Big(\delta_1 d_1 \wedge \delta_2 d_2 \wedge \upsilon_1 u_1 \wedge \upsilon_2 u_2 \\
&\qquad\qquad \wedge \Big(d' < \underline{f}\big((d_1, u_1), (d_2, u_2)\big)\Big)\Big) \\
\upsilon' u' &\Leftrightarrow j\exists(d_1, d_2, u_1, u_2 : \mathbb{Q}).\Big(\delta_1 d_1 \wedge \delta_2 d_2 \wedge \upsilon_1 u_1 \wedge \upsilon_2 u_2 \\
&\qquad\qquad \wedge \Big(\bar{f}\big((d_1, u_1), (d_2, u_2)\big) < u'\Big)\Big)
\end{aligned} \tag{4.21}$$

Indeed, this follows directly from Eq. (A.7).

We include one more technical lemma here, even though it will not be used again until Chap. 7. The reader can feel free to skip to Theorem 4.46.

Lemma 4.45 *Let ϕ_1, ϕ_2, ϕ_3 : `Prop` be such that $(\phi_1 \vee \phi_2) \Rightarrow \phi_3$, and let j_1, j_2, and j_3 be the corresponding closed modalities. Let $f : \bar{\underline{\mathbb{R}}}_{\text{pre}} \times \bar{\underline{\mathbb{R}}}_{\text{pre}} \to \bar{\underline{\mathbb{R}}}_{\text{pre}}$ be a function satisfying the four conditions Eq.* (4.20) *and the following "density"* *condition:*

$$\forall(c : \bar{\mathbb{R}}_{\text{pre}}).\, \exists(b_1, b_2 : \bar{\mathbb{R}}_{\text{pre}}).\, c \prec f(b_1, b_2).$$

Then there is an induced continuous function $F_{j_1,j_2,j_3} : \bar{\mathbb{R}}_{j_1} \times \bar{\mathbb{R}}_{j_2} \to \bar{\mathbb{R}}_{j_3}$ *and it preserves constants.*

Proof This follows from Propositions 4.44, A.41, and A.61. □

Addition

As mentioned above, the usual addition of improper intervals is defined coordinatewise, $(d_1, u_1) + (d_2, u_2) = (d_1 + d_2, u_1 + u_2)$, where $d_1, u_1, d_2, u_2 : \mathbb{Q}$ are rational numbers.

Theorem 4.46 *For any modality j, addition defines a continuous function* $+_j : \bar{\mathbb{R}}_j \times \bar{\mathbb{R}}_j \to \bar{\mathbb{R}}_j$*, which preserves constants. The formula for addition reduces to* $(\delta_1, \upsilon_1) + (\delta_2, \upsilon_2) = (\delta', \upsilon')$*, where*

$$\begin{aligned} \delta' q &\Leftrightarrow j\exists(q_1, q_2 : \mathbb{Q}).\, \delta_1 q_1 \wedge \delta_2 q_2 \wedge (q < q_1 + q_2) \\ \upsilon' q &\Leftrightarrow j\exists(q_1, q_2 : \mathbb{Q}).\, \upsilon_1 q_1 \wedge \upsilon_2 q_2 \wedge (q_1 + q_2 < q) \end{aligned} \tag{4.22}$$

A similar statement holds when $\bar{\mathbb{R}}_j$ *is replaced with any of the other Dedekind j-numeric types.*

Proof Working in the predomain $\mathbb{R}_{\text{pre}} = (\mathbb{Q}, <)$, it is clear that if $q_1 < q_2$ then $q_1 + q' \leq q_2 + q'$ and $q' + q_1 \leq q' + q_2$ for any q'. It is also clear that if $q' < q_1 + q_2$ then there exists some $q_1' < q_1$ such that $q' < q_1' + q_2$ and similarly for q_2. Thus the conditions of Proposition 4.44 are satisfied, and this defines an addition operation for eight of the ten Dedekind j-numeric types. It is easy to see that Eq. (4.22) agrees with Eq. (4.21).

It remains to check that if x_1 and x_2 are j-located, then so is their sum. So choose rationals $d < u$; we want to show $j((d < x_1 + x_2) \vee (x_1 + x_2 < u))$. By the arithmetic locatedness of x_1 (Proposition 4.26), we have

$$j\exists(d_1, u_1 : \mathbb{Q}).\, (d_1 < x_1 < u_1) \wedge (0 < u_1 - d_1 < u - d).$$

Then $d - d_1 < u - u_1$ so there exists d_2, u_2 such that $d - d_1 < d_2 < u_2 < u - u_1$, and hence $d < d_1 + d_2$ and $u_1 + u_2 < u$. By locatedness of x_2, we have $j(d_2 < x_2 \vee x_2 < u_2)$, hence

$$\begin{aligned} j\big((\exists(d_1, d_2 : \mathbb{Q}).\, (d_1 < x_1) \wedge (d_2 < x_2) \wedge (d < d_1 + d_2)) \vee \\ (\exists(u_1, u_2 : \mathbb{Q}).\, (x_1 < u_1) \wedge (x_2 < u_2) \wedge (u_1 + u_2 < u))\big) \end{aligned}$$

□

Subtraction

In Example A.43 we defined approximable mappings for the difference, the max, and the product of two improper intervals. The formula for subtraction reduces to $(\delta_1, \upsilon_1) -_j (\delta_2, \upsilon_2) = (\delta', \upsilon')$, where

$$\begin{aligned} \delta' q &\Leftrightarrow j\exists(d, u : \mathbb{Q}).\, \delta_1 d \wedge \upsilon_2 u \wedge (q < d - u) \\ \upsilon' q &\Leftrightarrow j\exists(d, u : \mathbb{Q}).\, \upsilon_1 u \wedge \delta_1 d \wedge (u - d < q). \end{aligned} \tag{4.23}$$

Remark 4.47 Note that this notion of subtraction extends the classical one for the set $\mathbb{IR}$ of proper intervals, where one thinks of $[d, u]$ as representing the set $\{x \mid d \leq x \leq u\}$. Then the difference is given by

$$[d_1, u_1] - [d_2, u_2] = [d_1 - u_2, u_1 - d_2],$$

which is the set of all pairwise differences. However, note that $\mathbb{IR}$ *does not form a group*, and neither does $\mathbb{IR}$! Subtraction is defined but is not inverse to addition.

Often the classical extended interval domain $\bar{\mathbb{R}}$ is used to correct this problem, embedding $\mathbb{IR}$ into something that does form a group, where, for example, the additive inverse of $[d, u]$ is $[-d, -u]$, sending proper intervals to improper intervals. However this operation is not constructively well-behaved; there does not seem to be an analogous operation on $\bar{\mathbb{R}}$ in a topos. Thus we use the definition given in Eq. (4.23), as it is the most convenient for us, and simply supply the above warning. Note that subtraction is inverse to addition for real numbers: $(\mathbb{R}_j, +_j, 0)$ forms an abelian group for any j, as shown in Corollary 4.29.

Maximum

The formula for the max function given in Example A.43 reduces to $\max_j((\delta_1, \upsilon_1), (\delta_2, \upsilon_2)) = (\delta', \upsilon')$, where

$$\begin{aligned} \delta' q &\Leftrightarrow j\exists(q_1, q_2 : \mathbb{Q}).\, \delta_1 q_1 \wedge \delta_2 q_2 \wedge (q < \max(q_1, q_2)) \\ \upsilon' q &\Leftrightarrow j\exists(q_1, q_2 : \mathbb{Q}).\, \upsilon_1 q_1 \wedge \upsilon_2 q_2 \wedge (\max(q_1, q_2) < q). \end{aligned} \tag{4.24}$$

However, we can do better.

Proposition 4.48 *We have the following equivalences:*

$$q < \max_j(x_1, x_2) \Leftrightarrow j(q < x_1 \vee q < x_2) \text{ and } \max_j(x_1, x_2) < q \Leftrightarrow x_1 < q \wedge x_2 < q$$

Proof Begin with the first statement. For the converse direction, suppose $q < x_1$, the case $q < x_2$ being similar. Then by roundedness $j\exists q_1.\, q < q_1 < x_1$, so for any $q_2 < x_2$ we have $q < \max(q_1, q_2)$. For the forwards direction, suppose $q <$

$\max_j(x_1, x_2)$ in the sense of Eq. (4.24). Then there exists q_1, q_2 with $q_1 < x_1$ and $q_2 < x_2$ and $q < \max(q_1, q_2)$. Then either $q < q_1$ or $q < q_2$, and we are done. The second statement is similarly straightforward, but uses that $j(P \wedge Q) = jP \wedge jQ$. □

Multiplication

The formula for multiplication is more involved, but again it comes directly from Example A.43. It is given by $(\delta_1, \upsilon_1) *_j (\delta_2, \upsilon_2) = (\delta', \upsilon')$, where

$$
\begin{aligned}
\delta' q \Leftrightarrow\ & j\exists(d_1, d_2, u_1, u_2 : \mathbb{Q}).\ \delta_1 d_1 \wedge \delta_2 d_2 \wedge \upsilon_1 u_1 \wedge \upsilon_2 u_2 \wedge \\
& \qquad q < \max(d_1^+ d_2^+, u_1^- u_2^-) - \max(u_1^+ d_2^-, u_2^+ d_1^-) \\
\upsilon' q \Leftrightarrow\ & j\exists(d_1, d_2, u_1, u_2 : \mathbb{Q}).\ \delta_1 d_1 \wedge \delta_2 d_2 \wedge \upsilon_1 u_1 \wedge \upsilon_2 u_2 \wedge \\
& \qquad \max(u_1^+ u_2^+, d_1^- d_2^-) - \max(d_1^+ u_2^-, u_1^- d_2^+) < q
\end{aligned}
\tag{4.25}
$$

where $q^+ = \max(q, 0)$ and $q^- = \max(-q, 0)$ for any $q : \mathbb{Q}$. It may be helpful to note that $q = q^+ - q^-$.

The following lemma—which simplifies the above multiplication formula in the case of positive proper intervals—is straightforward to prove.

Lemma 4.49 *Suppose (δ_1, υ_1) and (δ_2, υ_2) are each j-disjoint. If they are also positive, i.e. $\delta_1 0$ and $\delta_2 0$, then their product is $(\delta_1, \upsilon_1) *_j (\delta_2, \upsilon_2) = (\delta', \upsilon')$ where*

$$
\begin{aligned}
\delta' q &\Leftrightarrow j\exists(d_1, d_2 : \mathbb{Q}).\ \delta_1 d_1 \wedge \delta_2 d_2 \wedge (q < d_1 * d_2) \wedge (0 < d_1) \wedge (0 < d_2) \\
\upsilon' q &\Leftrightarrow j\exists(u_1, u_2 : \mathbb{Q}).\ \upsilon_1 u_1 \wedge \upsilon_2 u_2 \wedge (u_1 * u_2 < q)
\end{aligned}
$$

Theorem 4.50 *For any modality j, multiplication defines a continuous function $*_j : \bar{\mathbb{R}}_j \times \bar{\mathbb{R}}_j \to \bar{\mathbb{R}}_j$, which preserves constants. A similar statement holds when $\bar{\mathbb{R}}_j$ is replaced with any of the other two-sided j-numeric types, $\mathbb{IR}_j$, $\mathbb{R}_j$, $\bar{\mathbb{R}}_j^\infty$, $\mathbb{IR}_j^\infty$, or $\mathbb{R}_j^\infty$.*

Proof The first statement follows from Proposition 4.44 and from Example A.43, where it is shown that all of the functions whose composite is Eq. (4.25) are approximable mappings between predomains, e.g. $(q \mapsto q^+) : \bar{\mathbb{R}}_{\text{pre}} \to \bar{\mathbb{R}}_{\text{pre}}$. The statements for $\mathbb{IR}_j$, $\bar{\mathbb{R}}_j^\infty$, and $\mathbb{IR}_j^\infty$ follow similarly, replacing $\bar{\mathbb{R}}_{\text{pre}}$ with $\mathbb{IR}_{j,\text{pre}}$, $\bar{\mathbb{R}}_{\text{pre}}^\infty$, and $\mathbb{IR}_{j,\text{pre}}^\infty$ respectively. It remains to prove the result for $\mathbb{R}_j$ and $\mathbb{R}_j^\infty$. For this it suffices to show that if $x_1, x_2 : \mathbb{R}_j^\infty$ are located then so is $y := x_1 *_j x_2$. Take any rationals $d < u$.

As a first step, we prove $j(d < y \vee y < u)$ under two preliminary assumptions: that $0 < x_1$ and that $0 < d$. By the multiplicative locatedness of x_1 (Proposition 4.27)

$$j\exists(d_1, u_1 : \mathbb{Q}).\, 0 < d_1 < x_1 < u_1 \wedge \left(\frac{u_1}{d_1} < \frac{u}{d}\right).$$

Choose $q, q' : \mathbb{Q}$ such that $u_1 d < q < q' < u d_1$ and define $d_2 := \frac{q}{d_1 u_1}$ and $u_2 := \frac{q'}{d_1 u_1}$. So $d < d_1 d_2$, $d_2 < u_2$, and $u_1 u_2 < u$, and by locatedness of x_2, we have $j(d_2 < x_2 \vee x_2 < u_2)$. Hence,

$$j\big((\exists(d_1, d_2 : \mathbb{Q}).\, (d_1 < x_1) \wedge (d_2 < x_2) \wedge (d < d_1 d_2)) \vee \\ (\exists(u_1, u_2 : \mathbb{Q}).\, (x_1 < u_1) \wedge (x_2 < u_2) \wedge (u_1 u_2 < u))\big)$$

giving $j(d < y \vee y < u)$ as desired.

As a second step we retain the assumption $0 < x_1$, but we drop the second assumption and consider general $d < u$. Either $d < 0$ or $0 < u$. If $0 < u$, then there exists d' with $\max(0, d) < d' < u$, so $j(d' < y \vee y < u)$ by the first step, and hence $j(d < y \vee y < u)$. If $d < 0$ then $-u < -d$ and, just as before, there exists d' with $\max(0, -u) < d' < -d$, so $j(d' < -y \vee -y < -d)$ by the first step, so $j(d < y \vee y < u)$.

Finally, dropping all preliminary assumptions, take any located x_1 and x_2; we will show $d < y \vee y < u$ for $y := x_1 *_j x_2$. Let $m := \max(-d, u)$, so since $d < 0 \vee 0 < u$ we have $0 < m$, and hence by locatedness we have

$$(-1 < x_1 \vee x_1 < 0) \wedge (0 < x_1 \vee x_1 < 1) \wedge (-m < x_2 \vee x_2 < 0) \wedge (0 < x_2 \vee x_2 < m).$$

This implies $(x_1 < 0) \vee (0 < x_1) \vee (x_2 < 0) \vee (0 < x_2) \vee (|x_1| < 1 \wedge |x_2| < m)$. The first four cases are covered by the second step, above. In the last case $-1 < x_1 < 1$ implies $-|x_2| < y < |x_2|$, and $|x_2| < m$ implies $|x_2| < -d \vee |x_2| < u$, hence $(d < -|x_2| < y) \vee (y < |x_2| < u)$. □

Division

Division is just multiplication by a reciprocal. We next define a reciprocal function $r \mapsto 1/r$ for positive j-local reals $r > 0$. It is easily extended to a reciprocal for negative j-reals $r < 0$ by $r \mapsto -(1/(-r))$.

Consider the predomain $(\mathbb{Q}^{>0}, <)$, its opposite $(\mathbb{Q}^{>0}, >)$, and the product $(\bar{\mathbb{R}}^{>0})_{\text{pre}}$. There is a function $\bar{\mathbb{R}}^{>0}_{\text{pre}} \to \bar{\mathbb{R}}^{>0}_{\text{pre}}$ sending (d, u) to $(1/u, 1/d)$, and it induces an approximable mapping by Proposition A.39. Thus for any $(\delta, \upsilon) : \bar{\mathbb{R}}^{>0}_j$, define the reciprocal to be $(\delta', \upsilon') : \bar{\mathbb{R}}^{>0}_j$, where for a positive rational $q > 0$ we put

$$\delta' q \Leftrightarrow j\exists(u : \mathbb{Q}).\, \upsilon u \wedge (q * u < 1) \qquad \text{and} \qquad \upsilon' q \Leftrightarrow j\exists(d : \mathbb{Q}).\, \delta d \wedge (1 < q * d) \tag{4.26}$$

Lemma 4.51 $(0 < q < 1/x) \Leftrightarrow x < 1/q$, *for any* $x : \bar{\underline{\mathbb{R}}}_j$ *and* $q : \mathbb{Q}$.

Proof Since x is j-closed, it suffices to show $j(x < 1/q)$. By definition, $(0 < q < 1/x) \Leftrightarrow j\exists(u : \mathbb{Q}).\,(x < u) \wedge (q * u < 1)$, and we finish with a chain of equivalences

$$j\exists(u : \mathbb{Q}).\,(x < u) \wedge (q * u < 1) \;\Leftrightarrow\; j\exists(u : \mathbb{Q}).\,(x < u) \wedge (u < 1/q) \;\Leftrightarrow\; j(x < 1/q).$$

□

Proposition 4.52 *If* $x : \bar{\underline{\mathbb{R}}}_j^{>0}$ *is* j*-disjoint, then* $1/x$ *is* j*-disjoint; and if* x *is* j*-located, then* $1/x$ *is* j*-located. Moreover, for any positive real* $r : \mathbb{R}_j^{>0}$ *we have* $r * r' = 1$*, where* $r' := (\delta', \upsilon')$ *is as in Eq.* (4.26).

Proof Let (δ', υ') be the reciprocal of x. For j-disjointness, suppose $0 < q$ and $\delta' q \wedge \upsilon' q$. Then $j\exists d, u.\,\delta d \wedge \upsilon u \wedge q * u < 1 < q * d$, and we obtain $j\bot$. For j-locatedness, suppose $q_1 < q_2$. Then $1/q_2 < 1/q_1$, so $\delta(1/q_2) \vee \upsilon(1/q_1)$, and the result follows by Lemma 4.51.

For the final claim, $1 : \mathbb{R}_j$ is the j-local real represented by $1 : \mathbb{Q}$; see Eq. (4.12). We first show that if x is j-disjoint then $q < x * (1/x) \Rightarrow q < 1$:

$$\begin{aligned} q < x * (1/x) &\Leftrightarrow j\exists q_1, q_2.\,(0 < q_1 < x) \wedge (0 < q_2 < 1/x) \wedge (q < q_1 * q_2) \\ &\Leftrightarrow j\exists q_1, q_2.\,(0 < q_1 < x) \wedge (x < 1/q_2) \wedge (q < q_1 * q_2) \\ &\Rightarrow j\exists q_1, q_2.\,(0 < q_1 < 1/q_2) \wedge q < q_1 * q_2 \\ &\Leftrightarrow j(q < 1) \end{aligned} \tag{4.27}$$

The first equivalence is Lemma 4.49, the second is Lemma 4.51 and Remark 4.9, the implication is by j-disjointness, and the fourth is an easy calculation. It remains to prove that if $r : \mathbb{R}_j$ is real then $q < 1 \Rightarrow q < x * (1/x)$; assume $q < 1$. By Proposition 4.27, we have

$$j\exists q_1, q_2.\,(0 < q_1 < r < q_2) \wedge (q < q_1/q_2),$$

which is equivalent to (4.27) when $x := r$, so we obtain $q < r * (1/r)$. □

One must be careful to note that for numeric types other than $\mathbb{R}$, the distributive law does not hold. For example, if $x = [0, 1]$, $y = [1, 1]$, and $z = [-1, -1]$, then

$$\begin{aligned} x * (y + z) = [0, 1] * ([1, 1] + [-1, -1]) &= [0, 0] \\ &\neq [-1, 1] = [0, 1] + [-1, 0] \\ &= x * y + x * z. \end{aligned}$$

However, it is well-known that the real numbers form a commutative ring in any topos.

Proposition 4.53 *The type* $\mathbb{R}_j$ *of* j*-local real numbers forms a commutative ring satisfying the additional axiom*

$$\forall(r : \mathbb{R}_j).\big(j(r < 0 \vee r > 0) \Leftrightarrow j\exists(r' : \mathbb{R}_j).\, r *_j r' = 1\big). \tag{4.28}$$

Proof In any topos, including the subtopos $\mathcal{E}_j$, the Dedekind real numbers object R forms a residue field, in particular a commutative ring $(R, 0, +, 1, *)$ satisfying $\forall(r : R).\,(r < 0 \vee r > 0) \Leftrightarrow \exists(r' : R).\, r * r' = 1$. A constructive proof can be found in [Joh02, Proposition D.4.7.10]. □

Proposition 4.54 *For any morphism of modalities* $j' \Rightarrow j$*, there is an induced continuous map*

$$j : \mathbb{R}_{j'} \to \mathbb{R}_j$$

defined by sending (δ, υ) *to* $(j\delta, j\upsilon)$*, and it preserves constants, arithmetic, and inequalities. The analogous statement holds for each of the other nine Dedekind numeric types.*

Proof The continuity of the above-defined j follows from Proposition A.47. More explicitly, one can check that each of the j'-axioms for δ (or υ) immediately implies the corresponding j-axiom for $j\delta$ (or $j\upsilon$). The map is continuous because it is the restriction of a morphism of domains.

The fact that j preserves constants follows from Proposition A.59. The fact that it preserves arithmetic follows from Proposition A.50, because all arithmetic operations are induced by approximable mappings. For inequalities, suppose $a <_{j'} b$ in the sense of Eq. (4.16), i.e. $j'\exists q.\, \upsilon_a q \wedge \delta_b q$. This implies $j\exists q.\, j\upsilon_a q \wedge j\delta_b q$, as desired. □

Chapter 5
Axiomatics

In Chap. 4, we explained a connection between toposes, type theory, and logic. We also discussed modalities and numeric types in an arbitrary topos. In the current chapter, we will lay out the signature—meaning the atomic types, atomic terms, and axioms—for our specific topos, $\mathcal{B}$. It turns out that our signature consists of no atomic types, one atomic term, and ten axioms.

The real numbers and other numeric objects are particularly relevant in our study of $\mathcal{B}$. We introduced Dedekind j-numeric objects in Definition 4.21, where the j signifies an arbitrary modality. We will use that generality in Chap. 7, but not in the axioms. That is, in the present chapter, none of our numeric objects will be decorated by a j.

Notation 5.1 Recall from Definition 4.21 that by $\bar{\underline{\mathbb{R}}}$, $\mathbb{R}$, and $\mathbb{R}^{\infty}$, we mean the Dedekind numeric types $\bar{\underline{\mathbb{R}}}_j$, $\mathbb{R}_j$, and $\mathbb{R}_j^{\infty}$, where $j = \mathrm{id}$ is the identity modality. These numeric types are called the *improper intervals*, the *real numbers*, and the *unbounded real numbers*, respectively.

One reason that these Dedekind numeric types are so important here is that `Time` is defined in terms of them. Semantically speaking, a behavior of type `Time` assigns to every duration ℓ some interval $[t_0, t_0 + \ell]$ of real numbers: the values shown on a given clock during that interval. In other words, a behavior of type `Time` is a "time window." We will introduce the type `Time` in Sect. 5.2, but we begin in Sect. 5.1 with a discussion of what we call constant types.

Note: Although we will write our proofs in typical mathematical style, it is important to make note of the fact that *every proof and definition in Chap. 5 can be formalized in the type theory of Chap. 4.*

P. Schultz, D. I. Spivak, *Temporal Type Theory*, Progress in Computer Science and Applied Logic 29, https://doi.org/10.1007/978-3-030-00704-1_5

5.1 Constant Types

The notion of "constant types" plays an important role in these axiomatics. The word "constant" does not really refer to the type itself, but to its intended semantics.

Rather than defining a property of constancy for types, we consider it an axiomatic property. Specifically, whenever in an axiom we say "for all constant types C", we regard the axiom as an axiom *schema*.

Recall that a proposition $P : \texttt{Prop}$ is called *decidable* if $P \vee \neg P$ holds. We call $P : X \to \texttt{Prop}$ a *decidable predicate* if $P(x)$ is decidable for all $x : X$. If P is a decidable predicate, we also refer to the subtype $\{x : X \mid P(x)\}$ as a *decidable subtype*.

Definition 5.2 (Constant Types) We define a *constant type* to be any member of the smallest collection of types containing $\texttt{1}$, $\mathbb{N}$, $\mathbb{Z}$, $\mathbb{Q}$, and $\mathbb{R}^{\infty}$, and that is closed under finite sums, finite products, function types, and decidable subtypes.

We will see in Corollary 5.11 that $\mathbb{R}$ is also a constant type. This is just a definition; the axioms throughout this chapter will provide all the associated type-theoretic meaning.

Remark 5.3 It may come as a bit of a surprise to experts in topos theory that the real numbers object $\mathbb{R}$ is semantically a constant sheaf in $\mathcal{B} = \mathsf{Shv}(S_{\mathbb{IR}/\rhd})$. We prove this in Corollary 6.7, during our discussion of soundness.

For the time being, we can only give some evidence for the claim, namely that the type of real numbers object in the slice topos $\mathsf{Shv}(S_{\mathbb{IR}})$ is constant (see Proposition 3.11). Indeed, for any topological space X, the internal sheaf of Dedekind reals is externally the sheaf of continuous real-valued functions on X, and these are all constant for $X = \mathbb{IR}$, as we saw in Corollary 2.38.

5.1.1 First Axioms for Constant Types

We now begin laying out the axioms and their more immediate consequences. Our first axiom says that the cuts δ, υ defining any real number are decidable. The soundness of Axiom 1 is proven in Proposition 6.16.

Axiom 1 $\forall((\delta, \upsilon) : \mathbb{R}^{\infty})(q : \mathbb{Q}).\, (\delta q \vee \neg\delta q) \wedge (\upsilon q \vee \neg\upsilon q)$.

We can now begin proving some propositions. As mentioned above, every proof in Chap. 5—though written in typical mathematical style—can be formalized in a standard type theory and higher-order logic, as outlined in Chap. 4.

Proposition 5.4 (Trichotomy) *For all $r_1, r_2 : \mathbb{R}^{\infty}$, either $r_1 < r_2$ or $r_1 = r_2$ or $r_1 > r_2$. In particular, $(r_1 = r_2) \vee \neg(r_1 = r_2)$.*

Proof Let $r = (\delta, \upsilon) := r_2 - r_1$; it suffices to show that $(\delta 0) \vee (0 = r) \vee (\upsilon 0)$. By Axiom 1, it suffices to show $0 = r$ assuming $\neg\delta 0$ and $\neg\upsilon 0$. It follows easily that for all $q : \mathbb{Q}$ we have $(q < 0) \Leftrightarrow \delta q$ and $(0 < q) \Leftrightarrow \upsilon q$, so $0 = r$. The second statement is clear because $(r < 0) \Rightarrow \neg(0 = r)$. □

The next axiom is referred to by [Joh02, C.1.1.16.e] as a "dual Frobenius rule." It also is related to the notion of "overtness" for locales; one may say that any constant object is overt. Kuratowski-finite objects in a topos also satisfy this condition; see [Joh02, Lemma D.5.4.6]. The soundness of Axiom 2 is proven in Proposition 6.17.

Axiom 2 For any constant type $\mathtt{C}$, any $P : \mathtt{Prop}$, and any $P' : \mathtt{C} \to \mathtt{Prop}$:

$$\big(\forall(c : \mathtt{C}).\, P \vee P'(c)\big) \Rightarrow \big(P \vee \forall(c : \mathtt{C}).\, P'(c)\big).$$

Note that the converse is straightforward to prove constructively, hence Axiom 2 implies $\big(\forall(c : \mathtt{C}).\, P \vee P'(c)\big) \Leftrightarrow \big(P \vee \forall(c : \mathtt{C}).\, P'(c)\big)$.

Remark 5.5 Axiom 2 is a co-distributivity property $(\bigwedge_c (P \vee P'(c)) = P \vee \bigwedge_c P'(c))$, with universal quantification over a constant type taking the role of infinite intersections.

We say that a type X is *inhabited* if it satisfies $\exists(x : X).\, \top$.

Proposition 5.6 *For any constant type* $\mathtt{C}$ *and decidable predicate* $P : \mathtt{C} \to \mathtt{Prop}$ *(satisfying* $\forall(c : \mathtt{C}).\, Pc \vee \neg Pc$*), we have*

$$\exists(c : \mathtt{C}).\, Pc \vee \forall(c : \mathtt{C}).\, \neg Pc. \tag{5.1}$$

It follows immediately that both $\exists(c : \mathtt{C}).\, Pc$ *and* $\forall(c : \mathtt{C}).\, Pc$ *are decidable. In particular, the proposition that* $\mathtt{C}$ *is inhabited, namely* $\exists(c : \mathtt{C}).\, \top$*, is decidable.*

Proof Because P is decidable, we have $\forall(c : \mathtt{C}).\, (\exists(c' : \mathtt{C}).\, Pc') \vee \neg Pc$, and Eq. (5.1) follows from Axiom 2. It follows immediately that $\exists(c : C).\, Pc$ is decidable; since $\neg P$ is decidable, it also follows immediately that $\forall(c : C).\, Pc$ is decidable. The final statement follows by taking $Pc = \bot$. □

Say that a type A *has decidable equality* if $\forall(a, a' : A).\, (a = a') \vee \neg(a = a')$ holds.

Proposition 5.7 *Let* A *and* B *be types. If* A *and* B *have decidable equality, then so do the types* $A + B$, $A \times B$, *and* $\{a : A \mid Pa\}$ *for any decidable predicate* $P : A \to \mathtt{Prop}$.

Proof The case of decidable subtypes is easiest, and we leave this case to the reader. Consider the case of finite products. Let A and B be types with decidable equality, and let $x, y : A \times B$ be terms. Then, we have $(\pi_A(x) = \pi_A(y)) \vee \neg(\pi_A(x) = \pi_A(y))$ and similarly for π_B. In the case $\pi_A(x) = \pi_A(y)$ and $\pi_B(x) = \pi_B(y)$, we get $x = (\pi_A(x), \pi_B(x)) = (\pi_A(y), \pi_B(y)) = y$. For the other three cases, suppose without loss of generality that $\neg(\pi_A(x) = \pi_A(y))$. Then, if $x = y$, it follows that $\pi_A(x) = \pi_A(y)$ giving $\bot$. Hence, $\neg(x = y)$.

For the case of finite sums, we make use of Remark 4.3, including the notations is_empty and is_sing. Let A and B be types with decidable equality, and let $x, y : A + B$ be terms. By (4.6), we can consider x to be a pair $(x_A, x_B) : (A \to \mathtt{Prop}) \times (B \to \mathtt{Prop})$ such that $(\text{is\_sing}(x_A) \wedge \text{is\_empty}(x_B)) \vee (\text{is\_empty}(x_A) \wedge \text{is\_sing}(x_B))$. We similarly let y be the pair (y_A, y_B).

First, consider the case $\text{is\_sing}(x_A) \wedge \text{is\_sing}(y_A)$. Then, there exist elements $a_x, a_y : A$ such that $x_A(a_x)$, $y_A(a_y)$, $\forall(a : A). x_A(a) \Rightarrow a = a_x$, and $\forall(a : A). x_Y(a) \Rightarrow a = a_y$. By decidability of A, we have $(a_x = a_y) \vee \neg(a_x = a_y)$. It is now easy to show that $a_x = a_y \Rightarrow x = y$ (using propositional extensionality) and $\neg(a_x = a_y) \Rightarrow \neg(x = y)$.

In the case $\text{is\_sing}(x_A) \wedge \text{is\_empty}(y_A)$, we have $x_A(a_x)$ and $\neg y_A(a_x)$, therefore $\neg(x = y)$. The other two cases are symmetric. □

Corollary 5.8 (Constant Types Have Decidable Equality) *If* $\mathtt{C}$ *is a constant type, then it has decidable equality.*

Proof The type 1 obviously has decidable equality, and one can prove by induction that $\mathbb{N}$ and $\mathbb{Z}$ do too. The type $\mathbb{Q}$ is a quotient of $\mathbb{N} \times \mathbb{Z}$, under which equality $(n, z) = (n', z')$ holds iff $n * z' = n' * z$, which is decidable. And, $\mathbb{R}^\infty$ has decidable equality by Proposition 5.4.

By Definition 5.2, it suffices to check that finite sums, finite products, function types, and decidable subtypes of types with decidable equality have decidable equality. This has been done already in Proposition 5.7 for all but function types. So, suppose that $\mathtt{C}$ and $\mathtt{C}'$ are constant types with decidable equality. Given $f, g : \mathtt{C} \to \mathtt{C}'$, take $P : \mathtt{C} \to \mathtt{Prop}$ to be the decidable predicate $P := \lambda(c : \mathtt{C}). fc = gc$. By Proposition 5.6, $\forall(c : \mathtt{C}). fc = gc$ is decidable, and the result follows by function extensionality, Axiom 0. □

Because constant types are closed under decidable subtypes (see Definition 5.2), we have the following corollary to Axiom 2.

Corollary 5.9 *For any constant type* $\mathtt{C}$*, any* $P : \mathtt{Prop}$*,* $P' : \mathtt{C} \to \mathtt{Prop}$*, and any* $Q : \mathtt{C} \to \mathtt{Prop}$ *such that* $\forall(c : \mathtt{C}). Q(c) \vee \neg Q(c)$*:*

$$\big(\forall(c : \mathtt{C}). Q(c) \Rightarrow (P \vee P'(c))\big) \Leftrightarrow \big(P \vee \forall(c : \mathtt{C}). Q(c) \Rightarrow P'(c)\big).$$

Recall from Definition 4.21 the terminology of Dedekind numeric objects, including the conditions called "roundedness, boundedness, disjointness, and locatedness." We will use these names freely in this section. For example, any $u : \mathbb{R}^\infty$ is rounded, meaning $\forall(q : \mathbb{Q}). q < u \Rightarrow \exists(q' : \mathbb{Q}). q < q' < u$.

Proposition 5.10 *Let* $d, u, r : \mathbb{R}^\infty$ *with* $d < u$*. Then,* $(d < r) \vee (r < u)$*.*

Proof By definition, $d < u \Leftrightarrow \exists(q : \mathbb{Q}). d < q < u$. By roundedness, there exists $q' : \mathbb{Q}$ with $q < q' < u$. Then by locatedness, $(q < r) \vee (r < q')$, which implies the result. □

Definition 5.2 says that $\mathbb{R}^\infty$ is constant; now, we can prove that $\mathbb{R}$ is constant also.

Corollary 5.11 *The type* $\mathbb{R}$ *of Dedekind reals is constant.*

Proof By Definition 4.21, $\mathbb{R} = \{(\delta, \upsilon) : \mathbb{R}^\infty \mid P(\delta, \upsilon)\}$, where $P : \mathbb{R}^\infty \to \texttt{Prop}$ is the boundedness condition: $P(\delta, \upsilon) = (\exists (q : \mathbb{Q}).\, \delta q) \wedge (\exists (q : \mathbb{Q}).\, \upsilon q)$. It suffices by Definition 5.2 to show that P is decidable. This follows from Proposition 5.6 and Axiom 1. □

A function $f : A \to B$ is an *internal surjection* if it satisfies $\forall (b : B).\, \exists (a : A).\, f(a) = b$. It is an *internal injection* if it satisfies $\forall (a, a' : A).\, (f(a) = f(a')) \Rightarrow (a = a')$. It is an *internal bijection* if it satisfies both. We will not need the following, but it provides an example of something we can now prove in the type theory.

Proposition 5.12 *There is an internal bijection* $\mathbb{R} \sqcup \mathtt{1} \sqcup \mathtt{1} \to \mathbb{R}^\infty$.

Proof Define the function to send $(\delta, \upsilon) : \mathbb{R}$ to $(\delta, \upsilon) : \mathbb{R}^\infty$, it sends the first copy of $\mathtt{1}$ to $(\bot, \top)$, or more precisely the pair $(\lambda(q : \mathbb{Q}).\, \bot, \lambda(q : \mathbb{Q}).\, \top)$, and similarly it sends the second copy of $\mathtt{1}$ to $(\top, \bot)$. This is clearly injective since $\mathbb{Q}$ is inhabited and the cuts for real numbers are bounded. The function is surjective because for every $(\delta, \upsilon) : \mathbb{R}^\infty$, we can decide if δ or υ is bounded or not by Proposition 5.6. If one is unbounded (i.e., $\bot$), then the other is $\top$ by locatedness. □

5.1.2 Interlude: How to Read the Axiomatics Section

Now that the reader has seen some axioms and propositions that follow, he or she may be wondering about how the authors chose the axioms and consequences they did. Like in any mathematical theory, the soundness of an axiom or result is deduced from firm principles, whereas the choice of which axioms and results to announce is decided by history and necessity. One is science, the other is art.

The consistency of all of our axioms is proven in Sect. 6.5, by showing that they are sound in the sheaf topos $\mathcal{B} = \mathsf{Shv}(S_{\mathbb{IR}/\rhd})$. The soundness of the lemmas, propositions, etc. in this section then follows from the soundness of higher-order logic for toposes; see, e.g., [LS88].

Note that there is little hope that a finite set of axioms would be complete for $\mathcal{B}$. Thus, we had to make choices, and the question one might ask is how we did so. Indeed, we carried many competing criteria in mind while choosing the ten axioms and the consequences we announce. We were guided throughout by the National Airspace System "safe-separation" case study, detailed in Sect. 8.5, which involves a combination of differential equations, discrete-time signaling, and time delays. We chose this example because it was sufficiently complex and "exterior" to ourselves as mathematicians. Moreover, when tackling it, we were forced to define derivatives with respect to our internal notion of $t : \texttt{Time}$, and we were able to prove the Leibniz property internally to the logic. Having done so, we gained confidence that our system is indeed fairly robust.

Our criteria roughly were that the axioms should:

- Be powerful enough to prove the "safe-separation property";
- Be powerful enough to prove other facts we thought "should be provable";
- Be written at a "low logical level," e.g., commutation of $\vee$ and $\wedge$;
- Capture "simple geometric facts" about the topos $\mathcal{B}$; and
- Be "not too numerous" in number.

For example, Axiom 2 turns out to be quite powerful, it is written at a low logical level, and it tells us about how constant objects behave in $\mathcal{B}$. However, sometimes the above criteria conflict with each other. For example, Axioms 3c and 3d are chosen for their power, even though they are logically a bit convoluted and the geometric facts they convey are not intuitive.

A reader who has understood the rules of higher-order constructive logic (see Sect. 4.1.2) will have no trouble with the proofs in this section: they are all simply logical deductions. We try to give hints about the semantics when possible, but our goal is build up enough theory to internally define derivatives and eventually succeed in proving the safety property. We now return to these axioms and their results.

5.2 Introducing `Time`

We gave a semantic definition of `Time` as an object in $\mathcal{B}$ in Sect. 3.4. Below, we will define a certain type—in our internal language—that we also call `Time`. We will prove that the first is a valid semantic meaning of the second in Lemma 6.10.

`Time` will be defined as a subtype of the extended intervals $\bar{\underline{\mathbb{R}}}$ (see Definition 4.21), classified by a predicate:

$$\text{unit\_speed} : \bar{\underline{\mathbb{R}}} \to \texttt{Prop}.$$

This is the one atomic term we need to define our temporal type theory. With it in hand, we define `Time` to be the following subtype (see Sect. 4.1.3):

$$\texttt{Time} := \{ (\delta, \upsilon) : \bar{\underline{\mathbb{R}}} \mid \text{unit\_speed}\,(\delta, \upsilon) \}. \tag{5.2}$$

Just like we reserved the letters q, r, etc. for variables of type rationals and reals (see Definition 4.21), we reserve the letters t, t', t_1, t_2, etc. for variables of type `Time`.

Notation 5.13 We use the usual notation for inequalities comparing times $t = (\delta_t, \upsilon_t) : \texttt{Time}$ with rationals $q : \mathbb{Q}$ and unbounded reals $r = (\delta_r, \upsilon_r) : \mathbb{R}^{\infty}$:

$$\begin{aligned} q < t &:= \delta_t q & \qquad r < t &:= \exists (q : \mathbb{Q}).\, \upsilon_r q \wedge \delta_t q \\ t < q &:= \upsilon_t q & \qquad t < r &:= \exists (q : \mathbb{Q}).\, \upsilon_t q \wedge \delta_r q \end{aligned} \tag{5.3}$$

Thus using our notation, the top right-hand equation could be rewritten $r < t \Leftrightarrow \exists q.\, r < q < t$. We also introduce the following notations for $d, u, r : \mathbb{R}^\infty$ and $t : \texttt{Time}$:

- $[d, u] \ll t := (d < t) \wedge (t < u)$,
- $t \,\#\, [d, u] := (t < d) \vee (u < t)$,
- $t \,\#\, r := t \,\#\, [r, r]$.

The symbol # is pronounced *apart*.

Note that $t \,\#\, 0$ means $(t < 0) \vee (0 < t)$ which is strictly stronger than $\neg(t = 0)$. Indeed, we will see in Proposition 5.15 that the latter is automatically true.

Axiom 3a ensures that Time is inhabited (it is equivalent to $\exists(t : \texttt{Time}).\top$ by Eq. (5.2)). Axiom 3b implies that the cuts are disjoint, but it is stronger than disjointness; it can be read as saying "if the time t is never greater than q, then it is always less than q," and dually. The soundness of Axioms 3a–3d is shown in Proposition 6.18.

Axiom 3

3a. $\exists((\delta, \upsilon) : \bar{\mathbb{R}}).\, \text{unit\_speed}\,(\delta, \upsilon)$.
3b. $\forall(q : \mathbb{Q})((\delta, \upsilon) : \texttt{Time}).\, (\neg\delta q \Leftrightarrow \upsilon q) \wedge (\neg\upsilon q \Leftrightarrow \delta q)$.
3c. $\forall(t : \texttt{Time})(P : \mathbb{Q} \to \texttt{Prop}).\, (\forall q.\, Pq \vee \neg Pq) \Rightarrow (\forall q.\, Pq \vee q < t) \Rightarrow \exists q.\, Pq \wedge q < t$.
3d. $\forall(t : \texttt{Time})(P : \mathbb{Q} \to \texttt{Prop}).\, (\forall q.\, Pq \vee \neg Pq) \Rightarrow (\forall q.\, Pq \vee t < q) \Rightarrow \exists q.\, Pq \wedge t < q$.

Axiom 3b can be rewritten less pedantically using (5.3):

$$\forall(q : \mathbb{Q})(t : \texttt{Time}).\, (\neg(q < t) \Leftrightarrow (t < q)) \wedge (\neg(t < q) \Leftrightarrow (q < t)).$$

The primary use of Axioms 3c and 3d will be to ensure that many axioms stated in terms of rationals in fact generalize to reals. For a first example, we can generalize Axiom 3b.

Proposition 5.14 *For all times t : Time and unbounded reals $r : \mathbb{R}^\infty$:*

$$\neg(t < r) \Leftrightarrow r < t. \qquad \textit{and} \qquad \neg(r < t) \Leftrightarrow t < r$$

Proof We prove $\neg(t < r) \Leftrightarrow r < t$. The left side is by definition $\neg(\exists q.\, (t < q) \wedge (q < r))$. This is constructively equivalent to $\forall q.\, \neg((t < q) \wedge (q < r))$. Using $(q < r) \vee \neg(q < r)$ from Axiom 1, this is equivalent to $\forall q.\, \neg(t < q) \vee \neg(q < r)$, which is equivalent to $\forall q.\, \neg(q < r) \vee (q < t)$ by Axiom 3b. Because $q < r$ is decidable by Axiom 1, we can apply Axiom 3c to get $\exists q.\, \neg(q < r) \wedge (q < t)$. Taking such a q, there exists a q' with $q < q' < t$ because t is rounded ($\texttt{Time} \subseteq \bar{\mathbb{R}}$), and because r is located we have $(r < q') \vee (q < r)$. Since $\neg(q < r)$, we have proven $\exists q'.\, (r < q') \wedge (q' < t)$, i.e., $r < t$, as desired. □

In particular, the axioms for Time imply that it is nowhere constant.

Proposition 5.15 $\forall(t : \texttt{Time})(r : \mathbb{R}^\infty).\,\neg(t = r)$.

Proof If $t = r$, then $\neg(t < r)$ and $\neg(r < t)$, which implies $\bot$ by Proposition 5.14. □

Proposition 5.16 says that if t is never between d and u, then it is either always less than d or always greater than u.

Proposition 5.16 *For all times* t : `Time` *and reals* $d, u : \mathbb{R}^\infty$, *if* $d < u$ *then* $\neg([d, u] \ll t) \Leftrightarrow (t \# [d, u])$.

Proof Let $d < u$. The backward direction follows from Axiom 3b, so consider the forward direction. Assume $\neg(\exists(d', u' : \mathbb{Q}).\, d < d' < t < u' < u)$, we want to show $(t < d) \vee (u < t)$. By Axioms 1 and 3b, the assumption is constructively equivalent to $\forall(d', u' : \mathbb{Q}).\,\neg(d < d' < t) \vee \neg(t < u' < u)$. Using Axiom 2, we can split this up into $(\forall d'.\,\neg(d < d' < t)) \vee (\forall u'.\,\neg(t < u' < u))$. By Proposition 5.14, this is equivalent to $(t < d) \vee (u < t)$. □

Proposition 5.17 *For all* t : `Time` *and* $r, r' : \mathbb{R}^\infty$, *there is an equivalence between* $t \# [r, r']$ *and* $\forall(q, q' : \mathbb{Q}).\,((r < q) \wedge (q' < r')) \Rightarrow t \# [q, q']$.

Proof The proposition is proven by the following series of equivalences:

$\forall(q, q' : \mathbb{Q}).\,((r < q) \wedge (q' < r')) \Rightarrow t \# [q, q']$	
$\forall(q, q' : \mathbb{Q}).\,\neg(r < q) \vee \neg(q' < r') \vee (t < q) \vee (q' < t)$	Axiom 1
$(\forall q.\,\neg(r < q) \vee (t < q)) \vee (\forall q'.\,\neg(q' < r') \vee (q' < t))$	Axiom 2
$(t < r) \vee (r' < t)$	Axioms 3c and 3d

□

Recall from Theorem 4.28 that, even without any axioms introduced in this chapter, $\mathbb{R}$ forms a group under $+$. The next axiom encodes the essential connection between `Time` and $\mathbb{R}$, namely that `Time` is an "$\mathbb{R}$-torsor." This simply means that we can add a real number to a time to get a new translated time, that given any two times we can compute the difference between them as a real number, and moreover that these two operations are inverse. The soundness of Axiom 4 is proven in Proposition 6.19.

Axiom 4 `Time` is an $\mathbb{R}$-torsor:

4a. $\forall(t : \texttt{Time})(r : \mathbb{R}).\, t + r \in \texttt{Time}$,
4b. $\forall(t_1, t_2 : \texttt{Time}).\,\exists!(r : \mathbb{R}).\, t_1 + r = t_2$.

Proposition 5.18 *For any* t : `Time` *and* $r, r' : \mathbb{R}$, *it is the case that* $(t + r < r') \Leftrightarrow (t < r' - r)$.

Proof It follows from definitions (Eqs. (5.3) and (4.14)) that $t + r < r'$ iff $\exists(q_1, q_2 : \mathbb{Q}).\,(t < q_1) \wedge (r < q_2) \wedge (q_1 + q_2 < r')$. This holds iff $\exists(q_1', q_2').\,(t < q_1 + q_2') \wedge (q_1' <$

$s)\wedge(r < -q_2')$, where $q_1' = q_1+q_2$ and $q_2' = -q_2$. The latter holds iff $t < r'+(-r)$; see Corollary 4.29. □

Proposition 5.19 Time *has decidable equality.*

Proof Combine Axiom 4, Theorem 4.28, and Corollary 5.8 applied to the constant type $\mathbb{R}$. □

The next axiom states that for any terms t : Time and $q : \mathbb{Q}$, the propositions $t < q$ and $q < t$ are each *coprime*. Semantically, this corresponds to the fact that the corresponding open subsets of the interval domain are filtered (inhabited, up-closed, and down-directed); see Proposition 2.24. The soundness of Axiom 5 is proven in Proposition 6.20.

Axiom 5 For all $d, u : \mathbb{Q}$, all t : Time, and all P, Q : Prop:

5a. $(t < u \Rightarrow (P \vee Q)) \Rightarrow \big((t < u \Rightarrow P) \vee (t < u \Rightarrow Q)\big)$.
5b. $(d < t \Rightarrow (P \vee Q)) \Rightarrow \big((d < t \Rightarrow P) \vee (d < t \Rightarrow Q)\big)$.

The converse of each statement is easy. We can immediately generalize Axiom 5 from $\mathbb{Q}$ to $\mathbb{R}$.

Proposition 5.20 *For all* $d, u : \mathbb{R}^\infty$*, all* t : Time*, and all* P, Q : Prop*:*

1. $(t < u \Rightarrow (P \vee Q)) \Leftrightarrow \big((t < u \Rightarrow P) \vee (t < u \Rightarrow Q)\big)$.
2. $(d < t \Rightarrow (P \vee Q)) \Leftrightarrow \big((d < t \Rightarrow P) \vee (d < t \Rightarrow Q)\big)$.

Proof The two statements are similar, and the backward direction of each is trivial, so we just consider the forward direction of the first. Choosing $u = (\delta, \upsilon) : \mathbb{R}^\infty$, we can rewrite the hypothesis by undwinding definitions (5.3) and using Axiom 5:

$$\begin{aligned}
\big((\exists q.\, t < q \wedge \delta q) \Rightarrow (P \vee Q)\big) &\Leftrightarrow \forall q.\, \big((t < q \wedge \delta q) \Rightarrow (P \vee Q)\big)\\
&\Leftrightarrow \forall q.\, \big(\delta q \Rightarrow (t < q \Rightarrow (P \vee Q))\big)\\
&\Leftrightarrow \forall q.\, \big(\delta q \Rightarrow ((t < q \Rightarrow P) \vee (t < q \Rightarrow Q))\big)
\end{aligned}$$

We can similarly rewrite the conclusion:

$$\begin{aligned}
&\big((\exists q.\, t < q \wedge \delta q) \Rightarrow P\big) \vee \big((\exists q.\, t < q \wedge \delta q) \Rightarrow Q\big)\\
&\quad\Leftrightarrow \big(\forall q.\, (t < q \wedge \delta q) \Rightarrow P\big) \vee \big(\forall q.\, (t < q \wedge \delta q) \Rightarrow Q\big)\\
&\quad\Leftrightarrow \forall q_1, q_2.\, \big((t < q_1 \wedge \delta q_1) \Rightarrow P\big) \vee \big((t < q_2 \wedge \delta q_2) \Rightarrow Q\big)
\end{aligned}$$

where the backward direction of the last iff is by two applications of Axiom 2.

We will now prove the desired (reformulated) implication. Let q_1 and q_2 be arbitrary rationals, and assume without loss of generality that $q_1 \le q_2$. By Axiom 1, we have $\delta q_2 \vee \neg\delta q_2$. If $\neg\delta q_2$, then $(t < q_2 \wedge \delta q_2) \Rightarrow Q$ vacuously. On the other hand, if δq_2, then by the hypothesis we have $(t < q_2 \Rightarrow P) \vee (t < q_2 \Rightarrow Q)$. In the second case, we're clearly done. In the first case, since we have $(t < q_1 \wedge \delta q_1)$ implies $t < q_2$ because $q_1 \leq q_2$, and this implies P. □

Lemma 5.21 *For all $d, u : \mathbb{R}^\infty$, and all t : `Time`, the proposition $[d, u] \ll t$ is coprime, i.e., for all P, Q : `Prop`:*

$$([d, u] \ll t \Rightarrow (P \vee Q)) \Leftrightarrow \big(([d, u] \ll t \Rightarrow P) \vee ([d, u] \ll t \Rightarrow Q)\big).$$

Proof For any R : `Prop`, $[d, u] \ll t \Rightarrow R$ is equivalent to $d < t \Rightarrow (t < u \Rightarrow R)$. The lemma then follows from two applications of Proposition 5.20. □

Lemma 5.22 *For any $d, u : \mathbb{R}^\infty$ and any t : `Time`, if $d < u$ then:*

a. $(t < u \Rightarrow d < t) \Leftrightarrow d < t$,
b. $(d < t \Rightarrow t < u) \Leftrightarrow t < u$,
c. $(t < u \Rightarrow t < d) \Leftrightarrow (t < d \vee u < t)$,
d. $(d < t \Rightarrow u < t) \Leftrightarrow (t < d \vee u < t)$.
If $d < r < u$ and $d < r' < u$, then
e. $([d, u] \ll t \Rightarrow t < r) \Leftrightarrow t \# [r, u]$,
f. $([d, u] \ll t \Rightarrow r < t) \Leftrightarrow t \# [d, r]$,
g. $([d, u] \ll t \Rightarrow t \# [r, r']) \Leftrightarrow t \# [r, r']$,
h. $(t < u \Rightarrow t \# [r, r']) \Leftrightarrow t \# [r, r']$,
i. $(d < t \Rightarrow t \# [r, r']) \Leftrightarrow t \# [r, r']$.
For any r, if $d < u$, then
j. $(t \# [r, d] \Rightarrow t < u) \Leftrightarrow t < u$,
k. $(t \# [u, r] \Rightarrow d < t) \Leftrightarrow d < t$.
If $(r < u) \wedge (r' < u)$ or $(d < r) \wedge (d < r')$ (making no assumption about the order of d, u), then:
l. $(t \# [u, d] \Rightarrow t \# [r, r']) \Leftrightarrow t \# [r, r']$.

Proof The reverse directions are all trivial.

For the forward direction of a., it suffices by Proposition 5.14 to show that $\neg(t < d)$: supposing $t < d$, then $t < u$ (since $d < u$), which by assumption implies $d < t$, and by Proposition 5.14 again this is a contradiction. The proof of b. is dual.

For the forward direction of c., it suffices by Proposition 5.16 to show that $\neg([d, u] \ll t)$: if $d < t < u$ then by assumption $t < d$, a contradiction (by Proposition 5.14). Similarly for d.

For the forward direction of e., it suffices by c. to show that $t < u \Rightarrow t < r$. Assuming $t < u$, we have $d < t \Rightarrow t < r$, which by b. implies $t < r$. Similarly for f.

For the forward direction of g., suppose $[d, u] \ll t \Rightarrow t \# [r, r']$. By Lemma 5.21, we have $([d, u] \ll t \Rightarrow t < r) \vee ([d, u] \ll t \Rightarrow r' < t)$, which by e. and f. implies $t \# [r, u] \vee t \# [d, r']$, or equivalently $t \# [r, r']$.

Finally, h. and i. are immediate from g., using just $([d, u] \ll t) \Rightarrow (t < u)$; j. follows easily from b., and likewise k. from a.; and l. is immediate from h. and i using:

$$(t \,\#\, [u,d] \Rightarrow t \,\#\, [r,r']) \Leftrightarrow \big((t < u \Rightarrow t \,\#\, [r,r']) \wedge (d < t \Rightarrow t \,\#\, [r,r'])\big).$$

□

We will also need a sort of infinitary version of Axiom 5. The soundness of Axiom 6 is proven in Proposition 6.21.

Axiom 6 Let $\mathtt{C}$ be an inhabited constant type. Then for all $t : \mathtt{Time}$ and $P : \mathtt{C} \to \mathtt{Prop}$:

6a. $\forall(u : \mathbb{R}^\infty).\, (t < u \Rightarrow \exists(c : \mathtt{C}).\, P(c)) \Rightarrow \big(\forall(u' : \mathbb{Q}).\, u' < u \Rightarrow \exists(c : \mathtt{C}).\, t < u' \Rightarrow P(c)\big)$.
6b. $\forall(d : \mathbb{R}^\infty).\, (d < t \Rightarrow \exists(c : \mathtt{C}).\, P(c)) \Rightarrow \big(\forall(d' : \mathbb{Q}).\, d < d' \Rightarrow \exists(c : \mathtt{C}).\, d' < t \Rightarrow P(c)\big)$.

Proposition 5.23 *The converse of Axiom 6a and the converse of Axiom 6b also hold.*

Proof This follows from the definition of $t < u$, i.e., that $\exists(u' : \mathbb{Q}).\, t < u' < u$. □

Remark 5.24 Note that Axiom 6 without the extra $u' < u$ clause, namely:

$$(t < u \Rightarrow \exists(c : \mathtt{C}).\, P(c)) \Rightarrow^{?} \exists(c : \mathtt{C}).\, t < u \Rightarrow P(c),$$

does not hold in our semantics. Indeed, let $\mathtt{C} := \mathbb{Q}$ and in the context of some t, u define $P(q) := t < q < u$. While it is true that $t < u$ implies $\exists q.\, t < q < u$, it is not the case that there exists q with $(t < u) \Rightarrow t < q < u$.

Proposition 5.25 *Let $\mathtt{C}$ be an inhabited constant type. Then for all $d, u : \mathbb{R}^\infty$, all $t : \mathtt{Time}$, and all $P : \mathtt{C} \to \mathtt{Prop}$, the following are equivalent:*

1. $t < u \Rightarrow \exists(c : \mathtt{C}).\, P(c)$,
2. $\forall(u' : \mathbb{Q}).\, u' < u \Rightarrow \exists(c : \mathtt{C}).\, t < u' \Rightarrow P(c)$,
3. $\forall(u' : \mathbb{R}).\, u' < u \Rightarrow \exists(c : \mathtt{C}).\, t < u' \Rightarrow P(c)$.

Likewise, the following are equivalent:

1. $d < t \Rightarrow \exists(c : \mathtt{C}).\, P(c)$,
2. $\forall(d' : \mathbb{Q}).\, d < d' \Rightarrow \exists(c : \mathtt{C}).\, d' < t \Rightarrow P(c)$,
3. $\forall(d' : \mathbb{R}).\, d < d' \Rightarrow \exists(c : \mathtt{C}).\, d' < t \Rightarrow P(c)$.

Proof $3 \Rightarrow 2$ is obvious, $2 \Leftrightarrow 1$ is Proposition 5.23, and $2 \Rightarrow 3$ is the definition of $u' < u$: there exists $u'' : \mathbb{Q}$ with $u' < u'' < u$. □

Lemma 5.26 *Let $\mathtt{C}$ be an inhabited constant type. Then for all $d, u : \mathbb{R}^\infty$, all $t : \mathtt{Time}$, and all $P : \mathtt{C} \to \mathtt{Prop}$, the following are equivalent:*

1. $[d, u] \ll t \Rightarrow \exists(c : \mathtt{C}).\, P(c)$,
2. $\forall(d', u' : \mathbb{Q}).\, (d < d' \wedge u' < u) \Rightarrow \exists(c : \mathtt{C}).\, [d', u'] \ll t \Rightarrow P(c)$,
3. $\forall(d', u' : \mathbb{R}).\, (d < d' \wedge u' < u) \Rightarrow \exists(c : \mathtt{C}).\, [d', u'] \ll t \Rightarrow P(c)$.

Proof $3 \Rightarrow 2 \Rightarrow 1$ is obvious, and $1 \Rightarrow 3$ follows from two applications of Proposition 5.25. □

As a sort of dual to Lemma 5.21 (saying $[d, u] \ll t$ is coprime), the next axiom states that for any $d \leq u$, the proposition $t \,\#\, [u, d]$ is *prime*. The soundness of Axiom 7 is proven in Proposition 6.22.

Axiom 7 For all $d, u : \mathbb{Q}$ with $d \leq u$ and all $t : \mathtt{Time}$, the proposition $t \,\#\, [u, d]$ is prime, i.e., for all propositions $P, Q : \mathtt{Prop}$:

$$\big((P \wedge Q) \Rightarrow t \,\#\, [u, d]\big) \Rightarrow \big((P \Rightarrow t \,\#\, [u, d]) \vee (Q \Rightarrow t \,\#\, [u, d])\big). \tag{5.4}$$

Again, the converse trivially holds, and the statement is immediately generalized from rationals to reals.

Proposition 5.27 *For all* $d, u : \mathbb{R}^{\infty}$ *with* $d \leq u$ *and all* $t : \mathtt{Time}$*, the proposition* $t \,\#\, [u, d]$ *is prime, i.e., Eq.* (5.4) *holds for all* $P, Q : \mathtt{Prop}$.

Proof Assuming $(P \wedge Q) \Rightarrow t \,\#\, [u, d]$, we want to prove $(P \Rightarrow t \,\#\, [u, d]) \vee (Q \Rightarrow t \,\#\, [u, d])$. Using Proposition 5.17, it follows that $P \Rightarrow t \,\#\, [u, d]$ is equivalent to $\forall (d', u' : \mathbb{Q}). ((d' < d) \wedge (u < u')) \Rightarrow (P \Rightarrow t \,\#\, [u', d'])$. This combined with Axiom 2 shows that the desired conclusion is equivalent to:

$$\forall (d_1, u_1, d_2, u_2 : \mathbb{Q}). ((d_1 < d) \wedge (u < u_1) \wedge (d_2 < d) \wedge (u < u_2)) \Rightarrow \big((P \Rightarrow t \,\#\, [u_1, d_1]) \vee (Q \Rightarrow t \,\#\, [u_2, d_2])\big). \tag{5.5}$$

Likewise, the assumption is equivalent to:

$$\forall (d', u' : \mathbb{Q}). ((d' < d) \wedge (u < u')) \Rightarrow ((P \wedge Q) \Rightarrow t \,\#\, [u', d']),$$

and applying Axiom 7 to this gives

$$\forall (d', u' : \mathbb{Q}). ((d' < d) \wedge (u < u')) \Rightarrow \big((P \Rightarrow t \,\#\, [u', d']) \vee (Q \Rightarrow t \,\#\, [u', d'])\big). \tag{5.6}$$

To prove Eq. (5.5), suppose we are given rationals $d_1 < d$, $u < u_1$, $d_2 < d$, and $u < u_2$. Let $d' := \max(d_1, d_2)$ and $u' := \min(u_1, u_2)$. From Eq. (5.6), we have $(P \Rightarrow t \,\#\, [u', d']) \vee (Q \Rightarrow t \,\#\, [u', d'])$, which implies the result. □

5.3 Important Modalities in Temporal Type Theory

Recall from Sect. 4.2 the notion of modalities $j : \mathtt{Prop} \to \mathtt{Prop}$ in a type theory and their relationship with toposes. In the current section, we will specifically be interested in modalities with interesting semantics in $\mathcal{B}$, and in the slice topos $\mathcal{B}/\mathtt{Time}$; we call the latter *temporal modalities*. A temporal modality is a map

j : `Time` $\to$ (`Prop` $\to$ `Prop`) such that $j(t)$—or as we will generally denote it, j^t—is a modality for all t : `Time`.

5.3.1 Definition of the Modalities $\downarrow$, @, $\uparrow$, and π

In Sect. 4.2.3, we discussed three sorts of modalities that arise from any given proposition U : `Prop`, namely the open modality $o(U)$, the closed modality $c(U)$, and the quasi-closed modality $q(U)$. Below, we will introduce some modalities that turn out to be useful in our type theory. Given $d, u : \mathbb{R}$, and t : `Time`, we will use the closed modality $c(t \# [u, d])$, the quasi-closed modality $q(t \# [u, d])$, and the open modality $o(t \ll [d, u])$; see Notation 5.13. These are denoted $\downarrow^t_{[d,u]}$, $@^t_{[d,u]}$, and $\uparrow^t_{[d,u]}$, respectively. We will also introduce one more modality that does not depend on d, u, or t.

Notation 5.28 ($\downarrow$, @, $\uparrow$, and π)
For t : `Time` and $d, u : \mathbb{R}^\infty$, recall that $t \# [u, d] = (u < t) \vee (t < d)$ and $[d, u] \ll t = (d < t) \wedge (t < u)$. Given P : `Prop`, we write:

- $\downarrow^t_{[d,u]} P := t \# [u, d] \vee P$,
- $@^t_{[d,u]} P := (P \Rightarrow t \# [u, d]) \Rightarrow t \# [u, d]$,
- $\uparrow^t_{[d,u]} P := ([d, u] \ll t) \Rightarrow P$,
- $\pi P := \forall(t : \texttt{Time}).\ @^t_{[0,0]} P$.

Note that π is the only one among the above list that is a modality on $\mathcal{B}$; the rest are temporal modalities, i.e., they exist only in the context of some t : `Time`. One can pronounce the notation πP as "Pointwise, P holds," which means that P holds at every 0-length point. Note that we will not use the symbol π to indicate projections in this book, so the symbol is not overloaded if we use it to denote this modality.

When there is only one t : `Time` in the context, we often drop it from the notation, e.g., writing $\downarrow_{[d,u]}$ rather than $\downarrow^t_{[d,u]}$. One can pronounce the notation $@_{[d,u]} P$ as "At $[d, u]$, P holds." The authors often pronounce the notation $\uparrow_{[d,u]} P$ as "In the interval $[d, u]$, P holds" and $\downarrow_{[d,u]} P$ as "Seeing the interval $[d, u]$, P holds."

When $d = u$, we often simplify the notation by writing $\downarrow_d := \downarrow_{[d,d]}$ and $@_d := @_{[d,d]}$. For example, with this notation, we have $\pi P := \forall(t : \texttt{Time}).\ @_0 P$.

Remark 5.29 The modalities in Notation 5.28 correspond semantically to geometrically interesting subtoposes. We will discuss this in detail in Sect. 6.4, but we give the idea for the time being.

For any $d, u : \mathbb{R}^\infty$ with $d \le u$, there is an associated point $[d, u] \in \mathbb{IR}$; see Sect. 2.3.1. This point defines at least three subsets of $\mathbb{IR}$, an open subspace $\uparrow[d, u] \subseteq \mathbb{IR}$, a closed subspace $\downarrow[d, u]$, and the one-point subspace $\{[d, u]\}$, which in fact is never closed; in fact, the closure of $\{[d, u]\}$ is $\downarrow[d, u]$. As our notation hopefully makes obvious, the subspaces $\uparrow[d, u]$, $\downarrow[d, u]$, and $\{[d, u]\}$ correspond to the modalities $\uparrow_{[d,u]}$, $\downarrow_{[d,u]}$, and $@_{[d,u]}$ respectively; see also Sect. 4.2.3.

In Notation 5.28, the above modalities were defined for any $d, u : \mathbb{R}^\infty$, regardless of order. When $u < d$ or either is $\pm\infty$, the above "subspace" explanation does not apply exactly, but it can be massaged a bit. That is, for any $d, u \in \mathbb{R}^\infty$, we can consider the subspace $\downarrow[d, u] \subseteq \mathbb{IR}$ defined as follows:

$$\downarrow[d, u] := \{[d', u'] \in \mathbb{IR} \mid (d' \leq d) \wedge (u \leq u')\} \subseteq \mathbb{IR} \tag{5.7}$$

In other words, if $d \leq u$, then $\downarrow[d, u]$ denotes the set of intervals containing $[d, u]$, whereas if $u < d$, then $\downarrow[d, u]$ denotes the set of intervals which non-trivially intersect $[u, d]$.

It remains to consider the semantics of the pointwise modality π. It defines one of the "four relevant toposes" discussed early on, in Sect. 1.2.3. In the context of t : `Time`, the π-modality corresponds to the subspace $\mathbb{R} \subseteq \mathbb{IR}$.

5.3.2 Interlude on the Semantics of $\downarrow$, @, $\uparrow$, and π

We briefly explain the semantics of the $\downarrow_{[d,u]}$ and $@_{[d,u]}$ modalities for $d \leq u$ in $\mathbb{R}$ using Dyck paths; see Definition 2.33. On any length ℓ, a proposition P corresponds to an open subspace of $\uparrow[0, \ell]$, and can hence be drawn as a Dyck path: P is true at $[d, u]$ iff $[d, u]$ is under the Dyck path.

We first explain in a few steps how the modalities $\downarrow_{[d,u]}$ and $@_{[d,u]}$ operate. We begin by drawing two arbitrary propositions P, Q dashed, each together with the proposition $t \,\#\, [u, d] = (t < u) \vee (d < t)$ dotted.

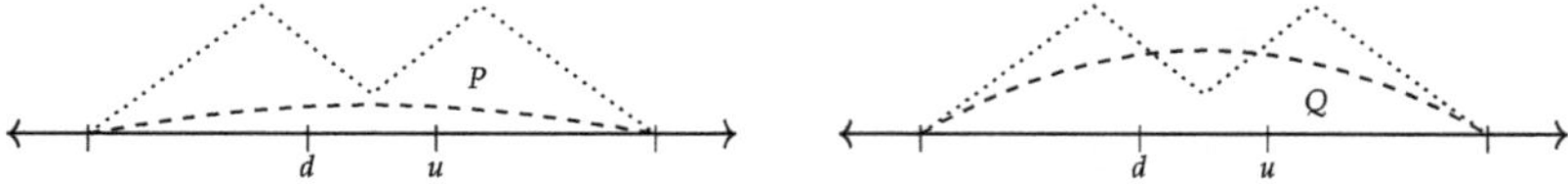

Next, we draw the propositions $\downarrow_{[d,u]}P$ and $\downarrow_{[d,u]}Q$.

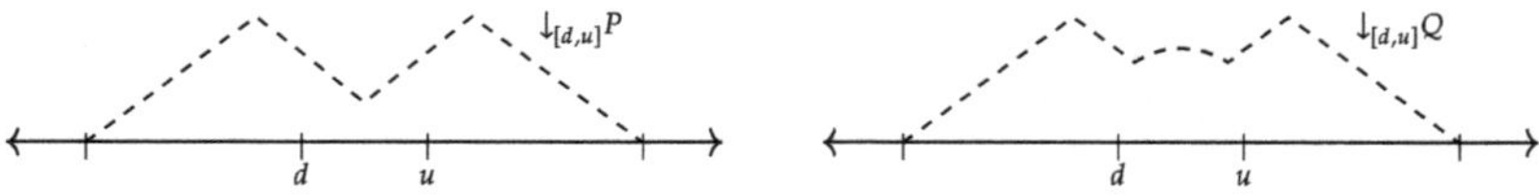

Next, we draw the propositions $P \Rightarrow t \,\#\, [u, d]$ and $Q \Rightarrow t \,\#\, [u, d]$, on our way to the @-modality:

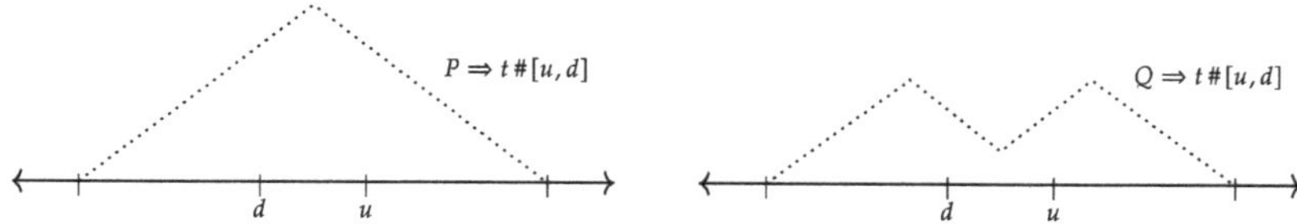

We are finally ready to draw the propositions $@_{[d,u]}P$ and $@_{[d,u]}Q$:

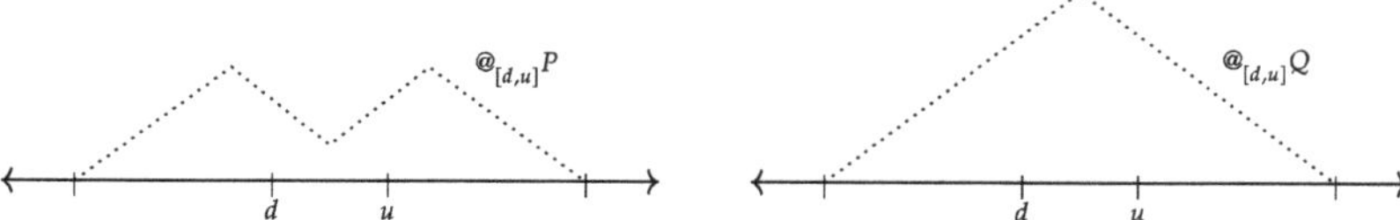

The picture on the left is also that of $@_{[d,u]}\bot = t\#[u,d] = \downarrow_{[d,u]}\bot$. That on the right is $\top$. We can see that applying the $@_{[d,u]}$ to any P : `Prop` returns one or the other of these, based on whether the point $[d,u]$ is strictly below the Dyck path P or not.

Next, on the left we draw $d < t < u$ dotted and P dashed. On the right, we draw $\Uparrow_{[d,u]}P$.

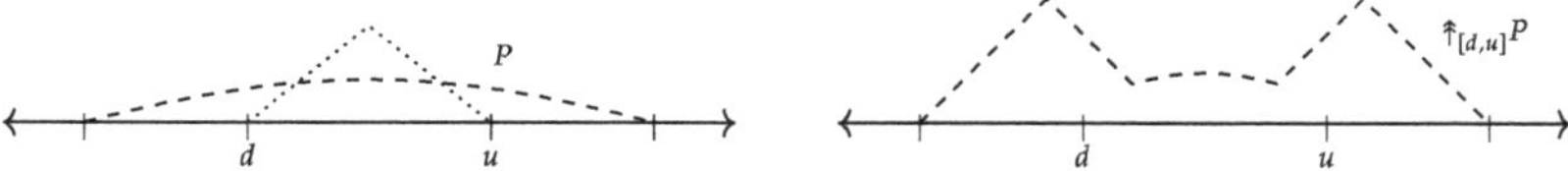

Finally, we draw a proposition P on the left and πP on the right. Thinking in terms of curves, one can see that π removes the subtlety of "height," replacing a Lipschitz curve P with the maximal one that is zero at the same places as P:

5.3.3 *First Properties of the Modalities* $\downarrow$, @, $\Uparrow$, *and* π

Before introducing any new axioms in terms of our modalities, we explore some basic consequences of their definitions. We break Sect. 5.3.3 into subsections; the first subsection is about π. The rest are about $\downarrow$, @, and $\Uparrow$ in various capacities, in particular regarding:

- Disjunction,
- Implication,
- Existential quantification,
- A De Morgan property,
- Decidability, and
- Containment and disjointness.

First Facts About π

Proposition 5.30 *For any* t : `Time` *and* P : `Prop`, *there is an equivalence* $\pi P \Leftrightarrow \forall(r : \mathbb{R}).\ @^t_r P$.

Proof By definition, $\pi P \Leftrightarrow \forall(t' : \texttt{Time}).\ @^{t'}_0 P$. By the torsor axiom, Axiom 4b, for any t' : `Time` there exists a unique $r : \mathbb{R}$ such that $t = t' + r$. Thus, $t' \# 0$ is equivalent to $t \# r$ by Proposition 5.18, and hence $\forall(t' : \texttt{Time}).\ @^{t'}_0 P$ is equivalent to $\forall(r : \mathbb{R}).\ @^t_r P$. □

Proposition 5.31 *Suppose given* P : `Prop` *and* t : `Time`. *Then for any* $a < b : \mathbb{R}^\infty$, *we have*

$$\big(\forall(r : \mathbb{R}).\, a < r < b \Rightarrow @^t_r P\big) \Rightarrow a < t < b \Rightarrow \pi P$$

Proof Assume the hypotheses. By definition of $a < t < b$ and by roundedness of t, there exists $a_1, a_2, b_1, b_2 : \mathbb{Q}$ such that $a < a_1 < a_2 < t$ and $t < b_2 < b_1 < b$. By Proposition 5.30, to prove πP it suffices to show $\big(\forall(r : \mathbb{R}).\ @^t_r P\big) \Rightarrow P$.

Choose $r : \mathbb{R}$; we want to prove $@^t_r P$. Because r is located by Definition 4.21 (5), we have either $a_1 < r$ or $r < a_2$ and either $b_2 < r$ or $r < b_1$. Thus, we have four cases, one of which is $a < r < b$ and the other three of which imply $t \# r$. In all cases, we obtain $@^t_r P$ as desired. □

Disjunction

For any modality j, it is easy to check that $(jP \vee jQ) \Rightarrow j(P \vee Q)$; the converse holds for $\downarrow$ and $\uparrow$, and most of the time for @.

Lemma 5.32 *For any* $d, u : \mathbb{R}^\infty$, *the modalities* $\downarrow_{[d,u]}$ *and* $\uparrow_{[d,u]}$ *commute with disjunction:*

$$\downarrow_{[d,u]}(P \vee Q) \Leftrightarrow \downarrow_{[d,u]}(P) \vee \downarrow_{[d,u]}(Q)$$
$$\uparrow_{[d,u]}(P \vee Q) \Leftrightarrow \uparrow_{[d,u]}(P) \vee \uparrow_{[d,u]}(Q).$$

When $d \le u$, *the modality* $@_{[d,u]}$ *also commutes with disjunction:*

$$@_{[d,u]}(P \vee Q) \Leftrightarrow @_{[d,u]}(P) \vee @_{[d,u]}(Q).$$

Proof The first statement is obvious by Notation 5.28, and the second statement is just a reformulation of Lemma 5.21. The third statement follows from Axiom 7, because $(P \vee Q) \Rightarrow (t \# [u, d]) \Leftrightarrow \big((P \Rightarrow (t \# [u, d])) \wedge (Q \Rightarrow (t \# [u, d]))\big)$. □

One may consider $\bot$ as nullary disjunction. We will see in Corollary 5.53, after adding some axioms, that $\pi\bot = \bot$. For the other modalities, we have the following lemma.

Lemma 5.33 *For all* $t : \mathtt{Time}$ *and* $r_1, r_2 : \mathbb{R}^\infty$, *we have* $\downarrow_{[r_2,r_1]}\bot \Leftrightarrow @_{[r_2,r_1]}\bot \Leftrightarrow t \,\#\, [r_1, r_2]$. *Assuming* $r_1 < r_2$, *we have*

$$\downarrow_{[r_2,r_1]}\bot \Leftrightarrow @_{[r_2,r_1]}\bot \Leftrightarrow \Uparrow_{[r_1,r_2]}\bot.$$

Proof The first statement follows directly from the definitions, Notation 5.28. The second that $t \,\#\, [r_1, r_2] \Leftrightarrow \neg([r_1, r_2] \ll t)$ is Proposition 5.16. □

Implication

For any modality j, it is easy to check that $j(P \Rightarrow Q) \Rightarrow (P \Rightarrow jQ)$ holds. The reader can verify immediately from Notation 5.28 that the converse holds for $\Uparrow_{[d,u]}$, for any $d, u : \mathbb{R}^\infty$:

$$\Uparrow_{[d,u]}(P \Rightarrow Q) \Leftrightarrow (P \Rightarrow \Uparrow_{[d,u]}Q). \tag{5.8}$$

Existential Quantification

For each of the modalities, $\Uparrow$, $\downarrow$, $@$, and π, we will prove some form of commutation with existential form of commutation with existential quantifiers. We will say something about $\downarrow$ and $\Uparrow$ now and give the Propositions 5.62 and 5.72.

Lemma 5.34 *Let* X *be an inhabited type, and suppose given* $t : \mathtt{Time}$ *and* $P : X \to \mathtt{Prop}$*:*

$$\downarrow^t_{[d,u]}(\exists(x : X).\, P(x)) \Leftrightarrow \exists(x : X).\, \downarrow^t_{[d,u]}P(x)$$

Proof $\downarrow^t_{[d,u]}P(x)$ is shorthand for a disjunction, $t \,\#\, [u, d] \vee P(x)$. For any type X, predicate $P : X \to \mathtt{Prop}$, and proposition $Q : \mathtt{Prop}$, one shows directly that $\exists(x : X).\, Q \vee P(x)$ implies $Q \vee \exists(x : X).\, P(x)$. The converse is easy too but requires X to be inhabited. □

The following is a direct consequence of Proposition 4.16 and Lemma 5.34.

Corollary 5.35 *For any inhabited type X, the inclusion* $\eta_{\downarrow[d,u]} : X \to \mathsf{sh}_{\downarrow[d,u]}(X)$, *of X into its* $\downarrow_{[d,u]}$*-sheafification, is surjective.*

The best we can do for $\Uparrow$ was already given in Lemma 5.26. It says that for an inhabited constant type $\mathtt{C}$:

$$\Uparrow_{[d,u]}(\exists(c : \mathbb{C}).\, P(c)) \Leftrightarrow \big(\forall(d', u' : \mathbb{R}).\, (d < d' \wedge u' < u) \Rightarrow \exists(c : \mathbb{C}).\, \Uparrow_{[d',u']} P(c)\big). \tag{5.9}$$

Remark 5.36 Equation (5.9) cannot be strengthened to avoid the quantification over d', u'. To see this, take as a context t : Time and recall that t is rounded. Then, the basic idea is that $t < 0 \Rightarrow \exists(q : \mathbb{Q}).\, t < q < 0$ holds but $\exists(q : \mathbb{Q}).\, t < 0 \Rightarrow t < q < 0$ does not.

More precisely, consider the predicate $P : \mathbb{Q} \to \texttt{Prop}$ given by:

$$P(q) := (t < q) \wedge (q < 0).$$

It is easy to prove that $\Uparrow^t_{[-\infty,0]}\exists(q : \mathbb{Q}).\, P(q)$, whereas one can show that $\exists(q : \mathbb{Q}).\, \Uparrow^t_{[-\infty,0]} P(q)$ is not sound in our semantics.

De Morgan

Semantically speaking, Lemma 5.37 says that the De Morgan laws hold in the closed subtopos $\downarrow[d, u]$; see [Joh02, Section D4.6].

Lemma 5.37 *Choose $d, u : \mathbb{R}^{\infty}$, t : Time, and P : Prop. If $d \leq u$, then:*

$$@_{[d,u]}P \vee (P \Rightarrow @_{[d,u]}\bot).$$

Proof Apply Proposition 5.27 to the implication $(P \wedge (P \Rightarrow t\#[u, d])) \Rightarrow t\#[u, d]$. □

Perhaps more memorably, the proposition $@P$ is either equal to $@\top$ or to $@\bot$:

$$(@_{[d,u]}P = @_{[d,u]}\top) \vee (@_{[d,u]}P = @_{[d,u]}\bot).$$

Decidability

Recall the definition (4.8) of a proposition P being j-closed for a modality j. For any modality j, say that a j-closed proposition P is *j-decidable* if $P \vee (P \Rightarrow j\bot)$ holds.

Corollary 5.38 *Choose $d, u : \mathbb{R}^{\infty}$, t : Time, and P : Prop. Then:*

1. *P is $@_{[d,u]}$-closed iff it is $\downarrow_{[d,u]}$-closed and $\downarrow_{[d,u]}$-decidable.*
2. *If P is $@_{[d,u]}$-closed, then it is $@_{[d,u]}$-decidable.*

Proof The second statement follows from the first because $\downarrow_{[d,u]}\bot \Leftrightarrow @_{[d,u]}\bot$ (Lemma 5.33).

Clearly, if $@_{[d,u]}P \Rightarrow P$, then P is $\downarrow_{[d,u]}$-closed, and it is $\downarrow_{[d,u]}$-decidable by Lemma 5.37. Conversely, suppose $\downarrow_{[d,u]}P \Rightarrow P$ and $P \vee (P \Rightarrow \downarrow_{[d,u]}\bot)$ and

$@_{[d,u]}P$; we want to show P. We may assume $P \Rightarrow \downarrow_{[d,u]}\bot$, so applying $@_{[d,u]}$ we get $@_{[d,u]}\downarrow_{[d,u]}\bot$, which implies $\downarrow_{[d,u]}\bot$ by Lemma 5.33, and $\downarrow_{[d,u]}\bot$ implies $\downarrow_{[d,u]}P$ and hence P. □

Containment and Disjointness

There are a number of containment and disjointness relations among our modalities. We begin with Proposition 5.39, which is just a matter of unwinding notation (5.28).

Proposition 5.39 *The following hold for any* P : `Prop`, t : `Time`, *and* $d_1, d_2, u_1, u_2 : \mathbb{R}^\infty$:

1. *If* $d_1 \le d_2$ *and* $u_2 \le u_1$, *then* $\Uparrow_{[d_1,u_1]}P \Rightarrow \Uparrow_{[d_2,u_2]}P$.
2. *If* $d_1 \le d_2$ *and* $u_2 \le u_1$, *then* $\downarrow_{[d_2,u_2]}P \Rightarrow \downarrow_{[d_1,u_1]}P$.
3. $\downarrow_{[d_1,u_1]}\downarrow_{[d_2,u_2]}P \Leftrightarrow \downarrow_{[\min(d_1,d_2),\max(u_1,u_2)]}P$.
4. $\Uparrow_{[d_1,u_1]}\Uparrow_{[d_2,u_2]}P \Leftrightarrow \Uparrow_{[\max(d_1,d_2),\min(u_1,u_2)]}P$.
5. $\downarrow_{[d_1,u_1]}P \Rightarrow @_{[d_1,u_1]}P$.

Proposition 5.40 *Suppose* $d < u : \mathbb{R}^\infty$. *Then,* $@_{[u,d]}P \Leftrightarrow \Uparrow_{[d,u]}\neg\neg P$.

Proof Suppose $@_{[u,d]}P$, or equivalently $(P \Rightarrow @_{[u,d]}\bot) \Rightarrow @_{[u,d]}\bot$. Then, $\neg P \Rightarrow \Uparrow_{[d,u]}\bot$ by Lemma 5.33, giving $\Uparrow_{[d,u]}\neg\neg P$ by Eq. (5.8). Conversely, suppose $\Uparrow_{[d,u]}\neg\neg P$ and $(P \Rightarrow @_{[u,d]}\bot)$. Again by Lemma 5.33 and Eq. (5.8), $P \Rightarrow \Uparrow_{[d,u]}\bot$, so $\Uparrow_{[d,u]}\neg P$, giving $\Uparrow_{[d,u]}\bot$ and hence $@_{[u,d]}\bot$ as desired. □

Lemma 5.41 $@_{[d_1,u_1]}P \Rightarrow \downarrow_{[d_1,u_1]}@_{[d_2,u_2]}P$ *holds for any* $d_1 \le d_2 \le u_2 \le u_1$ *and* P : `Prop`.

Proof Lemma 5.37 gives $(P \Rightarrow t\,\#\,[u_2,d_2]) \vee @_{[d_2,u_2]}P$. The result follows since $P \Rightarrow t\,\#\,[u_2,d_2]$ implies $P \Rightarrow t\,\#\,[u_1,d_1]$, which, assuming $@_{[d_1,u_1]}P$, implies $t\,\#\,[u_1,d_1]$. □

Lemmas 5.42 and 5.43 semantically tell us when a point is in, or is disjoint from, various open and closed subtoposes; see Sect. 4.2.3.

Lemma 5.42 *For any* P : `Prop` *and* $d_1, d_2, u_1, u_2 : \mathbb{R}^\infty$:

1. *If* $d_1 \le d_2 \wedge u_2 \le u_1$, *then* $\downarrow_{[d_2,u_2]}P \Rightarrow @_{[d_1,u_1]}P$.
2. *If* $d_1 < d_2 \le u_2 < u_1$, *then* $\Uparrow_{[d_1,u_1]}P \Rightarrow @_{[d_2,u_2]}P$.

Proof The first is two applications of Proposition 5.39, and the second follows easily from Lemma 5.22(g). □

Lemma 5.43 *For any* $d_1, d_2, u_1, u_2 : \mathbb{R}^\infty$ *with* $d_2 \le u_2$:

1. *If* $d_1 < d_2$ *or* $u_2 < u_1$, *then* $@_{[d_2,u_2]}\downarrow_{[d_1,u_1]}\bot$.
2. *If* $d_1 \le d_2$ *or* $u_2 \le u_1$, *then* $\Uparrow_{[d_2,u_2]}@_{[d_1,u_1]}\bot$.

Proof For the first, we assume $d_1 < d_2$ or $u_2 < u_1$ and need to show $(t\,\#\,[u_1,d_1] \Rightarrow t\,\#\,[u_2,d_2]) \Rightarrow t\,\#\,[u_2,d_2]$, but this is just Lemma 5.22(l). For the second, assume

$(d_1 \le d_2) \vee (u_1 \le u_2)$. We need to show $d_2 < t < u_2 \Rightarrow t \,\#\, [u_1, d_1]$, but this is direct. □

Proposition 5.44 *Let $d_1, u_1, d_2, u_2 : \mathbb{R}^\infty$. Then for any P : `Prop`:*

$$@_{[d_1,u_1]}\downarrow_{[d_2,u_2]}P \Leftrightarrow \big(((d_1 \le d_2) \wedge (u_2 \le u_1)) \Rightarrow @_{[d_1,u_1]}P\big).$$
$$\Leftrightarrow \big((d_2 < d_1) \vee (u_1 < u_2) \vee @_{[d_1,u_1]}P\big)$$

If $d_2 \le u_2$, then

$$: \Uparrow_{[d_1,u_1]}@_{[d_2,u_2]}P \Leftrightarrow \big(((d_1 < d_2) \wedge (u_2 < u_1)) \Rightarrow @_{[d_2,u_2]}P\big)$$
$$\Leftrightarrow \big((d_2 \le d_1) \vee (u_1 \le u_2) \vee @_{[d_2,u_2]}P\big).$$

Proof $@_{[d_1,u_1]}\downarrow_{[d_2,u_2]}P$ implies $((d_1 \le d_2) \wedge (u_2 \le u_1)) \Rightarrow @_{[d_1,u_1]}P$ by Lemma 5.42. This implies $(d_2 < d_1) \vee (u_1 < u_2) \vee @_{[d_1,u_1]}P$ by trichotomy (5.4). This implies $@_{[d_1,u_1]}\downarrow_{[d_2,u_2]}P$ by Lemma 5.43.

For the second statement, suppose $d_2 \le u_2$. Then, $\Uparrow_{[d_1,u_1]}@_{[d_2,u_2]}P$ implies $((d_1 < d_2) \wedge (u_2 < u_1)) \Rightarrow @_{[d_2,u_2]}P$ by Lemma 5.42. This implies $(d_2 \le d_1) \vee (u_1 \le u_2) \vee @_{[d_2,u_2]}P$ by trichotomy. This implies $\Uparrow_{[d_1,u_1]}@_{[d_2,u_2]}P$ by Lemma 5.43. □

Proposition 5.45 *If $d_1 \le d_2 \le u_2 \le u_1$, then for any P : `Prop`:*

$$@_{[d_1,u_1]}@_{[d_2,u_2]}P \Leftrightarrow \downarrow_{[d_1,u_1]}@_{[d_2,u_2]}P.$$

If $d_1 < d_2 \le u_2$ or $d_2 \le u_2 < u_1$, then:

$$@_{[d_2,u_2]}@_{[d_1,u_1]}P \Leftrightarrow \top.$$

Proof The first statement follows directly from Lemma 5.41. For the second, it suffices to show $@_{[d_2,u_2]}@_{[d_1,u_1]}\bot$, which reduces to $@_{[d_2,u_2]}t \,\#\, [u_1, d_1]$. This follows directly from Lemma 5.22(l). □

The next proposition looks at the intersection of $\Uparrow[d, u]$ and $\downarrow[r_1, r_2]$.

Proposition 5.46 *Choose $d, u, r, r' : \mathbb{R}^\infty$. If $r \le d$ or $u \le r'$, then:*

$$\Uparrow_{[d,u]}\downarrow_{[r,r']}P \Leftrightarrow \top.$$

If $d < r$ and $r' < u$, then:

$$\Uparrow_{[d,u]}\downarrow_{[r,r']}\bot \Leftrightarrow (t < d) \vee (t < r') \vee (u < t) \vee (r < t)$$
$$\Leftrightarrow t \,\#\, [\max(d, r'), \min(u, r)]$$

In particular, if $d < r < u$ *and* $d < r' < u$*, then:*

$$\Uparrow_{[d,u]}\Downarrow_{[r,r']}P \Leftrightarrow \Downarrow_{[r,r']}\Uparrow_{[d,u]}P.$$

Proof The first claim is easy: $\Uparrow_{[d,u]}\Downarrow_{[r,r']}\bot$ is $d < t < u \Rightarrow (t < r' \vee r < t)$, which is obvious. For the second claim, Lemma 5.21 says that $\Uparrow_{[d,u]}\Downarrow_{[r,r']}\bot$ is equivalent to $(d < t < u \Rightarrow t < r') \vee (d < t < u \Rightarrow r < t)$. Assume the first case; since either $d < r'$ or $r' \leq d$, we may assume $d < r'$ in which case we apply Lemma 5.22(e). The second case is similar.

For the third claim, the left-hand side is equivalent to $\Uparrow_{[d,u]}((\Downarrow_{[r,r']}\bot) \vee P)$ and the right-hand side is equivalent to $(\Downarrow_{[r,r']}\bot) \vee (\Uparrow_{[d,u]}P)$. The result follows from the second claim and Lemma 5.21. □

Proposition 5.47 says that the open set $\Uparrow[d, u]$ is covered by those way above it.

Proposition 5.47 *Choose* t : `Time`, P : `Prop`, *and* $d, u : \mathbb{R}^{\infty}$. *If* $u \leq d$, *then* $\Uparrow^{t}_{[d,u]}P = \top$. *If* $d < u$, *then:*

$$\Uparrow^{t}_{[d,u]}P \Leftrightarrow \big(\forall(d', u' : \mathbb{Q}).\, d < d' < u' < u \Rightarrow \Uparrow^{t}_{[d',u']}P\big).$$

Proof Since `Time` $\subseteq \bar{\underline{\mathbb{R}}}$ by Eq. (5.2), any t : `Time` is a pair of disjoint, rounded cuts. Disjointness gives the first statement and roundedness gives the second. See also Eq. (5.3). □

5.4 Remaining Axiomatics

We have now laid out some of the basic consequences that follow from the definition of our modalities $\Downarrow$, @, $\Uparrow$, and π. Because these correspond to semantically interesting subtoposes (see Remark 5.29), we can begin to record facts about these subtoposes internally to the type theory.

For example, Proposition 5.44 internally characterizes exactly which points are in the subtoposes of the form $\Downarrow[d, u]$ and $\Uparrow[d, u]$. We want to prove internally that those subtoposes are the union of those points they contain. This will be done in Propositions 5.58 and 5.57, which follows from our next axiom. Axiom 8 can be read as saying that the entire topos is covered by the single point subtoposes $\{\,[d, u]\,\}$.

The soundness of Axiom 8 is proven in Proposition 6.23.

Axiom 8 For all P: `Prop` and t : `Time`:

$$(\forall(d, u : \mathbb{Q}).\, d < u \Rightarrow @_{[d,u]}P) \Rightarrow P.$$

Of course, it follows directly that $(\forall(d, u : \mathbb{Q}).\, d \leq u \Rightarrow @_{[d,u]}P) \Rightarrow P$.

Corollary 5.48 *For all* $t : \mathtt{Time}$*, we have*

$$\neg(\forall(d, u : \mathbb{Q}).\, d < u \Rightarrow t \,\#\, [u, d]) \qquad \text{and} \qquad \neg(\forall(q : \mathbb{Q}).\, t \,\#\, q)$$

Proof For the first statement, simply take $P = \bot$ in Axiom 8. The second statement follows from the first, since $d < q < u$ and $t \,\#\, q$ implies $t \,\#\, [u, d]$. □

Recall from Eq. (5.2) that Time is a subtype of the type of improper intervals $\bar{\underline{\mathbb{R}}}$. Recall from Definition 4.21 that an improper interval $t = (\delta, \upsilon)$ is said to be located if $d < u$ implies $\delta d \vee \upsilon u$ for any $d, u : \mathbb{Q}$. Since $t \,\#\, [u, d]$ is defined as $(d < t) \vee (t < u) = \delta t \vee \upsilon u$, Corollary 5.48 reads directly as the statement "a time t (as a pair of cuts) is not located." If times were located, they would be real numbers, which we've seen are constant in the semantics. Hence, the corollary can be thought of as saying that time is not constant.

We have shown that $t : \mathtt{Time}$ is decidedly *not* located; however, the next proposition shows that, relative to the π modality, t *is* located, i.e., time is "pointwise located."

Proposition 5.49

$$\forall(t : \mathtt{Time})(d, u : \mathbb{R}^{\infty}).\, d < u \Rightarrow \pi(t \,\#\, [u, d]).$$

Proof By Proposition 5.30, this is equivalent to:

$$\forall(t : \mathtt{Time})(d, u : \mathbb{R}^{\infty}).\, d < u \Rightarrow \forall(r : \mathbb{R}).\, @_r(t \,\#\, [u, d]).$$

Taking arbitrary t, d, u, and r, with $d < u$, we have $(r < u) \vee (d < r)$ by Proposition 5.10. The result then follows from Lemma 5.22(1). □

We internally express the semantic fact that the topos $\mathsf{Shv}(\mathbb{IR})$ is the union of the closed subtoposes corresponding to $\downarrow[q, q]$ for $q : \mathbb{Q}$ in Proposition 5.50. Note that this proposition is a statement about subtoposes, or equivalently, about sub-locales. Indeed, it *does not* imply the (decidedly false) external statement that the topological space $\mathbb{IR}$ is the union of its closed subspaces of the form $\downarrow[q, q]$ for $q : \mathbb{Q}$.

Proposition 5.50 *The following holds for any proposition* $P : \mathtt{Prop}$ *and time* $t : \mathtt{Time}$*:*

$$\big(\forall(q : \mathbb{Q}).\, \downarrow_q P\big) \Rightarrow P.$$

Proof Choose $t : \mathtt{Time}$ and assume $\forall(q : \mathbb{Q}).\, \downarrow_q P$. This implies $P \vee \forall(q : \mathbb{Q}).\, t \,\#\, q$ by Axiom 2, and the conclusion follows from Corollary 5.48. □

Corollary 5.51 *The following holds for any proposition* $P : \mathtt{Prop}$*:*

$$\big(\forall(t : \mathtt{Time}).\, \downarrow_0^t P\big) \Rightarrow P.$$

Proof This follows from Proposition 5.50 and Axiom 4, because given t : `Time` and $\forall(t'$: `Time`$).\ \downarrow_0^{t'} P$, we obtain $\forall(r : \mathbb{R}).\ \downarrow_r^t P$, which implies P. □

Lemma 5.52 *Let P be a decidable proposition, i.e., where $P \vee \neg P$ holds. Then:*

$$@_{[d,u]}P \Leftrightarrow \downarrow_{[d,u]}P \qquad \textit{and} \qquad \pi P \Leftrightarrow P.$$

Proof The converses to both statements are obvious (see Proposition 5.39), so we proceed to show $@_{[d,u]}P \Rightarrow \downarrow_{[d,u]}P$. Assume $P \vee \neg P$. Trivially, $P \Rightarrow \downarrow_{[d,u]}P$, so suppose $\neg P$. Then, $@_{[d,u]}P$ implies $@_{[d,u]}\bot$, which is equivalent to $\downarrow_{[d,u]}\bot$, which implies $\downarrow_{[d,u]}P$.

Since now $@_0 P \Rightarrow \downarrow_0 P$, we have $\pi P \Rightarrow P$ by Proposition 5.50. □

Corollary 5.53 *$\bot$ is decidable, so $\pi\bot = \bot$ and we have $\pi P \Rightarrow \neg\neg P$ for all P :* `Prop`.

Corollary 5.54 *If X is a type with decidable equality, then it is π-separated.*

Proposition 5.55 *The cuts defining any t :* `Time` *are π-closed. That is, for all $q : \mathbb{Q}$, we have $\pi(t < q) \Rightarrow t < q$ and $\pi(q < t) \Rightarrow q < t$.*

Proof Choose t : `Time` and $q : \mathbb{Q}$ and suppose $\pi(t < q)$. By Axiom 3b, it suffices to show $\neg(q < t)$, so assume $q < t$. Then, we have $\pi(\bot)$, which implies $\bot$ by Lemma 5.52. □

Remark 5.56 In Eq. (5.2), `Time` was defined as a subtype of the extended interval type $\bar{\underline{\mathbb{R}}}$, but in fact it can also be identified with a subtype of the variable reals `Time` $\subseteq \mathbb{R}_\pi$. Indeed, consider the cuts $\delta, \upsilon : \mathbb{Q} \to$ `Prop` defining any $t = (\delta, \upsilon)$: `Time`. By definition, δ is down-closed and υ is up-closed, and both are rounded and bounded. It follows they are down-/up-closed (1), j-rounded (2), and j-bounded (3) for any j. The fact that they are π-closed (0) was shown in Proposition 5.55. The fact that they are π-disjoint (4) was shown in Axiom 3b. Finally, the fact that they are π-located (5) was shown in Proposition 5.49. See also Remark 7.23

In Propositions 5.57 and 5.58, we prove statements that reflect facts about the subspaces $\downarrow[d, u]$ and $\uparrow[d, u]$ as collections of points in $\mathbb{IR}$.

Proposition 5.57 *For any $d, u : \mathbb{R}^\infty$ and P :* `Prop`, *we have*

$$\big(\forall(d', u' : \mathbb{Q}).\ (d' \leq d \wedge u \leq u') \Rightarrow @_{[d',u']}P\big) \Leftrightarrow \downarrow_{[d,u]}P.$$

Proof The backward implication follows from Lemma 5.42. By Proposition 5.44, the left-hand side is equivalent to $\forall(d', u' : \mathbb{Q}).\ @_{[d',u']}\downarrow_{[d,u]}P$, which by Axiom 8 implies $\downarrow_{[d,u]}P$. □

For any reals $d < u$, we have $\uparrow[d, u] = \bigcup_{d<d'\leq u'<u}[d', u']$ by Proposition 2.23. This is expressed internally by Proposition 5.58.

Proposition 5.58 *For any* $d, u : \mathbb{R}^\infty$ *and* $P : \mathtt{Time} \to \mathtt{Prop}$*, we have*

$$\big(\forall(d', u' : \mathbb{Q}).\, d < d' \le u' < u \Rightarrow @_{[d',u']}P\big) \Leftrightarrow \Uparrow_{[d,u]}P.$$

Proof By Proposition 5.44, the left-hand side is equivalent to:

$$\forall(d', u' : \mathbb{Q}).\, d' \le u' \Rightarrow \Uparrow_{[d,u]}@_{[d',u']}P,$$

which is equivalent to:

$$d < t < u \Rightarrow \forall(d', u' : \mathbb{Q}).\, d' \le u' \Rightarrow @_{[d',u']}P,$$

and finally, applying Axiom 8 this is equivalent to $d < t < u \Rightarrow P$. □

Corollary 5.59 *For all* $t : \mathtt{Time}$ *and* $P : \mathtt{Prop}$*, we have*

$$\big(\forall(q : \mathbb{Q}).\, \Uparrow_{[-q,q]}P\big) \Rightarrow P.$$

Proof Choose t and P and suppose $\forall(q : \mathbb{Q}).\, \Uparrow_{[-q,q]}P$. By Proposition 5.50, it suffices to show $\forall(q' : \mathbb{Q}).\, \downarrow_{[q',q']}P$, so choose q'. By Proposition 5.57, it is enough to show that for any rationals $d' \le q' \le u'$ we have $@_{[d',u']}P$. Let $q = \max(-d'-1, u'+1)$, so $-q < d'$ and $u < q$. Then, $\Uparrow_{[-q,q]}P$ implies $@_{[d',u']}P$ by Proposition 5.58, and we are done. □

Corollary 5.60 *For all* $t : \mathtt{Time}$*,* $P : \mathtt{Prop}$*, and* $r_1, r_2 : \mathbb{R}$*, we have*

$$\big(\forall(q_1, q_2 : \mathbb{Q}).\, (q_1 < r_1) \Rightarrow (r_2 < q_2) \Rightarrow \Uparrow_{[q_1,q_2]}P\big) \Rightarrow P.$$

Proof Assume the hypothesis. If $q : \mathbb{Q}$ is a rational such that $\max(-r_1, r_2) < q$, then $\Uparrow_{[-q,q]}P$. Let q_0 be one such rational. Then for any $q : \mathbb{Q}$, either $q \le q_0$ or $q_0 < q$. In either case, we have $\Uparrow_{[-q,q]}P$ by Proposition 5.39, so we are done by Corollary 5.59. □

Proposition 5.61 *Let* $\mathtt{C}$ *be an inhabited constant type. Then for any* $P : \mathtt{Prop}$ *and* $t : \mathtt{Time}$*:*

$$\big(\forall(q : \mathbb{Q}).\, \exists(c : \mathtt{C}).\, \Uparrow_{[-q,q]}P\big) \Rightarrow \exists(c : \mathtt{C}).\, Pc.$$

Proof Assume the hypothesis. By Corollary 5.59, it suffices to show that $\Uparrow_{[-q,q]}\exists(c : \mathtt{C}).\, P(c)$ holds for any $q : \mathbb{Q}$. By Lemma 5.26, this is equivalent to $\forall(d, u : \mathbb{Q}).\, (-q < d \wedge u < q) \Rightarrow \exists(c : \mathtt{C}).\, \Uparrow_{[d,u]}P(c)$. But by hypothesis, we have some $c : \mathtt{C}$ with $\Uparrow_{[-q,q]}P(c)$, and this implies $\Uparrow_{[d,u]}P(c)$ by Proposition 5.39. □

Axiom 9 says that if a proposition P holds at a point, then it holds in some open neighborhood of the point. The converse of Axiom 9 holds, as we will see in Proposition 5.65. The soundness of Axiom 9 is proven in Proposition 6.24.

Axiom 9 For all times t : Time, propositions P : Prop, and $d, u : \mathbb{R}$ with $d \leq u$:

$$@_{[d,u]}P \Rightarrow \downarrow_{[d,u]}\exists(d', u' : \mathbb{Q}).\, d' < d \leq u < u' \wedge \Uparrow_{[d',u']}P.$$

In Sect. 5.3.3, page 103, we said that all of our modalities commute with existential quantification in one way or another. In Proposition 5.61, we give the limited sense in which the @-modality commutes with existential quantification.

Proposition 5.62 *For any constant type* C*, any* t : Time *and* $P : \mathtt{C} \to \mathtt{Prop}$*, and any* $d, u : \mathbb{R}$ *with* $d \leq u$*, if* $\exists(c : \mathtt{C}).\, \top$*, then:*

$$@^t_{[d,u]}(\exists(c : \mathtt{C}).\, P(c)) \Leftrightarrow \exists(c : \mathtt{C}).\, @^t_{[d,u]}P(c).$$

Proof The converse is obvious (see Eq. (4.7)), so we prove the forward direction. By Axiom 9, $@_{[d,u]}(\exists(c : \mathtt{C}).\, P(c))$ is equivalent to:

$$\downarrow_{[d,u]}\exists(d', u' : \mathbb{Q}).\, (d' < d) \wedge (u < u') \wedge \Uparrow_{[d',u']}\exists(c : \mathtt{C}).\, P(c)$$

which is equivalent to:

$$\downarrow_{[d,u]}\exists(d', u' : \mathbb{Q}).\, (d' < d) \wedge (u < u') \wedge \forall(d'', u'' : \mathbb{Q}).\, d' < d'' < u'' < u' \\ \Rightarrow \exists(c : \mathtt{C}).\, \Uparrow_{[d'',u'']}P(c)$$

by Lemma 5.26. Then by choosing $d' < d'' < d$ and $u < u'' < u'$, and using the fact (see, e.g., Proposition 5.58) that $\Uparrow_{[d'',u'']}P(c)$ implies $@_{[d,u]}P(c)$, this implies $\downarrow_{[d,u]}\exists(c : \mathtt{C}).\, @_{[d,u]}P(c)$. It is easy to see that this is equivalent to $\exists(c : \mathtt{C}).\, @_{[d,u]}P(c)$ as desired. □

The following is a direct consequence of Propositions 4.16 and 5.62.

Corollary 5.63 *For any inhabited type* X*, the inclusion* $\eta_{@[d,u]} : X \to \mathsf{sh}_{@[d,u]}(X)$*, of* X *into its* $@_{[d,u]}$*-sheafification, is surjective.*

Lemma 5.64 *For any reals* $d \leq r_1 \leq u$ *and* $d \leq r_2 \leq u$*, the following are equivalent:*

1. $\downarrow_{[r_1,r_2]}\exists(d', u' : \mathbb{Q}).\, (d' < d \leq u < u') \wedge \Uparrow_{[d',u']}P.$
2. $\exists(d', u' : \mathbb{Q}).\, (d' < d \leq u < u') \wedge \Uparrow_{[d',u']}\downarrow_{[r_1,r_2]}P.$

Proof It is easy to prove $\exists(d', u' : \mathbb{Q}).\, d' < d \leq u < u'$. Hence, 1. is equivalent to:

$$\exists(d', u' : \mathbb{Q}).\, (d' < d \leq u < u') \wedge \downarrow_{[r_1,r_2]}\Uparrow_{[d',u']}P.$$

By Proposition 5.46, this is equivalent to 2. □

Proposition 5.65 *The following are equivalent, for any* t : Time*,* P : Prop*, and* $d, u : \mathbb{R}$ *with* $d \leq u$*:*

1. $@_{[d,u]}P.$

2. $\downarrow_{[d,u]}\exists(d', u' : \mathbb{Q}).\,(d' < d \le u < u') \wedge \Uparrow_{[d',u']}P$.
3. $\exists(d', u' : \mathbb{Q}).\,(d' < d \le u < u') \wedge \forall(d'', u'' : \mathbb{Q}).\,((d' < d'' \le d) \wedge (u \le u'' < u')) \Rightarrow @_{[d'',u'']}P$.

Proof $1 \Rightarrow 2$ is Axiom 9, and the converse $2 \Rightarrow 1$ follows from Lemma 5.42.

We show $2 \Leftrightarrow 3$ by the chain of equivalences:

$$\begin{aligned}
&\downarrow_{[d,u]}\exists(d', u' : \mathbb{Q}).\,(d' < d \le u < u') \wedge \Uparrow_{[d',u']}P\\
\Leftrightarrow\quad &\exists(d', u' : \mathbb{Q}).\,(d' < d \le u < u') \wedge \Uparrow_{[d',u']}\downarrow_{[d,u]}P\\
\Leftrightarrow\quad &\exists(d', u' : \mathbb{Q}).\,(d' < d \le u < u')\\
&\qquad \wedge \forall(d'', u'' : \mathbb{Q}).\,d' < d'' \le u'' < u' \Rightarrow @_{[d'',u'']}\downarrow_{[d,u]}P\\
\Leftrightarrow\quad &\exists(d', u' : \mathbb{Q}).\,(d' < d \le u < u')\\
&\qquad \wedge \forall(d'', u'' : \mathbb{Q}).\,d' < d'' \le u'' < u' \Rightarrow d'' \le d \Rightarrow u \le u'' \Rightarrow @_{[d'',u'']}P\\
\Leftrightarrow\quad &\exists(d', u' : \mathbb{Q}).\,(d' < d \le u < u')\\
&\qquad \wedge \forall(d'', u'' : \mathbb{Q}).\,d' < d'' \le d \le u \le u'' < u' \Rightarrow @_{[d'',u'']}P
\end{aligned}$$

where the first three equivalences are by Lemma 5.64, Proposition 5.58, and Proposition 5.44, respectively. □

Proposition 5.66 *For any P :* `Prop`*, we have*

$$\pi P \Leftrightarrow \forall(t : \texttt{Time}).\,\exists(q : \mathbb{Q}).\,(0 < q) \wedge (-q < t < q \Rightarrow P).$$

Proof By definition, $\pi P \Leftrightarrow \forall(t : \texttt{Time}).\,@_0 P$, which by Axiom 9 and Corollary 5.51 is equivalent to:

$$\forall(t : \texttt{Time}).\,\exists(d, u : \mathbb{Q}).\,(d < 0 < u) \wedge (d < t < u \Rightarrow P)$$

and the result follows by taking $q = \min(-d, u)$. □

Proposition 5.67 *Suppose P :* `Prop` *is π-closed. Then, we have*

$$\forall(t : \texttt{Time})(r_1, r_2 : \mathbb{R}).\,r_1 < r_2 \Rightarrow (t \mathbin{\#} [r_2, r_1] \Rightarrow P) \Rightarrow P.$$

Proof Suppose $\pi P \Rightarrow P$ and choose t and $r_1 < r_2$ with $t \mathbin{\#} [r_2, r_1] \Rightarrow P$; then, $(t < r_2) \Rightarrow P$ and $(r_1 < t) \Rightarrow P$. It suffices by Proposition 5.31 to show $\forall(r : \mathbb{R}).\,@^t_r P$. Choose r and suppose $P \Rightarrow t \mathbin{\#} r$; we want to show $t \mathbin{\#} r$. Since r is located, either $r_1 < r$ or $r < r_2$, so we may apply Lemma 5.22(1). □

Corollary 5.68 *Suppose P :* `Prop` *is π-closed. Then, we have*

$$\forall(t : \texttt{Time})(r_1, r_2, r_3, r_4 : \mathbb{R}).\,r_1 < r_2 < r_3 < r_4 \Rightarrow \Uparrow_{[r_1,r_3]}P \Rightarrow \Uparrow_{[r_2,r_4]}P \Rightarrow \Uparrow_{[r_1,r_4]}P.$$

Proof Given t, reals $r_1 < r_2 < r_3 < r_4$, and the assumptions $\Uparrow_{[r_1,r_3]}P \Rightarrow \Uparrow_{[r_2,r_4]}P$ and $r_1 < t < r_4$, we directly obtain $t < r_3 \Rightarrow P$ and $r_2 < t \Rightarrow P$; in other words, $t \# [r_3, r_2] \Rightarrow P$. Thus, we obtain P as desired from Proposition 5.67. □

The following axiom immediately implies the converses to Proposition 5.67 and Corollary 5.68. The two together roughly say that π-closed sheaves are those for which behaviors compose. The soundness of Axiom 10 is proven in Proposition 6.25.

Axiom 10 For any proposition $P : \texttt{Prop}$, if:

$$\forall(t : \texttt{Time})(q_1, q_2 : \mathbb{Q}).\, q_1 < q_2 \Rightarrow (t \# [q_2, q_1] \Rightarrow P) \Rightarrow P$$

then $\pi P \Rightarrow P$.

Corollary 5.69 *For any proposition $P : \texttt{Prop}$, if:*

$$\begin{aligned}&\forall(t : \texttt{Time})(q_1, q_2, q_3, q_4 : \mathbb{Q}).\, q_1 < q_2 < q_3 < q_4 \\ &\quad \Rightarrow \Uparrow_{[q_1,q_3]}P \Rightarrow \Uparrow_{[q_2,q_4]}P \Rightarrow \Uparrow_{[q_1,q_4]}P\end{aligned}$$

then $\pi P \Rightarrow P$.

Proof Assume the hypothesis; to show $\pi P \Rightarrow P$, it suffices to show that the hypothesis of Axiom 10 holds. So, choose $t : \texttt{Time}$, rationals $q_2, q_3 : \mathbb{Q}$ with $q_2 < q_3$, and suppose $t \# [q_3, q_2] \Rightarrow P$. Then, $t < q_3 \Rightarrow P$ and $q_2 < t \Rightarrow P$. It follows that for any q_1, q_2, if $q_1 < q_2$ and $q_3 < q_4$ then $\Uparrow_{[q_1,q_4]}P$. Thus, we obtain P by Corollary 5.60. □

Proposition 5.70 *If $\texttt{C}$ is a constant type, then it is a π-sheaf.*

Proof $\texttt{C}$ is π-separated by Corollaries 5.54 and 5.8. To show it is a π-sheaf, we use Proposition 4.12. That is, we assume that $\phi : \texttt{C} \to \texttt{Prop}_\pi$ satisfies πP, where $P := \exists(c : \texttt{C}).\, \forall(c' : \texttt{C}).\, (\phi c' \Leftrightarrow c = c')$ is the "local singleton condition," and we prove P. By Axiom 10, it suffices to show $(t \# [q_2, q_1] \Rightarrow P) \Rightarrow P$ for any $t : \texttt{Time}$ and rationals $q_1 < q_2$.

Assuming $t \# [q_2, q_1] \Rightarrow P$, we can use Proposition 5.25 to choose intermediate rationals $q_1 < q_1' < q_2' < q_2$ such that the following hold:

$$\begin{aligned}&\exists(c_1 : \texttt{C}).\, (q_1' < t) \Rightarrow \forall(c' : \texttt{C}).\, (\phi c' \Leftrightarrow c_1 = c') \\ &\exists(c_2 : \texttt{C}).\, (t < q_2') \Rightarrow \forall(c' : \texttt{C}).\, (\phi c' \Leftrightarrow c_2 = c')\end{aligned} \tag{5.10}$$

By decidable equality, either $c_1 = c_2$ or $\neg(c_1 = c_2)$. In the first case, we obtain $t \# [q_2', q_1'] \Rightarrow \forall(c' : \texttt{C}).\, (\phi c' \Leftrightarrow c = c_1)$, so we are done by Proposition 5.67, because $\forall(c' : \texttt{C}).\, (\phi c' \Leftrightarrow c = c_1)$ is π-closed. In the second case, using $(q_1' < t < q_2') \Rightarrow c_1 = c_2$, we have $t \# [q_1', q_2']$ by Proposition 5.16. It follows that $t < q_2'$ or $q_1' < t$, and either way we use obtain $\exists(c : \texttt{C}).\, \forall(c' : \texttt{C}).\, (\phi c' \Leftrightarrow c = c')$ from Eq. (5.10). □

We will need the following lemma later.

Lemma 5.71 *For any* $t : \texttt{Time}$, $P, Q : \texttt{Prop}_\pi$, *the statement* $(t < 0 \wedge P) \vee (0 < t \wedge Q)$ *is* π*-closed.*

Proof Choose $t' : \texttt{Time}$ and $q_1 < q_2$ and assume $t' \# [q_2, q_1] \Rightarrow \big((t < 0 \wedge P) \vee (0 < t \wedge Q)\big)$; by Axiom 10, it suffices to show $\big((t < 0 \wedge P) \vee (0 < t \wedge Q)\big)$. Using Axiom 4, let $t' = t + r$ for $r : \mathbb{R}$, and let $q_1' := q_1 - r$ and $q_2' := q_2 - r$. By Proposition 5.20, we have $(t \# [q_2', q_1'] \Rightarrow (t < 0 \wedge P)) \vee (t \# [q_2', q_1'] \Rightarrow (0 < t \wedge Q))$. Since $t < 0$ and $0 < t$ are π-closed by Proposition 5.55, we are done by Proposition 5.67. □

A common application of Axiom 10 is to use the following proposition, which says that the subtopos $\mathcal{B}_\pi$ is "proper," see, e.g., [Gie+03, Def. VI-6.20].

Proposition 5.72 *The inclusion* $\texttt{Prop}_\pi \subseteq \texttt{Prop}$ *is continuous, i.e., preserves directed joins indexed by constant types.*

In detail, let D be a constant type and let $(D, \leq)$ *be a directed preorder, i.e., one satisfying:*

$$\exists(d : D).\, \top \qquad \textit{and} \qquad \forall(d_1, d_2 : D).\, \exists(d' : D).\, (d_1 \leq d') \wedge (d_2 \leq d').$$

Let $P : D \to \texttt{Prop}$ *be monotonic* ($\forall(d_1, d_2).\, d_1 \leq d_2 \Rightarrow Pd_1 \Rightarrow Pd_2$) *and* π*-closed* ($\forall(d : D).\, \pi\, Pd \Rightarrow Pd$). *Then:*

$$\big(\pi \exists(d : D).\, Pd\big) \Rightarrow \big(\exists(d : D).\, Pd\big).$$

Proof To set up an application of Axiom 10, let $t : \texttt{Time}$, let $q_1 < q_2$ be rationals, and suppose $t \# [q_2, q_1] \Rightarrow \exists(d : D).\, Pd$. By Proposition 5.25, this is equivalent to the conjunction of Eqs. (5.11) and (5.12):

$$\forall(q_1' : \mathbb{Q}).\, (q_1 < q_1') \Rightarrow \exists(d : D).\, q_1' < t \Rightarrow Pd \tag{5.11}$$

$$\forall(q_2' : \mathbb{Q}).\, (q_2' < q_2) \Rightarrow \exists(d : D).\, t < q_2' \Rightarrow Pd \tag{5.12}$$

Choose any rationals q_1' and q_2' satisfying $q_1 < q_1' < q_2' < q_2$. Then, applying (5.11) gives us a d_1 such that $q_1' < t \Rightarrow P(d_1)$ and applying (5.12) gives us a d_2 such that $t < q_2' \Rightarrow P(d_2)$. By directedness of D, there is a d such that $d_1 \leq d$ and $d_2 \leq d$, so by monotonicity of P we have $t \# [q_2', q_1'] \Rightarrow Pd$. This implies $P(d)$ by Proposition 5.67. □

Chapter 6
Semantics and Soundness

In this chapter, we prove that the temporal type theory, developed in Chaps. 4 and 5 is sound in the topos $\mathcal{B}$. To do so, we begin in Sect. 6.1 by recalling the Kripke–Joyal semantics by which to interpret type-theoretic formulas in the topos $\mathcal{B}$. Then in Sect. 6.3 we discuss the sheaf of real numbers and `Time`. We then proceed to our main goal: proving that our type signature—i.e., the one atomic predicate symbol and the ten axioms presented in Chap. 5—are sound in $\mathcal{B}$. This is done in Sect. 6.5, which begins with a Table 6.2 summarizing the type signature.

6.1 Categorical Semantics

When proving that a type signature is sound in a topos $\mathcal{E}$, one must give $\mathcal{E}$-interpretations to each type, term, and predicate in the type theory. To do so, one first assigns an object of $\mathcal{E}$ to each atomic type and a morphism of $\mathcal{E}$ to each atomic term. Recall that predicates in the type theory are identified with terms of type `Prop`. One thus assigns a morphism, whose codomain is the subobject classifier $\Omega_{\mathcal{E}}$, to each atomic predicate. Using categorical semantics (see, e.g., [Jac99]), the remaining types and terms are then recursively assigned interpretations, as briefly discussed in Sect. 4.1. The remaining part of the signature is the set of axioms, which are in particular predicates, so the remaining part of the work is to prove that each axiom is interpreted as *true* in $\mathcal{E}$. More precisely, the morphism $X \to \Omega_{\mathcal{E}}$ assigned to each must factor through $\texttt{true}\colon 1 \to \Omega_{\mathcal{E}}$.

We hope that the informal introduction in Sect. 4.1 is enough to give the reader a basic understanding of how to interpret the recursively constructed types and terms in the topos, e.g., product types and arrow types are sent to product objects and exponential objects, projections are sent to projections, etc. The least straightforward part, and the part we will be focusing on, is the $\mathcal{E}$-semantics of the logical connectives and quantifiers, so we explain this in more detail below.

P. Schultz, D. I. Spivak, *Temporal Type Theory*, Progress in Computer Science and Applied Logic 29, https://doi.org/10.1007/978-3-030-00704-1_6

The usual way to express the semantic interpretation of a type X or term $a : A, b : B \vdash e : C$, is using *Church brackets*, e.g., $[\![X]\!]$ or $[\![e]\!] \colon [\![A]\!] \times [\![B]\!] \to [\![C]\!]$ in $\mathcal{E}$. At certain points in this section, it is typographically more convenient to simply write $X \in \mathcal{E}$ or $A \times B \to C$, i.e., to drop the Church brackets. We hope this will not cause too much confusion.

We now recall the Kripke–Joyal sheaf semantics as it relates to our particular topos $\mathcal{B} = \mathsf{Shv}(S_{\mathbb{IR}/\rhd})$, see, e.g., [MM92, Theorem VI.7.1] for the general case. The definition of $S_{\mathbb{IR}/\rhd}$ is given in Definition 3.5, but see also Remark 3.6, which says that since the object $0 \in S_{\mathbb{IR}/\rhd}$ has an empty covering family, we may assume $\ell > 0$ for each object $\ell \in S_{\mathbb{IR}/\rhd}$.

Given a sheaf X, a length $\ell > 0$, a section $\alpha \in X(\ell)$, and a predicate $\phi : X \to \texttt{Prop}$, one can use forcing notation and write $\ell \Vdash \phi(\alpha)$ to mean that α is a section in the subobject determined by ϕ, or equivalently that $\phi(\alpha)$ factors through $\top : 1 \to \texttt{Prop}$.

For any morphism $\langle r, s\rangle \colon \ell' \to \ell$ in $\mathbb{IR}^{\mathrm{op}}_{/\rhd}$ and section $\alpha \in X(\ell)$ we denote the restriction by $\alpha\big|_{\langle r,s\rangle}$. An object ℓ in $S_{\mathbb{IR}/\rhd}$ has only one non-trivial covering family, namely the set of all wavy arrows[1] $\langle r, s\rangle \colon \ell' \rightsquigarrow \ell$ in $\mathbb{IR}^{\mathrm{op}}_{/\rhd}$. Thus predicates ϕ will always satisfy the following two rules:

Monotonicity: If $\ell \Vdash \phi(\alpha)$, then $\ell' \Vdash \phi(\alpha\big|_{\langle r,s\rangle})$ for all $\langle r, s\rangle \colon \ell' \to \ell$.
Local character: If $\ell' \Vdash \phi(\alpha\big|_{\langle r,s\rangle})$ for all wavy arrows $\langle r, s\rangle \colon \ell' \rightsquigarrow \ell$, then $\ell \Vdash \phi(\alpha)$.

The usual Kripke–Joyal sheaf semantics simplifies in our topos $\mathcal{B}$, in particular for the semantics of $\phi \vee \psi$ and $\neg\phi$. We will explain the simplification in Remark 6.1. A summary of the Kripke–Joyal semantics for $\mathcal{B}$ is shown in Table 6.1. For typographical reasons, we do not use Church brackets in this section.

Remark 6.1 Table 6.1 summarizes the Kripke–Joyal semantics in the topos $\mathcal{B} = \mathsf{Shv}(S_{\mathbb{IR}/\rhd})$, where a couple of things simplify, namely the rules for $\ell \Vdash \phi_1(\alpha) \vee \phi_2(\alpha)$ and $\ell \Vdash \neg\phi_1(\alpha)$, where $\alpha \in X(\ell)$. The usual rule for these are

$\ell \Vdash \phi_1(\alpha) \vee \phi_2(\alpha)$ iff, for each $\langle r, s\rangle \colon \ell' \rightsquigarrow \ell$, either $\ell' \Vdash \phi_1\big|_{\langle r,s\rangle}$ or $\ell' \Vdash \phi_2\big|_{\langle r,s\rangle}$.
$\ell \Vdash \neg\phi(\alpha)$ iff the empty family is a cover for ℓ.

Table 6.1 The Kripke–Joyal semantics in the case of $\mathcal{B} = \mathsf{Shv}(S_{\mathbb{IR}/\rhd})$

(i) $\ell \Vdash \phi_1(\alpha) \wedge \phi_2(\alpha)$ iff $\ell \Vdash \phi_1(\alpha)$ and $\ell \Vdash \phi_2(\alpha)$;
(ii) $\ell \Vdash \phi_1(\alpha) \vee \phi_2(\alpha)$ iff $\ell \Vdash \phi_1(\alpha)$ or $\ell \Vdash \phi_2(\alpha)$;
(iii) $\ell \Vdash \phi_1(\alpha) \Rightarrow \phi_2(\alpha)$ iff for all $\langle r, s\rangle \colon \ell' \to \ell$, $\ell' \Vdash \phi_1(\alpha\big|_{\langle r,s\rangle})$ implies $\ell' \Vdash \phi_2(\alpha\big|_{\langle r,s\rangle})$;
(iv) $\ell \Vdash \neg\phi(\alpha)$ iff for all $\langle r, s\rangle \colon \ell' \to \ell$, it is not the case that $\ell' \Vdash \phi(\alpha\big|_{\langle r,s\rangle})$;
(v) $\ell \Vdash \exists(y : Y).\,\phi(\alpha, y)$ iff, for each wavy arrow $\langle r, s\rangle \colon \ell' \rightsquigarrow \ell$, there exists $\beta \in Y(\ell')$ such that $\ell' \Vdash \phi(\alpha\big|_{\langle r,s\rangle}, \beta)$;
(vi) $\ell \Vdash \forall(y : Y).\,\phi(\alpha, y)$ iff, for all $\langle r, s\rangle \colon \ell' \to \ell$ and $\beta \in Y(\ell')$, we have $\ell' \Vdash \phi(\alpha\big|_{\langle r,s\rangle}, \beta)$.

[1] Recall from Definition 3.5 that a wavy arrow $\langle r, s\rangle \colon \ell' \rightsquigarrow \ell$ in $\mathbb{IR}^{\mathrm{op}}_{/\rhd}$ is an interval inclusion that is strict on both sides, i.e., where $r > 0$ and $s > 0$.

Table 6.2 Summary of notation and axioms from Chap. 5

Atomic symbols and other notation:

- $\mathbb{Q}$ is the usual type of rational numbers and $\texttt{Prop}$ is the type of truth values.
- $\bar{\underline{\mathbb{R}}}^{\infty}$ is the type of pairs $(\delta, \upsilon) : (\mathbb{Q} \to \texttt{Prop}) \times (\mathbb{Q} \to \texttt{Prop})$ that satisfy the following:
  - $\forall (q, q' : \mathbb{Q}).\, (q < q') \Rightarrow [(\delta q' \Rightarrow \delta q) \wedge (\upsilon q \Rightarrow \upsilon q')]$
  - $\forall (q : \mathbb{Q}).\, [\delta q \Rightarrow \exists (q' : \mathbb{Q}).\, (q < q') \wedge \delta q'] \wedge [\upsilon q \Rightarrow \exists (q' : \mathbb{Q}).\, (q' < q) \wedge \upsilon q']$

  $\bar{\underline{\mathbb{R}}} \subseteq \bar{\underline{\mathbb{R}}}^{\infty}$ is the type additionally satisfying

  - $\exists (q, q' : \mathbb{Q}).\, \delta q \wedge \upsilon q'$

  $\mathbb{R} \subseteq \bar{\underline{\mathbb{R}}}$ and $\mathbb{R}^{\infty} \subseteq \bar{\underline{\mathbb{R}}}^{\infty}$ are the subtypes additionally satisfying:

  - $\forall (q, q' : \mathbb{Q}).\, [(\delta q \wedge \upsilon q') \Rightarrow (q < q')] \wedge [(q < q') \Rightarrow (\delta q' \vee \upsilon q)]$

  For any $r = (\delta, \upsilon) : \mathbb{R}$, we write $q < r$ to denote δq and $r < q$ to denote υq. For any $(\delta', \upsilon') : \bar{\underline{\mathbb{R}}}$, we write $x + r$ to denote (δ'', υ'') with $\delta'' q'' \Leftrightarrow \exists (q, q').\, \delta q \wedge \delta' q' \wedge q'' < q + q'$ and $\upsilon'' q'' \Leftrightarrow \exists (q, q').\, \upsilon q \wedge \upsilon' q' \wedge q + q' < q''$
- $\texttt{C}$ represents any constant type; see Definition 5.2.
- unit_speed : $\bar{\underline{\mathbb{R}}} \to \texttt{Prop}$ is an atomic predicate symbol, and $\texttt{Time}$ is the subtype $\texttt{Time} = \{t : \bar{\underline{\mathbb{R}}} \mid \text{unit\_speed } t\}$. For any $t = (\delta, \upsilon) : \texttt{Time}$, we write $q < t$ to denote δq and $r < q$ to denote υq.
- For any $d, u : \mathbb{R}$, let $t \,\#\, [d, u]$ be shorthand for $(d < t) \vee (t < u)$. We define four modalities $\uparrow^t_{[d,u]}, \downarrow^t_{[d,u]}, @^t_{[d,u]}$, and π (see Lemma 4.6), given as follows:

$$\uparrow^t_{[d,u]} P \Leftrightarrow ((d < t < u) \Rightarrow P) \qquad \downarrow^t_{[d,u]} P \Leftrightarrow (t \,\#\, [d, u] \vee P)$$
$$@^t_{[d,u]} P \Leftrightarrow ((P \Rightarrow t \,\#\, [d, u]) \Rightarrow t \,\#\, [d, u]) \qquad \pi P \Leftrightarrow \forall (t : \texttt{Time}).\, @^t_{[0,0]} P$$

Axioms

Axiom 1	$\forall ((\delta, \upsilon) : \mathbb{R}^{\infty})(q : \mathbb{Q}).\, (\delta q \vee \neg \delta q) \wedge (\upsilon q \vee \neg \upsilon q)$
Axiom 2	$\forall (P : \texttt{Prop})(P' : \texttt{C} \to \texttt{Prop}).\, \big(\forall (c : \texttt{C}).\, P \vee P'(c)\big) \Rightarrow \big(P \vee \forall (c : \texttt{C}).\, P'(c)\big)$
Axiom 3a	$\exists (t : \texttt{Time}).\, \top$
Axiom 3b	$\forall (q : \mathbb{Q})(t : \texttt{Time}).\, (\neg (q < t) \Leftrightarrow (t < q)) \wedge (\neg (t < q) \Leftrightarrow (q < t))$
Axiom 3c	$\forall (t : \texttt{Time})(P : \mathbb{Q} \to \texttt{Prop}).\, (\forall q.\, Pq \vee \neg Pq) \Rightarrow (\forall q.\, Pq \vee q < t) \Rightarrow \exists q.\, Pq \wedge q < t$
Axiom 3d	$\forall (t : \texttt{Time})(P : \mathbb{Q} \to \texttt{Prop}).\, (\forall q.\, Pq \vee \neg Pq) \Rightarrow (\forall q.\, Pq \vee t < q) \Rightarrow \exists q.\, Pq \wedge t < q$
Axiom 4a	$\forall (t : \texttt{Time})(r : \mathbb{R}).\, t + r \in \texttt{Time}$
Axiom 4b	$\forall (t_1, t_2 : \texttt{Time}).\, \exists ! (r : \mathbb{R}).\, t_1 + r = t_2$
Axiom 5a	$\forall (t : \texttt{Time})(P, Q : \texttt{Prop})(q : \mathbb{Q}).\, (t < q \Rightarrow (P \vee Q)) \Rightarrow \big((t < q \Rightarrow P) \vee (t < q \Rightarrow Q)\big)$
Axiom 5b	$\forall (t : \texttt{Time})(P, Q : \texttt{Prop})(q : \mathbb{Q}).\, (q < t \Rightarrow (P \vee Q)) \Rightarrow \big((q < t \Rightarrow P) \vee (q < t \Rightarrow Q)\big)$
Axiom 6a	$\exists (c : \texttt{C}).\, \top \Rightarrow \forall (t : \texttt{Time})(u : \mathbb{R}^{\infty})(P : C \to \texttt{Prop}).\, (t < u \Rightarrow \exists (c : \texttt{C}).\, P(c)) \Rightarrow \big(\forall (u' : \mathbb{Q}).\, u' < u \Rightarrow \exists (c : \texttt{C}).\, t < u' \Rightarrow P(c)\big)$
Axiom 6b	$\exists (c : \texttt{C}).\, \top \Rightarrow \forall (t : \texttt{Time})(d : \mathbb{R}^{\infty})(P : C \to \texttt{Prop}).\, (d < t \Rightarrow \exists (c : \texttt{C}).\, P(c)) \Rightarrow \big(\forall (d' : \mathbb{Q}).\, d < d' \Rightarrow \exists (c : \texttt{C}).\, d' < t \Rightarrow P(c)\big)$
Axiom 7	$\forall (t : \texttt{Time})(P, Q : \texttt{Prop})(d, u : \mathbb{Q}).\, (d \leq u) \Rightarrow \big((P \wedge Q) \Rightarrow t \,\#\, [u, d]\big) \Rightarrow \big((P \Rightarrow t \,\#\, [u, d]) \vee (Q \Rightarrow t \,\#\, [u, d])\big)$
Axiom 8	$\forall (t : \texttt{Time})(P \colon \texttt{Prop}).\, \big(\forall (d, u : \mathbb{Q}).\, d < u \Rightarrow @^t_{[d,u]} P\big) \Rightarrow P$
Axiom 9	$\forall (t : \texttt{Time})(P \colon \texttt{Prop})(d, u : \mathbb{R}).\, d \leq u \Rightarrow @^t_{[d,u]} P \Rightarrow \downarrow^t_{[d,u]} \exists (d', u' : \mathbb{Q}).\, d' < d \leq u < u' \wedge \uparrow^t_{[d',u']} P$
Axiom 10	$\forall (P : \texttt{Prop}).\, \big(\forall (t : \texttt{Time})(q_1, q_2 : \mathbb{Q}).\, q_1 < q_2 \Rightarrow (t \,\#\, [q_2, q_1] \Rightarrow P) \Rightarrow P\big) \Rightarrow \pi P \Rightarrow P$

The only object of $S_{\mathbb{IR}/\rhd}$ covered by the empty family is $\ell = 0$, which we are not considering (as discussed in the opening remarks of Sect. 6.1). Thus the above rule for negation reduces to the one shown in Table 6.1 (iv). To prove that our simplification (ii) of the rule for implication is also valid, we will argue non-constructively. Clearly (ii) implies the above, so we want the converse.

So suppose that $\ell \Vdash \phi_1(\alpha) \vee \phi_2(\alpha)$ in the sense shown here, and suppose for contradiction that we have neither $\ell \Vdash \phi_1(\alpha)$ nor $\ell \Vdash \phi_2(\alpha)$. Then by local character, there is some $\langle r_1, s_1\rangle : \ell_1 \rightsquigarrow \ell$ and $\langle r_2, s_2\rangle : \ell_2 \rightsquigarrow \ell$ such that neither $\ell_1 \Vdash \phi_1(\alpha|_{\langle r_1,s_1\rangle})$ nor $\ell_2 \Vdash \phi_2(\alpha|_{\langle r_2,s_2\rangle})$. Let $r' := \max(r_1, r_2)$ and $s' := \min(s_1, s_2)$, giving $\langle r', s'\rangle : \ell' \rightsquigarrow \ell$. Then by assumption either $\ell' \Vdash \phi_1|_{\langle r',s'\rangle}$ or $\ell' \Vdash \phi_2|_{\langle r',s'\rangle}$, which is a contradiction by monotonicity.

The following lemma is easy to verify, but we prove it here because will use it quite often, and without mentioning it again.

Lemma 6.2 *Let Φ be a formula of the form $\forall(a_1 : A_1)\cdots(a_n : A_n).\,\phi(a_1,\ldots,a_n) \Rightarrow \psi(a_1,\ldots,a_n)$. Then Φ is sound iff, for every $\ell \in \mathbb{IR}_{/\rhd}$, every $\alpha_1 \in [\![A_1]\!](\ell)$ and so on through $\alpha_n \in [\![A_n]\!](\ell)$, if $\ell \Vdash \phi(\alpha_1,\ldots,\alpha_n)$ then $\ell \Vdash \psi(\alpha_1,\ldots,\alpha_n)$.*

Proof By definition, Φ is sound in the Kripke–Joyal semantics if $\ell \Vdash \Phi$ holds for all $\ell \in \mathbb{IR}_{/\rhd}$. This holds iff the following long formula does: For all $\ell \in \mathbb{IR}_{/\rhd}$,

$$
\begin{array}{ll}
\text{for all } \langle r_1, s_1\rangle : \ell_1 \to \ell & \text{and } a_1 \in [\![A_1]\!](\ell_1), \\
\text{for all } \langle r_2, s_2\rangle : \ell_2 \to \ell_1 & \text{and } a_2 \in [\![A_2]\!](\ell_2), \\
\vdots & \vdots \\
\text{for all } \langle r_n, s_n\rangle : \ell_n \to \ell_{n-1} & \text{and } a_n \in [\![A_n]\!](\ell_n), \text{ and} \\
\text{for all } \langle r_{n+1}, s_{n+1}\rangle : \ell_{n+1} \to \ell_n, & \text{if } \ell_{n+1} \Vdash \phi(a_1|_{\langle R_1,S_1\rangle},\ldots,a_n|_{\langle R_n,S_n\rangle}) \\
 & \quad \text{then } \ell_{n+1} \Vdash \psi(a_1|_{\langle R_1,S_1\rangle},\ldots,a_n|_{\langle R_n,S_n\rangle})
\end{array}
$$

where $R_k = r_{k+1} + \cdots + r_{n+1}$ and $S_k = s_{k+1} + \cdots + s_{n+1}$.

If the long formula above holds, then taking $r_i = 0 = s_i$ for each i, we have the desired conclusion: if $\ell \Vdash \phi(\alpha_1,\ldots,\alpha_n)$, then $\ell \Vdash \psi(\alpha_1,\ldots,\alpha_n)$. On the other hand, suppose it is the case that for every $\ell \in \mathbb{IR}_{/\rhd}$, if $\ell \Vdash \phi(\alpha_1,\ldots,\alpha_n)$, then $\ell \Vdash \psi(\alpha_1,\ldots,\alpha_n)$. Then in particular this holds for ℓ_{n+1} and $\alpha_i|_{\langle R_i,S_i\rangle}$, so the long formula above also holds. □

6.2 Constant Objects and Decidable Predicates

In order to prove that our axioms are sound, we need to interpret all of the types, terms, and predicates that appear in them. The most non-standard of these is the notion of constant types, so we record some of their semantic properties in this section.

6.2.1 *Constant Objects in $\mathcal{B}$*

For any sheaf topos $\mathcal{E}$, there is a unique geometric morphism $\Gamma : \mathcal{E} \leftrightarrows \mathbf{Set} : \mathsf{Cnst}$ to the category of sets, and an object $X \in \mathcal{E}$ is called *constant* if it is in the image of the left adjoint, Cnst. Because $\mathcal{B} = \mathsf{Shv}(S_{\mathbb{IR}/\rhd})$ is locally connected (Proposition 3.8), constant presheaves $(\ell \mapsto C) : \mathbb{IR}_{/\rhd} \to \mathbf{Set}$ are in fact sheaves. Thus for any constant type $\mathtt{C}$ and $\ell \in \mathbb{IR}_{/\rhd}$, we have $[\![\mathtt{C}]\!](\ell) = C$ and all the restriction maps are identity.

Morally, the converse should also hold: a type $\mathtt{C}$ in the type theory should be called constant if the corresponding sheaf $[\![\mathtt{C}]\!]$ is constant. However, in order to present a self-contained axiomatics, we gave a recursive formulation of the collection of constant types in Definition 5.2, namely as the smallest collection containing $\mathtt{1}$, $\mathbb{N}$, $\mathbb{Z}$, $\mathbb{Q}$, and $\mathbb{R}^{\infty}$, and that is closed under finite sums, finite products, exponentials, and decidable subtypes. In any sheaf topos, the objects $\mathbb{N}$, $\mathbb{Z}$, and $\mathbb{Q}$ are constant sheaves. We prove that $\mathbb{R}^{\infty}$ is constant in Corollary 6.7. Sums and products of constant sheaves are constant, as are complemented subobjects of constant sheaves. Exponentials of constant sheaves are constant by Corollary 3.9.

While we only consider this particular collection, the ideas in this section work in general. Later in Sect. 8.1.2 we will allow ourselves to add new constant types to the signature, and considering them as such in the axiomatics.

Exponentiating Constant Types For any sheaf X, the exponential $X^{\mathtt{C}}$ has a simple description: the set of sections $(X^{\mathtt{C}})(\ell)$ is simply the set of functions $C \to X(\ell)$. Indeed, $(X^{\mathtt{C}})(\ell) \cong \mathcal{B}(y(\ell), X^{\mathtt{C}}) \cong \mathcal{B}(\mathtt{C}, X^{y(\ell)}) \cong \mathbf{Set}(C, \Gamma(X^{y(\ell)})) \cong \mathbf{Set}(C, X(\ell))$, where Γ is the global sections functor, right adjoint to the constant sheaf functor $\mathbf{Set} \to \mathcal{B}$. Given a function $f : C \to X(\ell)$ and a morphism $\langle r, s\rangle : \ell' \to \ell$, the restriction is given by $(f\big|_{\langle r,s\rangle})(c) = (f(c))\big|_{\langle r,s\rangle}$.

In particular, for the type of predicates $\mathtt{Prop}^{\mathtt{C}}$, a section $\phi \in [\![\mathtt{Prop}^{\mathtt{C}}]\!](\ell)$ is just a function $\phi : C \to [\![\mathtt{Prop}]\!](\ell) = \Omega(\Uparrow[0, \ell])$, with restrictions given by $(\phi\big|_{\langle r,s\rangle})(c) = \phi(c) \cap \Uparrow[r, \ell - s]$.

6.2.2 *Pointwise Semantics of Predicates on Constant Types*

We just discussed the semantics of predicates on constant types, but there is another, dual, description of the set $[\![\mathtt{Prop}^{\mathtt{C}}]\!](\ell)$ which will be useful. Using the characterization of opens in a domain as continuous maps to the domain 2, we have

$$[\![\mathtt{Prop}^{\mathtt{C}}]\!](\ell) \cong \mathbf{Set}(C, \Omega(\Uparrow[0, \ell])) \cong \mathbf{Set}(C, \mathbf{Top}(\Uparrow[0, \ell], 2)) \cong \mathbf{Top}(\Uparrow[0, \ell], 2^{C}), \tag{6.1}$$

where 2^C is the powerset domain of C (see Example 2.15). Concretely, if $\phi : C \to \Omega(\Uparrow[0, \ell])$ is a function, then we write the corresponding continuous map as follows for any $0 < d \le u < \ell$:

$$[d,u] \mapsto (\phi_{[d,u]} \subseteq C), \qquad \text{given by } c \in \phi_{[d,u]} \text{ iff } [d,u] \in \phi(c). \tag{6.2}$$

By Proposition 2.16, continuity is simply the condition that $\phi_{[d,u]} = \bigcup_{[d',u'] \ll [d,u]} \phi_{[d',u']}$. From the perspective of Kripke–Joyal semantics, Eq. (6.2) is equivalent to the statement that $c \in \phi_{[d,u]}$ iff there exists an open neighborhood of $[d,u]$—a way-below subinterval $\langle r,s\rangle : \ell' \rightsquigarrow \ell$ with $r < d \leq u < \ell - s$—such that $\ell' \Vdash (\phi|_{\langle r,s\rangle})(c)$.

The reason this dual description of predicates on constant types is useful is that their semantics can often be understood "pointwise."

Proposition 6.3 *With notation as in Eq.* (6.2), *suppose given* $\phi, \phi' \in [\![\mathtt{Prop}^{\mathtt{C}}]\!](\ell)$ *and* $\psi \in [\![\mathtt{Prop}^{\mathtt{C}\times\mathtt{C}'}]\!](\ell) \cong [\![(\mathtt{Prop}^{\mathtt{C}})^{\mathtt{C}'}]\!](\ell)$. *Then for any* $[d,u] \in \Uparrow[0,\ell]$,

- $\top_{[d,u]} = C$
- $(\phi \wedge \phi')_{[d,u]} = \phi_{[d,u]} \cap \phi'_{[d,u]}$
- $\bot_{[d,u]} = \varnothing$
- $(\phi \vee \phi')_{[d,u]} = \phi_{[d,u]} \cup \phi'_{[d,u]}$
- $c \in (\exists(c' : \mathtt{C}).\, \psi(c'))_{[d,u]}$ *iff there exists* $c' \in C$ *such that* $(c, c') \in \psi_{[d,u]}$.

There is less control over implication and universal quantification. For any $[d,u] \in \Uparrow[0,\ell]$,

- $(\phi \Rightarrow \phi') = \top$ *iff* $\phi_{[d,u]} \subseteq \phi'_{[d,u]}$ *for all* $[d,u] \in \Uparrow[0,\ell]$
- $(\forall(c : \mathtt{C}).\, \phi(c)) = \top$ *iff* $\phi_{[d,u]} = C$ *for all* $[d,u] \in \Uparrow[0,\ell]$.

Proof We prove the existential case—which is the most difficult case—to give the idea. The remaining cases are proven similarly.

Consider a $\psi \in [\![\mathtt{Prop}^{\mathtt{C}\times\mathtt{C}'}]\!](\ell)$ and a point $[d,u] \in \Uparrow[0,\ell]$, and let $\psi' := \exists(c' : \mathtt{C}).\, \psi(c')$, i.e., $\psi'(c) = \exists(c' : \mathtt{C}).\, \psi(c,c')$. Unwinding the definition in Eq. (6.2), or more precisely its Kripke–Joyal semantic interpretation, we find that $c \in \psi'_{[d,u]}$ if and only if there exists a $\langle r,s\rangle : \ell' \rightsquigarrow \ell$ such that $r < d \leq u < \ell - s$, and such that for all $\langle r',s'\rangle : \ell'' \rightsquigarrow \ell'$, there exists a $c' \in C$ such that $\ell'' \Vdash (\psi|_{\langle r+r', s+s'\rangle})(c,c')$. But note that we can assume this ℓ'' is itself an open neighborhood of $[d,u]$, i.e., that $r + r' < d \leq u \leq \ell - (s+s')$.

Hence, the above simplifies to the following condition: $c \in \psi'_{[d,u]}$ if and only if there exists a $\langle r,s\rangle : \ell' \rightsquigarrow \ell$ such that $r < d$ and $u < \ell - s$, and such that there exists a $c' \in C$ such that $\ell' \Vdash (\psi|_{\langle r,s\rangle})(c,c')$. In other words, $c \in \psi'_{[d,u]}$ if and only if there exists a $c' \in C$ such that $(c,c') \in \psi_{[d,u]}$. □

The class of first-order formulas generated by $\top$, $\wedge$, $\bot$, $\vee$, and $\exists$—in particular, not including $\Rightarrow$ and $\forall$—are typically referred to as *coherent formulas*. If ϕ, ψ are coherent formulas, then for any sequence of variables $x_1, \ldots, x_n$, a statement of the form $\forall x_1, \ldots x_n.\, \phi \Rightarrow \psi$ is called a *coherent axiom*. Proposition 6.3 says that the fragment of the type theory consisting of constant types and coherent axioms has a particularly simple pointwise semantics.

Remark 6.4 Suppose $\phi : \mathtt{C} \to \mathtt{Prop}$ satisfies the additional condition $\forall(c : \mathtt{C}).\, \phi c \vee \neg(\phi c)$, i.e., ϕ is a decidable predicate. It is easy to check that this is equivalent to the requirement that $\phi_{[d,u]} = \phi_{[d',u']}$, as subsets of C, for all $[d, u], [d', u'] \in \Uparrow[0, \ell]$. In other words, we can identify ϕ with a subset of C.

6.3 Semantics of Dedekind Numeric Objects and Time

At this point, there is nothing stopping us from proving that each axiom from Chap. 5 is sound in $\mathcal{B}$. However, these axioms make repeated use of certain condensed definitions—namely, the Dedekind numeric types, the subtype Time, and certain modalities $j : \texttt{Prop} \to \texttt{Prop}$—the semantics of which it will be useful to unpack. We discuss the semantics of numeric types and Time in Sects. 6.3.1 and 6.3.2, and we discuss the semantics of the modalities in Sect. 6.4.

6.3.1 Semantics of Dedekind Numeric Objects

In Sect. 6.2 we discussed the semantics of predicates on constant sheaves. A particularly important case is that of Dedekind cuts, which are predicates $\mathbb{Q} \to \texttt{Prop}$. In this section we discuss the semantics of the types $\bar{\underline{\mathbb{R}}}$ and $\mathbb{R}$; see Definition 4.21. In particular we will show that the sheaf $[\![\mathbb{R}]\!]$ of Dedekind reals in $\mathcal{B} = \mathsf{Shv}(S_{\mathbb{IR}/\rhd})$ is constant.

Definition 6.5 (Sheaf of Continuous Functions) Let X be any topological space. Then we can define the *sheaf of continuous X-valued maps* $\mathrm{Fn}(X) \in \mathcal{B}$ as follows: sections are given by

$$\mathrm{Fn}(X)(\ell) := \mathbf{Top}(\Uparrow[0, \ell], X) \tag{6.3}$$

where $\Uparrow[0, \ell]$ is given the Scott topology, and for any $f \in \mathbf{Top}(\Uparrow[0, \ell], X)$ and any $\langle r, s\rangle : \ell \to \ell'$, the restriction $f\big|_{\langle r,s\rangle}$ is given by the composition

$$\Uparrow[0, \ell'] \cong \Uparrow[r, \ell - s] \hookrightarrow \Uparrow[0, \ell] \xrightarrow{f} X.$$

In particular, if P is any poset with directed joins, we define $\mathrm{Fn}(P)$ by equipping P with the Scott topology.

Hence $\mathrm{Fn}(P)(\ell)$ is the set of Scott-continuous functions $\Uparrow[0, \ell] \to P$, i.e., monotonic maps which preserve directed joins. The powerset $P = 2^{\mathbb{Q}}$ with its Scott topology (Remark 2.15) will be particularly relevant in the discussion below, because by Eqs. (6.1), (6.3) we can identify

$$[\![\mathbb{Q} \to \texttt{Prop}]\!] \cong \mathbf{Top}(\Uparrow[0, \ell], 2^{\mathbb{Q}}) = \mathrm{Fn}(2^{\mathbb{Q}})(\ell).$$

The Dedekind numeric types (for the trivial modality $j = \mathrm{id}$) are all constructed as subtypes of $\mathbb{Q} \to \texttt{Prop}$, or of $(\mathbb{Q} \to \texttt{Prop}) \times (\mathbb{Q} \to \texttt{Prop})$, namely those cut out by the axioms of Definition 4.21. One immediately checks that each of the Dedekind axioms is (equivalent to) a coherent axiom. Hence Proposition 6.3 allows us to use an alternative pointwise semantics to understand Dedekind cuts.

For any predicate $\delta\colon \mathbb{Q} \to \texttt{Prop}$, consider the corresponding continuous function $\Uparrow[0,\ell] \to 2^{\mathbb{Q}}$, sending $[d,u] \mapsto \delta_{[d,u]}$. If δ satisfies a coherent axiom ϕ such as $\forall(q, q' : \mathbb{Q}).\,(\delta q \wedge (q' < q)) \Rightarrow \delta q'$, then by Proposition 6.3 the subset $\delta_{[d,u]} \subseteq \mathbb{Q}$ will satisfy the set-theoretic property corresponding to ϕ, in this case being down-closed. This being true for each $[d,u]$ means that the subtype $[\![\{\mathbb{Q} \to \texttt{Prop} \mid \phi\}]\!]$ can be identified with the subsheaf of $\mathrm{Fn}(2^{\mathbb{Q}})$ corresponding to the subspace cut out by ϕ, e.g., $\{f : \mathbb{Q} \to 2 \mid f \text{ is down-closed}\} \subseteq 2^{\mathbb{Q}}$.

In this way, we can externalize the axioms defining each of the Dedekind numeric objects $\underline{\mathbb{R}}$, $\bar{\mathbb{R}}$, $\bar{\underline{\mathbb{R}}}$, $\mathbb{I}\ \mathbb{R}$, $\mathbb{R}$, $\underline{\mathbb{R}}^\infty$, $\bar{\mathbb{R}}^\infty$, $\bar{\underline{\mathbb{R}}}^\infty$, $\mathbb{I}\ \mathbb{R}^\infty$, and $\mathbb{R}^\infty$ from Definition 4.21. For example, $[\![\underline{\mathbb{R}}]\!]$ can be identified with the subsheaf of continuous functions $\Uparrow[0,\ell] \to 2^{\mathbb{Q}}$ whose image lies in the partially ordered *set*

$$\underline{\mathbb{R}} := \{D \subseteq \mathbb{Q} \mid D \text{ is nonempty, down-closed, and rounded}\}.$$

With the strength of classical logic, we can identify $\underline{\mathbb{R}}$ with $(\mathbb{R} \cup \{\infty\}, \leq)$, by sending a subset $D \subseteq \mathbb{Q}$ to its supremum in $\mathbb{R}$. Likewise, we can identify $\bar{\mathbb{R}}$ with $(\mathbb{R} \cup \{-\infty\}, \geq)$, $\bar{\underline{\mathbb{R}}}$ with $\underline{\mathbb{R}} \times \bar{\mathbb{R}}$, $\mathbb{I}\mathbb{R}$ with $\{(\underline{x}, \bar{x}) \in \underline{\mathbb{R}} \times \bar{\mathbb{R}} \mid \underline{x} \leq \bar{x}\}$, and $\mathbb{R}$ with the standard set of real numbers (having the discrete order). For the unbounded versions, we identify $\underline{\mathbb{R}}^\infty$ with $(\mathbb{R} \cup \{\infty, -\infty\}, \leq)$, $\bar{\mathbb{R}}^\infty$ with $(\mathbb{R} \cup \{\infty, -\infty\}, \geq)$, $\bar{\underline{\mathbb{R}}}^\infty$ with $\underline{\mathbb{R}}^\infty \times \bar{\mathbb{R}}^\infty$, $\mathbb{I}\mathbb{R}$ with $\{(\underline{x}, \bar{x}) \in \underline{\mathbb{R}}^\infty \times \bar{\mathbb{R}}^\infty \mid \underline{x} \leq \bar{x}\}$, and $\mathbb{R}^\infty$ with the $\mathbb{R} \cup \{\infty, -\infty\}$ with the discrete order.

Thus Proposition 6.3 implies the following very useful proposition.

Proposition 6.6 *There are isomorphisms of sheaves*

$$\begin{array}{ccccc}
[\![\underline{\mathbb{R}}]\!] \cong \mathrm{Fn}(\underline{\mathbb{R}}) & [\![\bar{\mathbb{R}}]\!] \cong \mathrm{Fn}(\bar{\mathbb{R}}) & [\![\bar{\underline{\mathbb{R}}}]\!] \cong \mathrm{Fn}(\bar{\underline{\mathbb{R}}}) & [\![\mathbb{I}\ \mathbb{R}]\!] \cong \mathrm{Fn}(\mathbb{I}\mathbb{R}) & [\![\mathbb{R}]\!] \cong \mathrm{Fn}(\mathbb{R}) \\
[\![\underline{\mathbb{R}}^\infty]\!] \cong \mathrm{Fn}(\underline{\mathbb{R}}^\infty) & [\![\bar{\mathbb{R}}^\infty]\!] \cong \mathrm{Fn}(\bar{\mathbb{R}}^\infty) & [\![\bar{\underline{\mathbb{R}}}^\infty]\!] \cong \mathrm{Fn}(\bar{\underline{\mathbb{R}}}^\infty) & [\![\mathbb{I}\ \mathbb{R}^\infty]\!] \cong \mathrm{Fn}(\mathbb{I}\mathbb{R}^\infty) & [\![\mathbb{R}^\infty]\!] \cong \mathrm{Fn}(\mathbb{R}^\infty)
\end{array}$$

where the sheaf of continuous maps $\mathrm{Fn}(P)$ *to a poset* P *is defined in Definition 6.5.*

From our work in Sect. 2.4, we have the following.

Corollary 6.7 *For any* $\ell \in \mathbb{I}\mathbb{R}_{/\rhd}$, *let* $(m, r) = (\frac{\ell}{2}, \frac{\ell}{2}) \in H$ *be the midpoint-radius coordinates of* $[0, \ell]$. *Then we have the following:*

$$\begin{aligned}
[\![\underline{\mathbb{R}}]\!](\ell) &\cong \{\, f\colon \Uparrow(m,r) \to \mathbb{R} \cup \{\infty\} \mid \\
&\qquad f \textit{ is lower semi-continuous and order-preserving}\,\} \\
[\![\bar{\mathbb{R}}]\!](\ell) &\cong \{\, f\colon \Uparrow(m,r) \to \mathbb{R} \cup \{-\infty\} \mid
\end{aligned}$$

$$
\begin{aligned}
& \qquad f \textit{ is upper semi-continuous and order-reversing} \,\} \\
\llbracket \bar{\underline{\mathbb{R}}} \rrbracket(\ell) &\cong \llbracket \underline{\mathbb{R}} \rrbracket(\ell) \times \llbracket \bar{\mathbb{R}} \rrbracket(\ell) \\
\llbracket \mathtt{I}\ \mathbb{R} \rrbracket(\ell) &\cong \{\, (\underline{f}, \bar{f}) \in \llbracket \bar{\underline{\mathbb{R}}} \rrbracket(\ell) \mid \forall x \in \Uparrow(m,r).\, \underline{f}(x) \leq \bar{f}(x) \,\} \\
\llbracket \mathbb{R} \rrbracket(\ell) &\cong \{\, (\underline{f}, \bar{f}) \in \llbracket \bar{\underline{\mathbb{R}}} \rrbracket(\ell) \mid \forall x \in \Uparrow(m,r).\, \underline{f}(x) = \bar{f}(x) \,\} \cong \mathbb{R} \\
\llbracket \underline{\mathbb{R}}^{\infty} \rrbracket(\ell) &\cong \{\, f \colon \Uparrow(m,r) \to \mathbb{R} \cup \{-\infty, \infty\} \mid \\
& \qquad f \textit{ is lower semi-continuous and order-preserving} \,\} \\
\llbracket \bar{\mathbb{R}}^{\infty} \rrbracket(\ell) &\cong \{\, f \colon \Uparrow(m,r) \to \mathbb{R} \cup \{-\infty, \infty\} \mid \\
& \qquad f \textit{ is upper semi-continuous and order-reversing} \,\} \\
\llbracket \bar{\underline{\mathbb{R}}}^{\infty} \rrbracket(\ell) &\cong \llbracket \underline{\mathbb{R}} \rrbracket(\ell) \times \llbracket \bar{\mathbb{R}} \rrbracket(\ell) \\
\llbracket \mathtt{I}\ \mathbb{R}^{\infty} \rrbracket(\ell) &\cong \{\, (\underline{f}, \bar{f}) \in \llbracket \bar{\underline{\mathbb{R}}} \rrbracket(\ell) \mid \forall x \in \Uparrow(m,r).\, \underline{f}(x) \leq \bar{f}(x) \,\} \\
\llbracket \mathbb{R}^{\infty} \rrbracket(\ell) &\cong \{\, (\underline{f}, \bar{f}) \in \llbracket \bar{\underline{\mathbb{R}}} \rrbracket(\ell) \mid \forall x \in \Uparrow(m,r).\, \underline{f}(x) = \bar{f}(x) \,\} \cong \mathbb{R}^{\infty}
\end{aligned}
$$

Proof This follows directly from Proposition 6.6 and Corollary 2.38. □

The following gives an alternate proof of Proposition 6.6. It is more technical, but it has the advantage that it proves that the semantics of our internal addition, multiplication, etc. is just pointwise addition, multiplication, etc.

Lemma 6.8 *Let B be a constant predomain, such that $\llbracket B \rrbracket$ is the constant sheaf on a set $\tilde{B}$. If the predomain order $\prec \colon B \times B \to \mathtt{Prop}$ is identified with a subset $\prec$ of $\tilde{B} \times \tilde{B}$ as in Remark 6.4, then $(\tilde{B}, \prec)$ is an external predomain, $\llbracket \mathrm{RId}(B) \rrbracket \cong \mathrm{Fn}(\mathrm{RId}(\tilde{B}))$, and $\llbracket \Omega(B) \rrbracket \cong \mathrm{Fn}(\Omega(\tilde{B}))$.*

If $H \colon B \to B'$ is a decidable approximable mapping between constant predomains, then $H \colon B \times B' \to \mathtt{Prop}$ can similarly be identified with a subset $\tilde{H} \subseteq \tilde{B} \times \tilde{B}'$. Then $\tilde{H}$ defines an approximable mapping between the external predomains $\tilde{B}$ and $\tilde{B}'$, such that the sheaf homomorphism $\llbracket \mathrm{RId}(H) \rrbracket \in \llbracket \mathrm{RId}(B) \to \mathrm{RId}(B') \rrbracket$ makes the diagram commute:

$$
\begin{array}{ccc}
\llbracket \mathrm{RId}(B) \rrbracket & \cong & \mathrm{Fn}(\mathrm{RId}(\tilde{B})) \\
{\scriptstyle \llbracket \mathrm{RId}(H) \rrbracket} \big\downarrow & & \big\downarrow {\scriptstyle \mathrm{Fn}(\mathrm{RId}(\tilde{H}))} \\
\llbracket \mathrm{RId}(B') \rrbracket & \cong & \mathrm{Fn}(\mathrm{RId}(\tilde{B}'))
\end{array}
$$

Proof This is a simple repeated application of Proposition 6.3 and Remark 6.4.

$(\tilde{B}, \prec)$ is an external predomain because the single axiom in Definition A.1 is equivalent to a pair of coherent axioms, while $\llbracket \mathrm{RId}(B) \rrbracket \cong \mathrm{Fn}(\mathrm{RId}(\tilde{B}))$ because the condition for $I \colon B \to \mathtt{Prop}$ to be a rounded ideal is equivalent to the collection of coherent axioms:

$$\exists(b : B).\, Ib$$
$$\forall(b, b' : B).\, Ib \wedge b' \prec b \Rightarrow Ib'$$
$$\forall(b_1, b_2 : B).\, Ib_1 \wedge Ib_2 \Rightarrow \exists(b' : B).\, b_1 \prec b' \wedge b_2 \prec b'$$

hence $I \in [\![B \to \mathtt{Prop}]\!](\ell)$ is in $[\![\mathrm{RId}(B)]\!](\ell)$ if and only if for all $[d, u] \in {\uparrow}[0, \ell]$, $I_{[d,u]} \in \mathrm{RId}(\tilde{B})$. Finally, the isomorphism $[\![B \to \mathtt{Prop}]\!](\ell) \cong \mathbf{Top}({\uparrow}[0, \ell], 2^{\tilde{B}})$ shows that this assignment is continuous, giving $[\![\mathrm{RId}(B)]\!] \cong \mathrm{Fn}(\mathrm{RId}(\tilde{B}))$. Likewise for $[\![\Omega(B)]\!] \cong \mathrm{Fn}(\Omega(\tilde{B}))$.

From Remark 6.4, if H is decidable, then it determines a subset $\tilde{H} \subset \tilde{B} \times \tilde{B}'$. The axioms in Definition A.28 are easily seen to be equivalent to a collection of coherent axioms, so H is an approximable mapping if and only if $\tilde{H}$ is an external approximable mapping. The commuting diagram follows from the fact that $\exists(b : B).\, Ib \wedge H(b, b')$ is a coherent formula, so for any $I \in [\![\mathrm{RId}(B)]\!](\ell)$ and any $[d, u] \in {\uparrow}[0, \ell]$,

$$\mathrm{RId}(H)(I)_{[d,u]} = \{\, b' \in \tilde{B}' \mid \exists(b \in \tilde{B}).\, b \in I_{[d,u]} \wedge (b, b') \in H_{[d,u]} \,\} = \mathrm{RId}(\tilde{H})(I_{[d,u]}).$$

□

Corollary 6.9 *The semantics of arithmetic and order on* $\bar{\mathbb{R}}$ *agrees with pointwise arithmetic and order on* $\mathrm{Fn}(\bar{\mathbb{R}})$.

In Sect. 4.3, we discussed the type of real numbers in various subtoposes, namely those corresponding to modalities $\uparrow$, $@$, $\downarrow$, and π. This level of understanding is not necessary for the purposes of verifying the soundness of our axioms, so we do not discuss it further now. However, we do need to better understand the semantics of these modalities, so we turn to that in Sect. 6.4. But first we discuss the semantics of `Time`.

6.3.2 *Semantics of* `Time`

Recall that `Time` was defined as a sheaf in Eq. (3.2), by

$$\mathtt{Time}(\ell) = \{\, (d, u) \in \mathbb{R}^2 \mid u - d = \ell \,\} \tag{6.4}$$

where for any subinterval $\langle r, s\rangle : \ell' \to \ell$, the restriction is given by $(d, u)\big|_{\langle r,s\rangle} = (d + r, u - s)$. This is an external definition; its internal counterpart is also denoted `Time` and was defined in Eq. (5.2) to be a subtype of $\bar{\mathbb{R}}$, the type of bounded rounded pairs of cuts $(\delta, \upsilon) : (\mathbb{Q} \to \mathtt{Prop})^2$. Thus we want to show that the sheaf `Time` is a candidate semantics for the type `Time`; this is the content of the following lemma.

Lemma 6.10 *The sheaf* `Time` *is isomorphic to a subsheaf of* $[\![\bar{\mathbb{R}}]\!]$.

Proof It suffices to construct a monomorphism $c\colon \texttt{Time} \to [\![\bar{\underline{\mathbb{R}}}]\!]$. For any $\ell \in \mathbb{IR}^{\mathrm{op}}_{/\rhd}$, we have from Proposition 6.6 that $[\![\bar{\underline{\mathbb{R}}}]\!](\ell) \cong \mathrm{Fn}(\bar{\underline{\mathbb{R}}}) \cong \mathrm{Fn}(\underline{\mathbb{R}}) \times \mathrm{Fn}(\bar{\mathbb{R}})$. Given any section $(d, u) \in \texttt{Time}(\ell)$, define $c_\ell(d, u) := (\underline{d}, \bar{u})$, where $\underline{d}\colon \Uparrow[0, \ell] \to \underline{\mathbb{R}}$ is the continuous map $\underline{d}([d', u']) = d + d'$, and $\bar{u}\colon \Uparrow[0, \ell] \to \bar{\mathbb{R}}$ is the continuous map $\bar{u}([d', u']) = u - (\ell - u')$.

The maps $\underline{d}$ and $\bar{u}$ are obviously monotonic (recall that $\bar{\mathbb{R}} = (\mathbb{R}, \geq))$ and continuous. It is also clear that the function $c_\ell\colon \texttt{Time}(\ell) \to [\![\bar{\underline{\mathbb{R}}}]\!](\ell)$ is injective for all ℓ, and that these functions commute with restrictions, thus defining a monomorphism $c\colon \texttt{Time} \to [\![\bar{\underline{\mathbb{R}}}]\!]$. □

For any ℓ and $t = (d_t, u_t) : \texttt{Time}(\ell)$ and $r \in \mathbb{R} = [\![\mathbb{R}]\!](\ell)$, we have

$$(\ell \Vdash r < t) \text{ iff } r \leq d_t \qquad \text{and} \qquad (\ell \Vdash t < r) \text{ iff } u_t \leq r. \tag{6.5}$$

Indeed, by Corollary 6.9 and the above proof, $\ell \Vdash r < t$ holds iff for all $d' \in \Uparrow[0, \ell]$ we have $r < d_t + d'$, which is equivalent to $r \leq d_t$.

Remark 6.11 We saw in Remark 5.56 that, internally, there is a containment of types $\texttt{Time} \subseteq \mathbb{R}_\pi$, and by definition there is a containment of types $\texttt{Time} = \{t : \bar{\underline{\mathbb{R}}} \mid \text{unit\_speed}\, t\} \subseteq \bar{\underline{\mathbb{R}}}$.[2] In Lemma 6.10 we showed that the sheaf Time is a valid semantics for this type, but in fact it is not the only one.

As the name of the atomic term "unit_speed" suggests, our axioms only constrain the semantics of the type $\texttt{Time} = \{t : \bar{\underline{\mathbb{R}}} \mid \text{unit\_speed}\, t\}$ to be a sheaf of "unit-speed" real-valued functions. More precisely, for any nonzero real number $0 \neq r \in \mathbb{R}$, there is a sheaf $\texttt{Time}_r$ given by $\texttt{Time}_r(\ell) := \{\, (d, u) \in \mathbb{R}^2 \mid u - d = \ell * r\}$.

This is significant when thinking about the semantics of Time. Our convention in Eq. (6.4) and throughout this book is $\texttt{Time} = \texttt{Time}_1$. From this point of view one should interpret each section of time as a *clock behavior*, counting upward like a stopwatch. While this is straightforward, it does have counter-intuitive aspects if we accidentally think of $t : \texttt{Time}$ as indicating a point in time, e.g., the point where the clock shows 0. This is often the natural interpretation of the "rough monotonicity" formula (1.1) or simply the monotonicity formula

$$\forall (t_1, t_2 : \texttt{Time}).\, t_1 \leq t_2 \Rightarrow f(t_1) \leq f(t_2). \tag{6.6}$$

Namely, it looks like $t_1 \leq t_2$ means t_1 comes before t_2. But in fact $t_1 \leq t_2$ means clock t_1 shows a value that is less than clock t_2, so its 0-point t_1 comes *after* that of t_2. Thus Eq. (6.6) actually says that f is *decreasing*!

Consider instead the sheaf $\texttt{Time}_{-1}$. Here it is appropriate to interpret $t : \texttt{Time}$ as indicating a *moment in time*, with sections acting as a count-down. Using $\texttt{Time}_{-1}$ as semantics, Eq. (6.6) indeed says that f is increasing.

One way to get the best of both worlds is to use $\texttt{Time}_1$, as we do, but to restrict oneself to terms in which at most one Time variable occurs, using the torsor

---

[2] We will see in Proposition 7.12 that there is also containment of types $\mathbb{R}_\pi \subseteq \bar{\underline{\mathbb{R}}}$.

axiom—Axiom 4—to convert all other occurrences to real numbers. For example, the above formula becomes

$$\forall(t : \texttt{Time})(r : \mathbb{R}).\, (0 \le r) \Rightarrow f(t) \le f(t + r),$$

and it has the expected semantics that f is increasing.

6.4 Semantics of the Modalities $\Uparrow$, $\downarrow$, @, and π

Recall that for any $d, u : \mathbb{Q}$ the temporal modalities $\Uparrow_{[d,u]}$, $\downarrow_{[d,u]}$, and $@_{[d,u]}$ are defined by

$$\Uparrow_{[d,u]} P := (d < t < u) \Rightarrow P \qquad\qquad \downarrow_{[d,u]} P := (t \,\#\, [u, d]) \vee P$$
$$@_{[d,u]} P := (P \Rightarrow t \,\#\, [u, d]) \Rightarrow t \,\#\, [u, d]$$

where $t \,\#\, [u, d]$ means $(t < u) \vee (d < t)$. The modality $\pi : \texttt{Prop} \to \texttt{Prop}$ is defined by

$$\pi P := \forall(t : \texttt{Time}).\, @_{[0,0]} P.$$

Proposition 6.12 *Suppose given $P \in [\![\texttt{Prop}]\!](\ell)$ and $t = (d_t, u_t) \in [\![\texttt{Time}]\!](\ell)$, for some ℓ, as well as $d, u \in \mathbb{R} = [\![\mathbb{R}]\!]$. Consider the open set $P \in \Omega(\Uparrow[d_t, u_t])$.*

1. *If $u \le d$, then $\ell \Vdash \Uparrow^t_{[d,u]} P$.*
2. *If $d < u$, then $\ell \Vdash \Uparrow^t_{[d,u]} P$ iff $\Uparrow[d, u] \cap \Uparrow[d_t, u_t] \subseteq P$.*
3. *$\ell \Vdash \downarrow^t_{[d,u]} P$ iff it is the case that $[d, u] \in \Uparrow[d_t, u_t]$ implies $P = \Uparrow[d_t, u_t]$.*
4. *Suppose $d \le u$. Then $\ell \Vdash @^t_{[d,u]} P$ iff $[d, u] \in \Uparrow[d_t, u_t]$ implies $[d, u] \in P$.*
5. *Suppose $u < d$. Then $\ell \Vdash @^t_{[d,u]} P$ iff $\Uparrow[u, d] \cap \Uparrow[d_t, u_t] \neq \varnothing$ implies $\Uparrow[u, d] \cap P \neq \varnothing$.*

Proof The meaning of $\ell \Vdash \Uparrow_{[d,u]} P$ is the following statement: for any $\langle r, s\rangle : \ell' \to \ell$, if $d \le d_t + r \le u_t - s \le u$, then $\ell' \Vdash P\big|_{\langle r,s\rangle}$. Since $\ell' > 0$, this statement is vacuously true if $u \le d$, in which case $\ell \Vdash \Uparrow_{[d,u]} P$; this proves claim 1. If $d < u$, the statement is true for all r, s iff it is true when $r := \max(0, d - d_t)$ and $s := \max(0, u_t - u)$, i.e., when $d_t + r = \max(d_t, d)$ and $u_t - s = \min(u_t, u)$. This is the case if $\Uparrow[d, u] \cap \Uparrow[d_t, u_t] \subseteq P$, proving claim 2.

By Eq. (6.5) we have $\ell \Vdash (d < t) \vee (t < u) \vee P$ iff either $d \le d_t$ or $u_t \le u$ or $\ell \Vdash P$. Claim 3 follows because $\ell \Vdash P$ iff $P = \Uparrow[d_t, u_t]$.

For claim 4, suppose $d \le u$. Then $\ell \Vdash (P \Rightarrow t \,\#\, [u, d]) \Rightarrow t \,\#\, [u, d]$ means, by contrapositive, that if $d_t < d \le u < u_t$ then there exists some $\langle r, s\rangle : \ell' \to \ell$ such that $d_t + r < d \le u < u_t - s$ and $\Uparrow[d_t + r, u_t - s] \subseteq P$. The hypothesis is equivalent to $[d, u] \in \Uparrow[d_t, u_t]$ and the conclusion is equivalent to $[d, u] \in P$.

For claim 5, suppose $u < d$. Then one can prove by cases that $(d_t < d) \wedge (u < u_t)$ is equivalent to $\Uparrow[u, d] \cap \Uparrow[d_t, u_t] \neq \varnothing$. Again by contrapositive, $\ell \Vdash @_{[d,u]}P$ means that if $\Uparrow[u, d] \cap \Uparrow[d_t, u_t] \neq \varnothing$ then there exists $\langle r, s\rangle : \ell' \to \ell$ such that $\Uparrow[u, d] \cap \Uparrow[d_t + r, u_t - s] \neq \varnothing$ and $\Uparrow[d_t + r, u_t - s] \subseteq P$. The conclusion is equivalent to saying there exists $d' < u'$ such that $\Uparrow[d', u'] \subseteq \Uparrow[u, d] \cap P$. This clearly implies that $\Uparrow[u, d] \cap P \neq \varnothing$, and the converse follows from the fact that $\mathbb{IR}$ has a basis consisting of intervals $\Uparrow[d', u']$ with $d' < u'$. □

Corollary 6.13 *Suppose given $P \in [\![\texttt{Prop}]\!](\ell)$ and $t = (d_t, u_t) \in [\![\texttt{Time}]\!](\ell)$, for some ℓ, as well as $d, u \in \mathbb{R} = [\![\mathbb{R}]\!]$ with $d \leq u$.*

1. *$\ell \Vdash \uparrow^t_{[d,u]}P$ iff either $d = u$ or $\ell' \Vdash P\big|_{\langle r,s\rangle}$, where $\langle r, s\rangle : \ell' \to \ell$ is given by $r := \max(0, d - d_t)$ and $s := \max(0, u_t - u)$.*
2. *$\ell \Vdash \downarrow^t_{[d,u]}P$ iff it is the case that $d_t < d \leq u < u_t$ implies $\ell \Vdash P$.*
3. *$\ell \Vdash @^t_{[d,u]}P$ iff it is the case that $d_t < d \leq u < u_t$ implies: there exists $\langle r, s\rangle : \ell' \to \ell$ with $d_t + r < d \leq u < u_t - s$, such that $\ell' \Vdash P\big|_{\langle r,s\rangle}$.*

Proof Follows directly from Proposition 6.12. □

Proposition 6.14 *Suppose given $P \in [\![\texttt{Prop}]\!](\ell)$ for some $\ell \in \mathbb{IR}$. Then $\ell \Vdash \pi P$ iff, for all $0 < a < \ell$ there exists $\langle r, s\rangle : \ell' \to \ell$ such that $r < a < \ell - s$ and $\ell' \Vdash P\big|_{\langle r,s\rangle}$.*

Proof $\ell \Vdash \pi P$ implies that for all $t \in \texttt{Time}(\ell)$ we have $@^t_{[0,0]}P$. By Corollary 6.13, this implies that for all $d < 0 < d + \ell$ there exists $\langle r, s\rangle : \ell' \to \ell$ with $0 \leq r < -d$ and $0 \leq s < d + \ell$, such that $\ell' \Vdash P\big|_{\langle r,s\rangle}$. We have the result by letting $a := -d$. □

Suppose $\texttt{C}$ is the constant sheaf on the set C. For any $\phi \in [\![\texttt{Prop}^{\texttt{C}}]\!]$, recall the pointwise-semantics notation $\phi_{[d,u]} \subseteq C$ from Eq. (6.2).

Proposition 6.15 *Choose $P \in [\![\texttt{Prop}^{\texttt{C}}]\!](\ell)$ and $t = (d_t, u_t) \in [\![\texttt{Time}]\!](\ell)$, and take $[d', u'] \in \Uparrow[d_t, u_t]$. For any $c \in C$ and $d, u \in \mathbb{R}$,*

- *$c \in (\downarrow_{[d,u]}P)_{[d',u']}$ iff $[d', u'] \sqsubseteq [d, u] \Rightarrow c \in P_{[d',u']}$.*
- *$c \in (@_{[d,u]}P)_{[d',u']}$ iff $[d', u'] \sqsubseteq [d, u] \Rightarrow c \in P_{[d,u]}$.*
- *$c \in (\pi P)_{[d',u']}$ iff $c \in P_{[q,q]}$ for all q with $d' \leq q \leq u'$.*

Moreover,

- *$P(c)$ is $\downarrow_{[d,u]}$-closed for all $c \in C$ iff, for all $[d', u'] \in \Uparrow[d_t, u_t]$, either $[d', u'] \sqsubseteq [d, u]$ or $P_{[d',u']} = C$.*
- *$P(c)$ is $@_{[d,u]}$-closed for all $c \in C$ iff, for all $[d', u'] \in \Uparrow[d_t, u_t]$, if $[d', u'] \sqsubseteq [d, u]$ then $P_{[d',u']} = P_{[d,u]}$, and if $[d', u'] \not\sqsubseteq [d, u]$, then $P_{[d',u']} = C$.*
- *$P(c)$ is π-closed for all $c \in C$ iff for all $[d', u'] \in \Uparrow[0, \ell]$, $P_{[d',u']} = \bigcap_{[d',u'] \sqsubseteq [q,q]} P_{[q,q]}$.*

Proof The first three are straightforward using Propositions 6.3 and 6.14 and Corollary 6.13. The second three immediately follow. □

6.5 Proof that Each Axiom Is Sound

We are now set up to prove—with little effort—that each axiom is sound in $\mathcal{B}$. See Table 6.2 for a summary of the axioms. In every proof in this section, we will use Lemma 6.2—often without mentioning it—to peel off the top-level universal quantifiers and the top-level implication (if any such are present).

Proposition 6.16 *Axiom 1, denoted ϕ below, is sound in $\mathcal{B}$:*

$$\phi := \forall((\delta, \upsilon) : \mathbb{R}^{\infty})(q : \mathbb{Q}).\, (\delta q \vee \neg\delta q) \wedge (\upsilon q \vee \neg\upsilon q).$$

Proof We must show that for every $\ell \in S_{\mathbb{IR}/\rhd}$, every pair of cuts $(\delta, \upsilon) \in \mathbb{R}(\ell)$ and every rational $q \in \mathbb{Q} = \mathbb{Q}(\ell)$, we have

$$\ell \Vdash (\delta q \vee \neg\delta q) \wedge (\upsilon q \vee \neg\upsilon q).$$

Note that $\mathbb{R}^{\infty}$ is the constant sheaf on the set $\mathbb{R}^{\infty}$ of (set-theoretic) extended real numbers by Corollary 6.7; hence we can identify (δ, υ) with an extended real number $r \in \mathbb{R} \cup \{-\infty, \infty\}$. Thus the above holds iff we have $\ell \Vdash (q < r \vee r \leq q) \wedge (r < q \vee q \leq r)$, which is true because $\mathbb{R} \cup \{-\infty, \infty\}$ is a linear order. □

Proposition 6.17 *Let $\mathtt{C}$ be the constant sheaf on some set C. Then Axiom 2, denoted ϕ below, is sound in $\mathcal{B}$:*

$$\phi := \forall(P : \mathtt{Prop})(P' : \mathtt{C} \to \mathtt{Prop}).\, \big(\forall(c : \mathtt{C}).\, P \vee P'(c)\big) \Rightarrow \big(P \vee \forall(c : \mathtt{C}).\, P'(c)\big).$$

Proof Choose $\ell \in S_{\mathbb{IR}/\rhd}$, and sections $P \in \mathtt{Prop}(\ell)$ and $P' \in \mathtt{Prop}^{\mathtt{C}}(\ell)$. Assume that $\ell \Vdash \forall(c : \mathtt{C}).\, P \vee P'(c)$ holds; we want to show $\ell \Vdash P \vee \forall(c : \mathtt{C}).\, P'(c)$. Arguing classically, suppose $\ell \nVdash P$. Then for each $c \in \mathtt{C}(\ell)$, we have $\ell \Vdash P'(c)$.

We want to prove $\ell \vdash \forall(c : C).\, P'(c)$, so choose $f : \ell' \to \ell$ and $c' \in \mathtt{C}(\ell')$. Since $\mathtt{C}$ is constant, $\mathtt{C}(\ell) = C = \mathtt{C}(\ell')$ for every ℓ'. Thus $\ell \Vdash P'(c')$, and it follows that $\ell' \Vdash P'(c')$ by monotonicity (page 116). □

Proposition 6.18 *Axiom 3a–3d, denoted ϕ_1, ϕ_2, ϕ_3, and ϕ_4 below, are sound in $\mathcal{B}$:*

$$\phi_1 := \exists(t : \mathtt{Time}).\, \top$$

$$\phi_2 := \forall(q : \mathbb{Q})(t : \mathtt{Time}).\, (\neg(q < t) \Leftrightarrow (t < q)) \wedge (\neg(t < q) \Leftrightarrow (q < t))$$

$$\phi_3 := \forall(t : \mathtt{Time})(P : \mathbb{Q} \to \mathtt{Prop}).\, (\forall q.\, Pq \vee \neg Pq) \Rightarrow (\forall q.\, Pq \vee q < t) \Rightarrow \exists q.\, Pq \wedge q < t$$

$$\phi_4 := \forall(t : \mathtt{Time})(P : \mathbb{Q} \to \mathtt{Prop}).\, (\forall q.\, Pq \vee \neg Pq) \Rightarrow (\forall q.\, Pq \vee t < q) \Rightarrow \exists q.\, Pq \wedge t < q.$$

Proof Refer to Sect. 6.3.2 for the semantics of Time. To prove $\ell \Vdash \phi_1$ it suffices to recognize that the set $\mathtt{Time}(\ell) = \{(d,u) \in \mathbb{R}^2 \mid u - d = \ell\}$ is nonempty for any ℓ. To prove $\ell \Vdash \phi_2$, take $q \in \mathbb{Q}$ and $t = (d,u) \in \mathtt{Time}(\ell)$; we want to show $\ell \Vdash \neg(q < t) \Leftrightarrow (t < q)$, the other conjunct being similar.

Recall from Lemma 6.10 that the semantics of $q < t$ and $t < q$ are given by Eq. (6.5). Thus the implication $(t < q) \Rightarrow (q < t) \Rightarrow \bot$ is easy: since $u - d = \ell \geq 0$, we have $d \leq u$ so it cannot be that both $q \leq d$ and $u \leq q$. For the converse, suppose that $\ell' \Vdash \neg(q < t)$, i.e., for all $\langle r, s\rangle : \ell' \to \ell$, if $\ell' > 0$, then it is not the case that $q \leq d + r$. We want to show $u \leq q$. Recall that $\ell = u - d$, by definition of the category $\mathbb{IR}_{/\rhd}$, so for any $r + \ell' + s = \ell$ with $r, s \geq 0$ and $\ell' > 0$ we have $d + r < q$. Consider the case $s = 0$, where $d + r = u - \ell'$. For any $\ell' > 0$ we get $u - \ell' < q$; this implies $u \leq q$ as desired.

The proofs of $\ell \Vdash \phi_3$ and $\ell \Vdash \phi_4$ are similar, so we only prove the former. Choose $t = (d,u)$ and a decidable predicate P; by Remark 6.4 we can identify P with a subset $P \subseteq \mathbb{Q}$. Assume $\ell \Vdash (\forall q.\, Pq \vee q < t)$. Then, in particular, for all $q \in \mathbb{Q}$ either $q \in P$ or $q \leq d$. We must show $\ell \Vdash \exists(q : \mathbb{Q}).\, Pq \vee q < d$, i.e., that for any wavy arrow $\langle r, s\rangle : \ell' \rightsquigarrow \ell$, there exists q such that $q \in P$ and $q \leq d + r$. Since by definition $0 < r$, simply take q to satisfy $d < q < d + r$. □

Proposition 6.19 *Axioms 4a and 4b, denoted ϕ_1 and ϕ_2 below, are sound in $\mathcal{B}$:*

$$\phi_1 := \forall(t : \mathtt{Time})(r : \mathbb{R}).\, t + r \in \mathtt{Time}$$

$$\phi_2 := \forall(t_1, t_2 : \mathtt{Time}).\, \exists(r : \mathbb{R}).\, (t_1 + r = t_2) \wedge \forall(r' : \mathbb{R}).\, (t_1 + r' = t_2) \Rightarrow r = r'.$$

Proof By Corollary 6.7 and Lemma 6.10, we can regard any $t = (d,u) \in \mathtt{Time}(\ell)$ and any $r \in \mathbb{R} = [\![\mathbb{R}]\!]$ as elements of $[\![\bar{\underline{\mathbb{R}}}]\!]$. That is, each is a pair of functions $U \to \mathbb{R}$, say $(\underline{d}, \bar{u})$ and $(\underline{r}, \bar{r})$, respectively, where $U = \{(a,b) \in H \mid (\ell/2, \ell/2) \ll (a,b)\}$. Here $\underline{d}(a,b) = d + a$, $\bar{u}(a,b) = u + b$, $\underline{r}(a,b) = r$, and $\bar{r}(a,b) = r$.

To prove $\ell \Vdash \phi_1$, take t and r as above. By Corollary 6.9, the sum $t + r$ is the pointwise sum $(a,b) \mapsto (d + a + r, u + b + r)$. But this is the function associated with the time $(d + r, u + r) \in \mathtt{Time}(\ell)$. To prove $\ell \Vdash \phi_2$, choose $t_1 = (d_1, u_1)$ and $t_2 = (d_1, u_1)$ in $\mathtt{Time}(\ell)$. Their pointwise difference $f := t_1 - t_2$ is the map $f(a,b) = (d_1 - d_2, u_1 - u_2)$, which is independent of a and b and hence a constant section of $[\![\bar{\underline{\mathbb{R}}}]\!](\ell)$. And since $u_1 - d_1 = \ell = u_2 - d_2$, we have $u_1 - u_2 = d_1 - d_2$, so f can be identified with an element $r \in [\![\mathbb{R}]\!](\ell)$, again by Corollary 6.7. □

Proposition 6.20 *Axiom 5a, denoted ϕ below, is sound in $\mathcal{B}$:*

$$\phi := \forall(t : \mathtt{Time})(P, Q : \mathtt{Prop})(q : \mathbb{Q}).\, (t < q \Rightarrow (P \vee Q))$$
$$\Rightarrow \big((t < q \Rightarrow P) \vee (t < q \Rightarrow Q)\big).$$

Axiom 5b, which is similar, is also sound in $\mathcal{B}$.

Proof The proofs of these two statements are similar so we prove the first that $\ell \Vdash \phi$. Choose $t = (d,u) \in \mathtt{Time}(\ell)$, $q \in \mathbb{Q}$, and $P, Q \in \mathtt{Prop}(\ell)$, and assume the hypothesis, $\ell \Vdash t < q \Rightarrow (P \vee Q)$. This means that for all $\langle r, s\rangle : \ell' \to \ell$, if $u - s \le q$, then either $\ell' \Vdash P\big|_{\langle r,s\rangle}$ or $\ell' \Vdash Q\big|_{\langle r,s\rangle}$.

If $q \le d$, then $\ell \Vdash (t < q) \Leftrightarrow \bot$ and the conclusion is vacuously true. If $u \le q$, then $\ell \Vdash t < q$ so $\ell \Vdash P \vee Q$, and again the conclusion follows easily. So suppose $d < q < u$. We want to show $\ell \Vdash t < q \Rightarrow P$ or $\ell \Vdash t < q \Rightarrow Q$.

Arguing classically, suppose that $\ell \Vdash t < q \Rightarrow P$ does not hold. Then there is some $\langle r, s\rangle : \ell_0 \to \ell$ for which $\ell_0 \Vdash u - s \le q$ but $\ell_0 \nVdash P\big|_{\langle r,s\rangle}$. Let $\ell' := \ell - s$; by monotonicity, we also have $\ell' \nVdash P\big|_{\langle 0,s\rangle}$. Thus by hypothesis we have $\ell' \Vdash Q\big|_{\langle 0,s\rangle}$. It follows that $\ell \Vdash t < q \Rightarrow Q$, as desired. □

Proposition 6.21 *Let* $\mathtt{C}$ *be the constant sheaf on some set* C. *Then Axiom 6a, denoted* ϕ *below, is sound in* $\mathcal{B}$.

$$\phi := \exists(c : \mathtt{C}).\, \top \Rightarrow \forall(t : \mathtt{Time})(u : \mathbb{R}^{\infty})(P : C \to \mathtt{Prop}).$$
$$(t < u \Rightarrow \exists(c : \mathtt{C}).\, P(c)) \Rightarrow \big(\forall(u' : \mathbb{Q}).\, u' < u \Rightarrow \exists(c : \mathtt{C}).\, t < u' \Rightarrow P(c)\big)$$

Axiom 6b, which is similar, is also sound in $\mathcal{B}$.

Proof The proofs of these two statements are similar so we prove the first that $\ell \Vdash \phi$. Suppose C is nonempty, choose $t = (d_t, u_t) \in \mathtt{Time}(\ell)$, $P \in \mathtt{Prop}^{\mathtt{C}}(\ell)$, and $u \in \mathbb{R} \cup \{\infty, -\infty\}$. If $u = \infty$ or $u = -\infty$, the claim is vacuously true, so assume $u \in \mathbb{R}$. Assume the hypothesis, $\ell \Vdash t < u \Rightarrow \exists(c : \mathtt{C}).\, P(c)$. If $u_t \le u$, then $\ell \Vdash t < u$ by Eq. (6.5) and the hypothesis directly implies the conclusion. If $u \le d_t$, then for any u', any choice of $c \in C$ will vacuously satisfy the conclusion. Thus we may assume $d_t < u < u_t$; we next unpack the hypothesis in this case.

Let $r_0 := 0$ and $s_0 := u_t - u$; then on the restriction $\langle r_0, s_0\rangle : \ell_0 \to \ell$, we have $\ell_0 \Vdash (t\big|_{\langle r_0,s_0\rangle} < u)$, so we also have $\ell_0 \Vdash \exists(c : \mathtt{C}).\, P\big|_{\langle r_0,s_0\rangle}(c)$. Thus for every $\langle r_1, s_1\rangle : \ell_1 \rightsquigarrow \ell_0$ there exists a $c \in C$ with $\ell_1 \Vdash P\big|_{\langle r_1,s_1\rangle}(c)$.

For the conclusion, take any $\langle r', s'\rangle : \ell' \to \ell$ and $u' < u$; we need to show $\ell' \Vdash \exists(c : \mathtt{C}).\, (t\big|_{\langle r',s'\rangle} < u') \Rightarrow P\big|_{\langle r',s'\rangle}(c)$. For this it suffices to take an arbitrary $\langle r'', s''\rangle : \ell'' \rightsquigarrow \ell'$ and show that there exists $c \in C$ such that $\ell'' \Vdash (t\big|_{\langle r'',s''\rangle} < u') \Rightarrow P\big|_{\langle r'',s''\rangle}(c)$. Let $r_1 := r' + r''$ and let $s_1 := u_t - u'$; then $r_1 > r_0$ and $s_1 > s_0$, so by the paragraph above, on the subinterval $\langle r_1, s_1\rangle : \ell_1 \rightsquigarrow \ell_0$, there exists $c \in C$ such that $\ell_1 \Vdash P\big|_{\langle r_1,s_1\rangle}(c)$. This is our candidate c.

To show $\ell'' \Vdash (t\big|_{\langle r'',s''\rangle} < u') \Rightarrow P\big|_{\langle r'',s''\rangle}(c)$, take any $\langle r''', s'''\rangle : \ell''' \to \ell''$ and assume the hypothesis, $u_t - s' - s'' - s''' < u' = u_t - s_1$. Then ℓ''' is a subinterval of ℓ_1 and we are done by monotonicity: $\ell_1 \Vdash P\big|_{\langle r_1,s_1\rangle}(c)$ implies $\ell''' \Vdash P\big|_{\langle r''',s'''\rangle}(c)$.

□

Proposition 6.22 *Axiom 7, denoted* ϕ *below, is sound in* $\mathcal{B}$:

$$\phi := \forall(t : \mathtt{Time})(P, Q : \mathtt{Prop})(d, u : \mathbb{Q}).\, (d \le u) \Rightarrow$$
$$\big((P \wedge Q) \Rightarrow t\,\#\,[u, d]\big) \Rightarrow \big((P \Rightarrow t\,\#\,[u, d]) \vee (Q \Rightarrow t\,\#\,[u, d])\big)$$

Proof To show $\ell \Vdash \phi$, choose $t = (d_t, u_t) \in \texttt{Time}(\ell)$, $P, Q \in \texttt{Prop}(\ell)$, and $d, u \in \mathbb{Q}$. Assume the hypothesis that $d \leq u$ and that $\ell \Vdash (P \wedge Q) \Rightarrow t\,\#\,[u, d]$. This means that for all $\langle r, s\rangle : \ell' \to \ell$, if $\ell' \Vdash P\big|_{\langle r,s\rangle}$ and $\ell' \Vdash Q\big|_{\langle r,s\rangle}$, then either $d \leq d_t + r$ or $u_t - s \leq u$.

Suppose $\ell \nVdash P \Rightarrow t\,\#\,[u, d]$. Then for some $\langle r', s'\rangle : \ell' \to \ell$ we have $\ell' \Vdash P\big|_{\langle r',s'\rangle}$ and $d_t + r' < d$ and $u < u_t - s'$. We want to show $\ell \Vdash Q \Rightarrow t\,\#\,[u, d]$, so suppose that for some $\langle r'', s''\rangle : \ell'' \to \ell$ we have $\ell'' \Vdash Q\big|_{\langle r'',s''\rangle}$ and for contradiction suppose that $d_t + r'' < d$ and $u < u_t - s''$.

Let $r_0 := \max(r', r'')$ and $s_0 := \max(s', s'')$, and let $\ell_0 := \ell - (r + s)$. Then $d_t + r_0 < d \leq u < u_t - s_0$ and $\ell = u_t - d_t$, so $0 < \ell_0$. We get a contradiction because $\ell_0 \Vdash P\big|_{\langle r_0,s_0\rangle}$ and $\ell_0 \Vdash Q\big|_{\langle r_0,s_0\rangle}$, which by hypothesis gives $d \leq d_t + r_0$ or $u_t - s_0 \leq u$. □

Proposition 6.23 *Axiom 8, denoted ϕ below, is sound in $\mathcal{B}$:*

$$\phi := \forall(t : \texttt{Time})(P : \texttt{Prop}).\, \big(\forall(d, u : \mathbb{Q}).\, d < u \Rightarrow @_{[d,u]}P\big) \Rightarrow P$$

Proof To show $\ell \Vdash \phi$, choose $t = (d_t, u_t) \in \texttt{Time}(\ell)$ and $P \in \texttt{Prop}(\ell)$, and assume the hypothesis. That is, for every $\langle r, s\rangle : \ell' \to \ell$ and $d, u \in \mathbb{Q}$ with $d < u$, we have $\ell' \Vdash @^{t|_{\langle r,s\rangle}}_{[d,u]} P\big|_{\langle r,s\rangle}$. It follows by Corollary 6.13 (3) that $\ell' \Vdash P\big|_{\langle r,s\rangle}$ for all $r, s > 0$, i.e., for all $\langle r, s\rangle : \ell' \rightsquigarrow \ell$. By local character, this implies $\ell \Vdash P$. □

Proposition 6.24 *Axiom 9, denoted ϕ below, is sound in $\mathcal{B}$.*

$$\forall(t : \texttt{Time})(P : \texttt{Prop})(d, u : \mathbb{R}).\, d \leq u \Rightarrow$$
$$@_{[d,u]}P \Rightarrow \downarrow_{[d,u]}\exists(d', u' : \mathbb{Q}).\, d' < d \leq u < u' \wedge \mathop{\uparrow}_{[d',u']}P$$

Proof To show $\ell \Vdash \phi$, choose $t = (d_t, u_t) \in \texttt{Time}(\ell)$, $P \in \texttt{Prop}(\ell)$, and $d \leq u$, and suppose $\ell \Vdash @_{[d,u]}P$ holds. The result follows from the three parts of Corollary 6.13. Indeed, by (2), the conclusion is vacuously true unless $d_t < d \leq u < u_t$, so we may assume it. But then by (3) there exists $\langle r, s\rangle : \ell' \to \ell$ with $d_t + r < d \leq u < u_t - s$ such that $\ell' \Vdash P\big|_{\langle r,s\rangle}$. Letting $d' := d_t + r$ and $u' := u_t - s$, the result follows from (1). □

Proposition 6.25 *Axiom 10, denoted ϕ below, is sound in $\mathcal{B}$.*

$$\phi := \forall(P : \texttt{Prop}).\, \big(\forall(t : \texttt{Time})(q_1, q_2 : \mathbb{Q}).\, q_1 < q_2 \Rightarrow (t\,\#\,[q_2, q_1] \Rightarrow P) \Rightarrow P\big)$$
$$\Rightarrow \pi P \Rightarrow P.$$

Proof To show $\ell \Vdash \phi$, choose $P : \texttt{Prop}(\ell)$ and suppose that $\ell \Vdash \pi P$ and the "covering hypothesis," that $\ell \Vdash \forall(t : \texttt{Time})(q_1, q_2 : \mathbb{Q}).\, q_1 < q_2 \Rightarrow (t\,\#\,[q_2, q_1] \Rightarrow P) \Rightarrow P$. We use local character to show $\ell \Vdash P$. That is, suppose given an arbitrary $\langle r', s'\rangle : \ell' \rightsquigarrow \ell$ with $r', s' > 0$; we will show $\ell' \Vdash P\big|_{\langle r',s'\rangle}$ using a compactness argument.

Proposition 6.14 says that for every $0 < a < \ell$ there exists an open neighborhood of a in $\mathbb{R}$ on which P holds, i.e., there exists $\langle r_a, s_a\rangle \colon \ell_a \to \ell$ with $r_a < a < \ell - s_a$ and $\ell_a \Vdash P\big|_{\langle r_a, s_a\rangle}$. The set of open intervals $\{(r_a, \ell - s_a) \subseteq \mathbb{R} \mid 0 \leq a \leq \ell\}$ covers the closed interval $[0, \ell] \subseteq \mathbb{R}$. Since this interval is compact, we can choose a finite subcover, say $\langle r_i, s_i\rangle \colon \ell_i \to \ell$ with $\ell_i \Vdash P\big|_{\langle r_i, s_i\rangle}$ for $0 \leq i \leq n$ and $n \in \mathbb{N}$. We may assume that none is strictly contained in any other and thus order them by left endpoint. Then we have $r_0 < r' < s' < s_n$ and $r_i < r_{i+1} < s_i < s_{i+1}$ for each $0 \leq i \leq n-1$.

For each $0 \leq i \leq n$, consider the restriction $\langle r_0, s_i\rangle \colon \ell'_i \to \ell$. We will prove by induction on i that $\ell'_i \Vdash P\big|_{\langle r_0, s_i\rangle}$. This is enough because if $\ell'_n \Vdash P\big|_{\langle r_0, s_n\rangle}$, then $\ell' \Vdash P\big|_{\langle r', s'\rangle}$.

We have that $\ell_i \Vdash P\big|_{\langle r_i, s_i\rangle}$ holds for each $0 \leq i \leq n$, so in particular the base case holds because $\ell'_0 = \ell_0$. Assume the result is true for some $0 \leq i < n$, i.e., that $\ell'_i \Vdash P\big|_{\langle r_0, s_i\rangle}$; we want to show $\ell'_{i+1} \Vdash P\big|_{\langle r_0, s_{i+1}\rangle}$. Since $r_{i+1} < s_i$, we may choose rationals $q_1, q_2 : \mathbb{Q}$ with $r_{i+1} < q_1 < q_2 < s_i$. By the covering hypothesis with $t = (r_0, s_{i+1}) \in \mathtt{Time}(\ell'_{i+1})$, it suffices to show $\ell'_{i+1} \Vdash (t \mathbin{\#} [q_2, q_1]) \Rightarrow P\big|_{\langle r_0, s_{i+1}\rangle}$.

So choose any $\langle r'', s''\rangle \colon \ell'' \to \ell'_{i+1}$ and assume $\ell'' \Vdash t\big|_{\langle r'', s''\rangle} \mathbin{\#} [q_2, q_1]$, i.e., either $s_{i+1} - s'' < q_2$ or $q_1 < r_0 + r''$; we want to show $\ell'' \Vdash P\big|_{\langle r_0 + r'', s_{i+1} + s''\rangle}$. Since $q_2 < s_i$ and $r_{i+1} < q_1$, we have in the first case that ℓ'' is contained in ℓ'_i and in the second case that ℓ'' is contained in ℓ_{i+1}. Either way, the result follows from monotonicity. □

Chapter 7
Local Numeric Types and Derivatives

In Sect. 4.3 we introduced ten Dedekind j-numeric types for an arbitrary modality j. Namely, we have the j-local lower and upper reals, improper and proper intervals, and real numbers, as well as their unbounded counterparts. They are denoted

$$\underline{\mathbb{R}}_j, \quad \bar{\mathbb{R}}_j, \quad \bar{\underline{\mathbb{R}}}_j, \quad \mathbb{IR}_j, \quad \mathbb{R}_j, \quad \underline{\mathbb{R}}_j^{\infty}, \quad \bar{\mathbb{R}}_j^{\infty}, \quad \bar{\underline{\mathbb{R}}}_j^{\infty}, \quad \mathbb{IR}_j^{\infty}, \quad \mathbb{R}_j^{\infty}.$$

However, our axiomatics (Chap. 5) were given using only three of these: $\bar{\underline{\mathbb{R}}}$, $\mathbb{R}^{\infty}$, and $\mathbb{R}$. Not only did we only use three of ten, we also did not make use of the extra generality afforded by the modality j, i.e., using only $j = \mathrm{id}$. In this section, we explore the more general j-numeric types, both type-theoretically and semantically in our topos $\mathcal{B}$.

Recall from Sect. 4.2.3 that for any topos $\mathcal{E}$, there is a one-to-one correspondence between subtoposes of $\mathcal{E}$ and modalities $j : \mathtt{Prop} \to \mathtt{Prop}$. In our setting, the modalities

$$\downarrow^t_{[d,u]}, \quad @^t_{[d,u]}, \quad \pi,$$

defined in Notation 5.28 thus correspond to particular subtoposes of $\mathcal{B}/\mathtt{Time}$ and $\mathcal{B}$. The Dedekind j-numeric types thus correspond to numeric objects in those subtoposes.

For example, the type $\mathbb{R}_\pi$ is semantically a sheaf on the translation-invariant interval site $S_{\mathbb{IR}/\rhd}$ (see Definition 3.5), but the "pointwise" modality π in some sense eliminates all of the nonzero-length points in $\mathbb{IR}$. Thus the sections of $\mathbb{R}_\pi$ can be identified with continuous real-valued functions on the ordinary real line $\mathbb{R} \subseteq \mathbb{IR}$. While perhaps it would be most consistent to call $\mathbb{R}_\pi$ the type of "pointwise reals," we generally refer to it as the type of *variable reals*. This is meant to evoke the idea that the sections of $\mathbb{R}_\pi$ are real numbers that vary in time, as opposed to the sections of $\mathbb{R}$ which are real numbers that remain constant.

P. Schultz, D. I. Spivak, *Temporal Type Theory*, Progress in Computer Science and Applied Logic 29, https://doi.org/10.1007/978-3-030-00704-1_7

In Sect. 7.1 we compare Dedekind j-numeric types for the modalities discussed above: $\downarrow$, $@$, and π. In Chap. 5 we also introduced and used the modality $\uparrow$, but we will not discuss the Dedekind $\uparrow$-numeric objects, except in Remark 7.1 where we explain our reasons for leaving it out. In Sect. 7.2 we discuss the semantics of the $\downarrow$-, $@$-, and π-numeric types; it is a short section because most of the work has already been done.

In Sect. 7.3 we define an internal notion of derivative and differentiability for variable reals $x : \mathbb{R}_\pi$. For example, if x is differentiable, then its derivative $\frac{d}{dt}(x) : \mathbb{R}_\pi$ is also a variable real. We prove internally that differentiation is linear, satisfies $\frac{d}{dt}(t) = 1$ for $t : \mathtt{Time}$, and satisfies the Leibniz rule

$$\frac{d}{dt}(x * y) = x * \frac{d}{dt}(y) + \frac{d}{dt}(x) * y$$

for differentiable $x, y : \mathbb{R}_\pi$. Finally, we prove that the semantics of the $\frac{d}{dt}$ operation really is that of the Newtonian derivative.

7.1 Relationships Between Various Dedekind j-Numeric Types

All proofs in Sect. 7.1 take place in the temporal type theory given in Chaps. 4 and 5.

Let j be a modality. Recall from Definition 4.21 that every j-numeric type consists of either one or two predicates $\mathbb{Q} \to \mathtt{Prop}_j$, where $\mathtt{Prop}_j$ denotes the type of j-closed predicates. For example, a lower real is a predicate $\delta : \mathbb{Q} \to \mathtt{Prop}$ satisfying $\forall(q : \mathbb{Q}).\ j\delta q \Rightarrow q$—i.e., j-closure in the sense of Definition 4.8—plus two other conditions.

There are many relationships between Dedekind j-numeric types for varying j, and some make sense in any topos. For example, let R_j denote any of the ten Dedekind j-numeric types from Definition 4.21. Proposition 4.54 says that a map of modalities $j' \Rightarrow j$ induces a map

$$j : \underline{\mathbb{R}}_{j'} \to \underline{\mathbb{R}}_j \tag{7.1}$$

which preserves arithmetic, inequalities, and constants. For example, in the case $R = \underline{\mathbb{R}}$, the above map sends $\delta : \mathbb{Q} \to \mathtt{Prop}_{j'}$ to the predicate $j\delta : \mathbb{Q} \to \mathtt{Prop}_j$. In this section, we focus our attention on just a few such modalities j, namely $\downarrow$, $@$, and π.

Remark 7.1 We will not discuss the $\uparrow$-numeric types, e.g., $\bar{\underline{\mathbb{R}}}_{\uparrow[d,u]}$, anywhere in the book. We omit them for a couple of reasons. First, none of them has seemed to come up in practice, at least to this point in the authors' investigations. Second, our work in this section and the next is to understand how the j-numeric types embed—both internally and externally—into their undecorated ($j = \mathrm{id}$) counterparts. Our proof technique for this, developed in Appendix A.3, works only in certain cases. These cases happen to include the $\downarrow$, @, and π modalities but not the $\uparrow$-modality.

Notation 7.2 Again let R_j denote any of the ten Dedekind j-numeric types from Definition 4.21. When $j = @^t_{[r,r]}$ for some real number $r : \mathbb{R}$ and $t : \texttt{Time}$, and when j' is any modality with $j' \Rightarrow j$, we have a special notation for the map $j : R_{j'} \to R_j$ from Proposition 4.54.

To make it explicit, suppose R_j is a two-sided numeric type and suppose $j \Rightarrow @_r$, e.g., $j = \text{id}$, $j = \downarrow_r$, or $j = \pi$. Then for $x = (\delta, \upsilon) : R$, we write $x^{@t}(r)$ to denote $@^t_{[r,r]}(x) : R_j$; it is the pair of cuts $(@^t_r\delta,\ @^t_r\upsilon)$. If t is clear from context, we can drop it as usual:

$$x^{@}(r) := (@_r\delta,\ @_r\upsilon). \tag{7.2}$$

We read $x^{@t}(r)$ as "the value of x at time $t = r$," or just "x at r." Note that this map $x \mapsto x^{@}(r)$ preserves inequalities, arithmetic, and constants by Proposition 4.54.

Of course the same idea works for the one-sided numeric types.

Example 7.3 Considering $t : \texttt{Time}$ as a variable real by Remark 5.56, then for any $r : \mathbb{R}$ we have

$$t^{@}(r) = \downarrow_r(r)$$

or explicitly $@^t_r(q < t) \Leftrightarrow \downarrow^t_r(q < r)$ and $@^t_r(t < q) \Leftrightarrow \downarrow^t_r(r < q)$ for all $q : \mathbb{Q}$. Indeed, for the first equivalence—the second being similar—the forward direction follows by unraveling the definitions (5.28) and by trichotomy $(q < r) \vee (q = r) \vee (r < q)$ (Proposition 5.4), and the reverse direction follows from the fact that $(q < r) \Rightarrow @^t_r(q < t)$, which is Lemma 5.22i.

Remark 7.4 Strictly speaking $\mathbb{R}_{@[d,u]}$ and later $\mathbb{R}_{\downarrow[d,u]}$ are *dependent types*: they are dependent on $t : \texttt{Time}$ and $d, u : \mathbb{R}$. Each is the subtype of a non-dependent type, $(\mathbb{Q} \to \texttt{Prop}) \times (\mathbb{Q} \to \texttt{Prop})$, cut out by propositions that depend on t, d, and u. See Sect. 4.1.4.

There are other relationships between the various Dedekind j-numeric types, beyond the one shown in Eq. (7.1). These relationships arise because several of the j-local Dedekind axioms do not actually depend on j. The results from the remainder of this section are all to this effect; they are summarized in Sect. 7.1.1. Readers who feel the urge to do so may skip ahead to that summary.

Proposition 7.5 *Let* $\phi : (\mathbb{Q} \to \texttt{Prop}) \times (\mathbb{Q} \to \texttt{Prop})$. *Then:*

1. *ϕ is disjoint iff it is π-disjoint.*
2. *ϕ is $\downarrow_{[d,u]}$-disjoint iff it is $@_{[d,u]}$-disjoint.*
3. *If ϕ is $\downarrow_{[d,u]}$-closed, then it is $\downarrow_{[d,u]}$-located iff it is located.*
4. *If ϕ is $@_{[d,u]}$-closed, then it is $@_{[d,u]}$-located iff it is located.*

Proof Statements 1 and 2 follow directly from the facts $\pi\bot \Leftrightarrow \bot$ and $\downarrow_{[d,u]}\bot \Leftrightarrow @_{[d,u]}\bot$, which are shown in Corollary 5.53 and Lemma 5.33. Statements 3 and 4 follow directly from the fact that $\downarrow_{[d,u]}$ and $@_{[d,u]}$ preserve disjunction, Lemma 5.32. □

Remark 7.6 Note that π-located does not imply located. Semantically, this is obvious: sections of $\mathbb{R}_\pi$ are arbitrary continuous functions whereas sections of $\mathbb{R}$ are constants; see Theorem 7.19 and Corollary 7.22.

Before proving similar results relating numeric types for varying modalities, we pause to recall the notion of j-constant numeric types from Definition 4.37. Namely, if R_j is a j-numeric type, then $\mathrm{c}R_j$ denotes the j-numeric subtype whose cuts are j-decidable. As a sanity check, we have said many times that the type $\mathbb{R}$ of real numbers is constant in $\mathcal{B}$ (as is $\mathbb{R}^\infty$), and we prove in Proposition 7.7 that the terminology matches.

Proposition 7.7 *There are internal bijections* $\mathrm{c}\mathbb{R} \cong \mathbb{R}$ *and* $\mathrm{c}\mathbb{R}^\infty \cong \mathbb{R}^\infty \cong \mathbb{R} \sqcup \{-\infty, \infty\}$.

Proof The first two isomorphisms are Axiom 1 and Corollary 5.11; the isomorphism $\mathbb{R}^\infty \cong \mathbb{R} \sqcup \{-\infty, \infty\}$ is shorthand for the isomorphism $\mathbb{R}^\infty \cong \mathbb{R} \sqcup 1 \sqcup 1$ provided in Proposition 5.12. □

Note that the analogous statement does not hold when $\mathbb{R}$ is replaced by other Dedekind numeric types. For example, there is no isomorphism $\mathrm{c}\underline{\mathbb{R}} \overset{?}{\cong} \underline{\mathbb{R}}$, and there should not be one because $\mathrm{c}\underline{\mathbb{R}}$ is semantically the constant sheaf on $\mathbb{R}\cup\{\infty\}$, whereas $\underline{\mathbb{R}}$ is semantically the sheaf of lower semi-continuous real-valued functions on $\mathbb{IR}$. Similarly, $\mathbb{R}_\pi$ is semantically the sheaf of continuous real-valued functions on $\mathbb{R}$, and $\mathrm{c}\mathbb{R}_\pi$ is the proper subsheaf of constant ones.

Proposition 7.8 *For any d, u there is an inclusion $\bar{\underline{\mathbb{R}}}_{@[d,u]} \subseteq \bar{\underline{\mathbb{R}}}_{\downarrow[d,u]}$, and we can identify its image with the subtype of constants, in the sense that the following diagram commutes:*

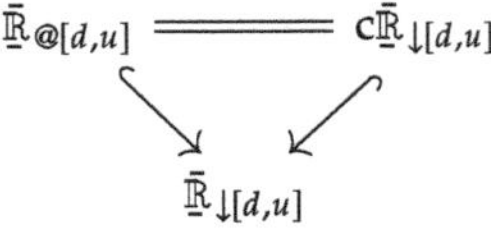

The inclusion preserves domain order and arithmetic. The same holds when $\bar{\underline{\mathbb{R}}}$ is replaced by any of the other nine Dedekind numeric types.

Proof We first prove that there is an inclusion; consider the case of $\delta : \underline{\mathbb{R}}$. Since δ is $@_{[d,u]}$-closed, it is also $\downarrow_{[d,u]}$-closed. To show δ is $\downarrow_{[d,u]}$-rounded, we have the chain of implications

$$\begin{aligned}\delta q &\Rightarrow @_{[d,u]}\exists q'.\,(q < q') \wedge \delta q'\\ &\Rightarrow \exists q'.\,@_{[d,u]}(q < q') \wedge \delta q'\\ &\Rightarrow \exists q'.\,\downarrow_{[d,u]}(q < q') \wedge \delta q' \Rightarrow \downarrow_{[d,u]}\exists q'.\,(q < q') \wedge \delta q',\end{aligned}$$

where the second implication is Proposition 5.62 and the third is Lemma 5.52. One proves similarly that $@_{[d,u]}\exists q\,.\,\delta q \Rightarrow \downarrow_{[d,u]}\exists q\,.\,\delta q$, so @-bounded implies $\downarrow$-bounded.

A similar argument works for $\bar{\mathbb{R}}$, $\bar{\underline{\mathbb{R}}}$, $\underline{\mathbb{R}}^\infty$, $\bar{\mathbb{R}}^\infty$, and $\bar{\underline{\mathbb{R}}}^\infty$. For $\mathbb{IR}$ and $\mathbb{IR}^\infty$ we need that disjointness is preserved, and for $\mathbb{R}$ and $\mathbb{R}^\infty$ we further need that locatedness is preserved; this is shown in Proposition 7.5.

To see that $\underline{\mathbb{R}}_{@[d,u]} = \mathrm{c}\underline{\mathbb{R}}_{\downarrow[d,u]}$, we simply apply Corollary 5.38: a predicate $\delta : \mathbb{Q} \to \texttt{Prop}$ is $@_{[d,u]}$-closed iff it is $\downarrow_{[d,u]}$-closed and $\downarrow_{[d,u]}$-constant. The same argument works for all Dedekind numeric types. Finally, the fact that the inclusion preserves arithmetic follows from Proposition A.60. □

We will see that the inclusion $\bar{\underline{\mathbb{R}}}_{@[d,u]} \subseteq \bar{\underline{\mathbb{R}}}_{\downarrow[d,u]}$ preserves inequalities in Proposition 7.16

Proposition 7.9 *For Dedekind reals, the inclusion $\mathbb{R}_{@[d,u]} \subseteq \mathbb{R}_{\downarrow[d,u]}$ from Proposition 7.8 is an isomorphism,*

$$\mathbb{R}_{@[d,u]} \cong \mathbb{R}_{\downarrow[d,u]}.$$

Proof Let $(\delta, \upsilon) \in \mathbb{R}_{\downarrow[d,u]}$. We have a retraction $@_{[d,u]} : \mathbb{R}_{\downarrow[d,u]} \to \mathbb{R}_{@[d,u]}$, so it suffices to show that δ and υ are already $@_{[d,u]}$-closed. The argument for δ and υ is similar, so we consider δ.

It suffices to show that for any $q : \mathbb{Q}$, we have $@_{[d,u]}(\delta q) \Rightarrow \downarrow_{[d,u]}(\delta q)$. Using $\delta q \Rightarrow \downarrow_{[d,u]}(\exists q'\,.\,(q < q') \wedge \delta q')$, we get $@_{[d,u]}(\exists q'\,.\,(q < q') \wedge \delta q')$. Then using Proposition 5.62 we get $\exists q'\,.\,@_{[d,u]}(q < q') \wedge @_{[d,u]}(\delta q')$. Take such a q'.

By locatedness, we have $\downarrow_{[d,u]}(\delta q \vee \upsilon q')$, i.e., $t \,\#\, [u, d] \vee \delta q \vee \upsilon q'$. The first two cases trivially imply $\downarrow_{[d,u]}(\delta q)$. In the last case, suppose $\upsilon q'$. We also have $@_{[d,u]}(\delta q')$, which by disjointness gives $@_{[d,u]}\bot$, which is equivalent to $t \,\#\, [u, d]$ and hence implies $\downarrow_{[d,u]}(\delta q)$. □

Proposition 7.10 *Suppose j is a closed modality (e.g., $j = \downarrow_{[d,u]}$). Then there is an inclusion $\bar{\underline{\mathbb{R}}}_j \subseteq \bar{\underline{\mathbb{R}}}$, which is a section of $j : \bar{\underline{\mathbb{R}}} \to \bar{\underline{\mathbb{R}}}_j$, and which preserves domain order, constants, arithmetic, and inequalities. An analogous statement holds when $\bar{\underline{\mathbb{R}}}$ is replaced with $\underline{\mathbb{R}}$, $\bar{\mathbb{R}}$, $\bar{\underline{\mathbb{R}}}^\infty$, $\underline{\mathbb{R}}^\infty$, or $\bar{\mathbb{R}}^\infty$.*

Proof The existence of the inclusion (which obviously preserves domain order) follows directly from Proposition A.51, and it is easy to check it is a section of j. The preservation of constants and arithmetic follows from Proposition A.61 where $j = \mathrm{id}$, and $j_1' = j_2' = j_3' = j$.

To prove that the inclusion preserves $<$, recall the open $U_< : \Omega(\bar{\underline{\mathbb{R}}}_{\mathrm{pre}} \times \bar{\underline{\mathbb{R}}}_{\mathrm{pre}})$ defined by $U_<(d, u, d', u') := (u < d')$ from Proposition 4.34, where we showed that for any $(x, y) \in \bar{\underline{\mathbb{R}}}_j \times \bar{\underline{\mathbb{R}}}_j \cong \mathrm{RId}_j(\bar{\underline{\mathbb{R}}}_{\mathrm{pre}} \times \bar{\underline{\mathbb{R}}}_{\mathrm{pre}})$, we have $(x, y) \models_j U_<$ iff $x <_j y$ by Proposition 4.34. Now by Proposition A.55, the inclusion $\bar{\underline{\mathbb{R}}}_j \subseteq \bar{\underline{\mathbb{R}}}$ is continuous, meaning $(I \models_j jU_<) \Leftrightarrow (I \models U_<)$. Thus $(x <_j y) \Leftrightarrow (x < y)$. □

Remark 7.11 (Mixed Arithmetic) In fact, the proof of Proposition 7.10 implies more. Let $\phi_1, \phi_2, \phi_3 : \texttt{Prop}$ be such that $(\phi_1 \vee \phi_2) \Rightarrow \phi_3$, let j_1', j_2', and j_3'

be the corresponding closed modalities. Let j be any modality such that $jP \Rightarrow (j_1'P \wedge j_2'P \wedge j_3'P)$ for all P. Then there is a sort of mixed arithmetic, preserving constants: for any arithmetic operation f (including $+$, max, $-$, $*$, etc.), we have the left-hand diagram

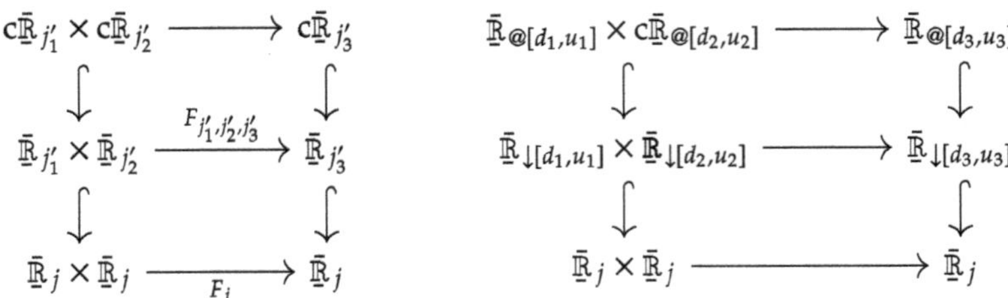

where F_j and $F_{j_1',j_2',j_3'}$ are as in Proposition 4.44 and Lemma 4.45. Again, this is a direct application of Proposition A.61. The right-hand diagram is the special case $j_i' = \downarrow_{[d_i,u_i]}$, and its commutativity follows from that and Proposition 7.8.

The statement of Proposition 7.10 holds when the closed modality j is replaced by π, though for different reasons.

Proposition 7.12 *There is an inclusion* $\bar{\underline{\mathbb{R}}}_\pi \subseteq \bar{\underline{\mathbb{R}}}$, *which preserves domain order and constants and is a section of* $\pi : \bar{\underline{\mathbb{R}}} \to \bar{\underline{\mathbb{R}}}_\pi$. *Thus we can make the identification*

$$\bar{\underline{\mathbb{R}}}_\pi = \{(\delta, \upsilon) : \bar{\underline{\mathbb{R}}} \mid \delta \text{ and } \upsilon \text{ are } \pi\text{-closed}\}.$$

Analogous statements hold when $\bar{\underline{\mathbb{R}}}$ *is replaced with any of the other numeric domains.*

Proof We first prove there is an inclusion $\underline{\mathbb{R}}^\infty_\pi \subseteq \underline{\mathbb{R}}^\infty$. Suppose given $\delta \in \underline{\mathbb{R}}^\infty_\pi$ that is π-rounded, i.e., $\forall(q_1 : \mathbb{Q}).\, \delta q_1 \Rightarrow \pi\exists(q_2 : \mathbb{Q}).\, q_1 < q_2 \wedge \delta q_2$; we need to show it is rounded. Choose q_1 with δq_1, and let $\mathsf{C} := \{q : \mathbb{Q} \mid q_1 < q\}$; it is a constant type (see Definition 5.2).[1] Thus we have $\pi\exists(q_2 : \mathsf{C}).\, \delta q_2$ and we want to remove the π. Because δ is down-closed and $(\mathbb{Q}, <)$ has binary meets, the following condition is satisfied:

$$\forall(q_1, q_2 : \mathsf{C}).\, \exists(q' : \mathsf{C}).\, (\phi q_1 \Rightarrow \phi q') \wedge (\phi q_2 \Rightarrow \phi q')$$

Thus by Proposition 5.72 and the fact that δ is π-closed, we obtain $\exists(q_2 : \mathsf{C}).\, \delta q_2$. This proves that δ is rounded and hence gives the inclusion $\underline{\mathbb{R}}^\infty_\pi \subseteq \underline{\mathbb{R}}^\infty$.

The proof that π-bounded implies bounded is similar, so we obtain the inclusion $\underline{\mathbb{R}}_\pi \subseteq \underline{\mathbb{R}}$; the analogous inclusions for $\bar{\mathbb{R}}^\infty$ and $\bar{\mathbb{R}}$, and those for $\bar{\underline{\mathbb{R}}}^\infty$ and $\bar{\underline{\mathbb{R}}}$ follow

[1] In fact, the type C is dependent on q_1; see Sect. 4.1.4.

directly. Finally, for $\mathbb{IR}^{\infty}$ and $\mathbb{IR}$, we need that π-disjoint implies disjoint, but this is shown in Proposition 7.5.

The inclusions obviously preserve domain order. They preserve constants because $\pi\bot \Leftrightarrow \bot$. □

Of course the analogue of Proposition 7.12 should not hold for the other two numeric objects, $\mathbb{R}_{\pi} \not\subseteq \mathbb{R}$ and $\mathbb{R}_{\pi}^{\infty} \not\subseteq \mathbb{R}^{\infty}$, since it does not hold semantically; see Theorem 7.19.

Remark 7.13 Note that the inclusion $\bar{\underline{\mathbb{R}}}_{\pi} \subseteq \bar{\underline{\mathbb{R}}}$ from Proposition 7.12 preserves *neither arithmetic nor inequalities*. For example, recall that $\mathtt{Time} \subseteq \mathbb{R}_{\pi} \subseteq \bar{\underline{\mathbb{R}}}_{\pi}$, so take $t : \mathtt{Time}$ and consider $t -_{\pi} t$ vs. $t - t$. Since $\mathbb{R}_{\pi}$ is a group, we have $t -_{\pi} t = 0$. However, $\bar{\underline{\mathbb{R}}}$ is not a group, so we do not expect $t - t =^{?} 0$ to hold. In fact given $\ell \in S_{\mathbb{IR}}$, the cuts for $(\delta, \upsilon) := t - t$ are $[\![\delta q]\!] \Leftrightarrow (q < -\ell)$ and $[\![\upsilon q]\!] \Leftrightarrow (\ell < q)$, which are decidedly nonzero since $\ell > 0$.

Similarly, consider t and $t + 1$, the latter of which equals $t +_{\pi} 1$ by Proposition 7.14. Then it is not the case that $t < t + 1$, whereas we can prove $t <_{\pi} t + 1$ using Proposition 5.66 and Axiom 4. Indeed, it suffices to show

$$\forall(r : \mathbb{R}).\, \exists(q : \mathbb{Q}).\, 0 < q \wedge \big(r - q < t < r + q \Rightarrow \exists(q' : \mathbb{Q}).\, t < q' < t + 1\big)$$

and for any r we can take $q := \frac{1}{2}$. Then assuming $r - q < t < r + q$, we can find q' with $t < q' < r + q$ by roundedness, and it follows that $q' < t + 1$.

To see that semantically $t <^{?} t + 1$ does not hold, consider a section of length $\ell \geq 1$. Then there is some $d \in \mathbb{R}$ such that $[\![t]\!] = (d, d + \ell) \in \mathtt{Time}(\ell)$, and $t + 1 = (d + 1, d + 1 + \ell)$. If $[\![\exists(q : \mathbb{Q}).\, t < q < t + 1]\!]$, then $d + \ell < q < d + 1$, which is false.

Despite Remark 7.13, we can use the fact that π is a proper modality (see Remark A.68) and the general results from Appendix A.3.4 to establish some useful special cases in which arithmetic and inequalities *are* preserved.

Proposition 7.14 *The inclusion $\bar{\underline{\mathbb{R}}}_{\pi} \subseteq \bar{\underline{\mathbb{R}}}$ commutes with adding constants and inequalities with constants; in other words the diagrams below commute:*

$$\begin{array}{ccccc} c\bar{\underline{\mathbb{R}}}_{\pi} \times \bar{\underline{\mathbb{R}}}_{\pi} & \hookrightarrow & \bar{\underline{\mathbb{R}}}_{\pi} \times \bar{\underline{\mathbb{R}}}_{\pi} & \xrightarrow{+_{\pi}} & \bar{\underline{\mathbb{R}}}_{\pi} \\ \| & & & & \downarrow \\ c\bar{\underline{\mathbb{R}}} \times \bar{\underline{\mathbb{R}}} & \hookrightarrow & \bar{\underline{\mathbb{R}}} \times \bar{\underline{\mathbb{R}}} & \xrightarrow[+]{} & \bar{\underline{\mathbb{R}}} \end{array} \qquad \begin{array}{ccccc} c\bar{\underline{\mathbb{R}}}_{\pi} \times \bar{\underline{\mathbb{R}}}_{\pi} & \hookrightarrow & \bar{\underline{\mathbb{R}}}_{\pi} \times \bar{\underline{\mathbb{R}}}_{\pi} & \xrightarrow{<_{\pi}} & \mathtt{Prop}_{\pi} \\ \| & & & & \downarrow \\ c\bar{\underline{\mathbb{R}}} \times \bar{\underline{\mathbb{R}}} & \hookrightarrow & \bar{\underline{\mathbb{R}}} \times \bar{\underline{\mathbb{R}}} & \xrightarrow[<]{} & \mathtt{Prop} \end{array}$$

and similarly when $c\bar{\underline{\mathbb{R}}}_{\pi} \times \bar{\underline{\mathbb{R}}}_{\pi}$ is replaced by $\bar{\underline{\mathbb{R}}}_{\pi} \times c\bar{\underline{\mathbb{R}}}_{\pi}$.

This doesn't hold for multiplication by constants in general, but it does on the subtype $\mathbb{IR}_{\pi} \subseteq \mathbb{IR}$; in other words, the following diagram commutes:

$$\begin{array}{ccccc} c\mathbb{IR}_\pi \times \mathbb{IR}_\pi & \hookrightarrow & \mathbb{IR}_\pi \times \mathbb{IR}_\pi & \xrightarrow{*_\pi} & \mathbb{IR}_\pi \\ \| & & & & \downarrow \\ c\mathbb{IR} \times \mathbb{IR} & \hookrightarrow & \mathbb{IR} \times \mathbb{IR} & \xrightarrow[*]{} & \mathbb{IR} \end{array}$$

Proof To show that the inclusion commutes with adding constants, it suffices by Propositions A.41 and A.72 to show that $+\colon \bar{\underline{\mathbb{R}}}_{\text{pre}} \times \bar{\underline{\mathbb{R}}}_{\text{pre}} \to \bar{\underline{\mathbb{R}}}_{\text{pre}}$ preserves meets in the second variable (and by in the first variable by symmetry). This requires showing that

$$(d, u) + ((d_1, u_1) \sqcap (d_2, u_2)) = ((d, u) + (d_1, u_1)) \sqcap ((d, u) + (d_2, u_2)),$$

which comes down to the fact that $d + \min(d_1, d_2) = \min(d + d_1, d + d_2)$, and likewise for max.

The statement about inequalities follows similarly from Proposition A.73. Recall the open $U_< := \{(d_1, u_1), (d_2, u_2) \mid u_1 < d_2\} \in \Omega(\bar{\underline{\mathbb{R}}}_{\text{pre}} \times \bar{\underline{\mathbb{R}}}_{\text{pre}})$, from Proposition 4.34, where it was shown that $x_1 <_j x_2$ iff $(x_1, x_2) \models_j U_<$, for $x_1, x_2 : \bar{\underline{\mathbb{R}}}_j$. Clearly $U_<$ is decidable, and it is filtered (i.e., closed under meets) in each variable, as witnessed by

$$(u_1 < d_2 \wedge u_1' < d_2) \Rightarrow \max(u_1, u_1') < d_2$$
$$(u_1 < d_2 \wedge u_1 < d_2') \Rightarrow u_1 < \min(d_2, d_2').$$

Finally, multiplication $*\colon \bar{\underline{\mathbb{R}}}_{\text{pre}} \times \bar{\underline{\mathbb{R}}}_{\text{pre}} \to \bar{\underline{\mathbb{R}}}_{\text{pre}}$ is not filtered. A counterexample is $((-1, -1) \sqcap (1, 1)) * (1, -1) = (-1, 1) * (1, -1) = (0, 0)$, while $((-1, -1) * (1, -1)) \sqcap ((1, 1) * (1, -1)) = (1, -1) \sqcap (1, -1) = (1, -1)$; see Eq. (A.6). However, restricted to $\mathbb{IR}_{j,\text{pre}}$ multiplication *is* filtered. In this case, one can check by cases on the signs of d_1 and d_2 that the formula for multiplication simplifies to $(d_1, u_1) * (d_2, u_2) = (d', u')$, where

$$d' := \min(d_1 d_2, d_1 u_2, u_1 d_2, u_1 u_2)$$
$$u' := \max(d_1 d_2, d_1 u_2, u_1 d_2, u_1 u_2).$$

Then it is a simple exercise, e.g., by cases on the signs of d', u', to show that

$$((d_1, u_1) \sqcap (d_2, u_2)) * (d', u') = ((d_1, u_1) * (d', u')) \sqcap ((d_2, u_2) * (d', u')).$$

□

We need a technical lemma before proving Proposition 7.16.

Lemma 7.15 *Suppose that B is a constant predomain (see Definition A.56), and $I \in \mathrm{RId}(B)$ and $U \in \Omega(B)$ are such that $@_{[d,u]} I \Rightarrow I$ and $@_{[d,u]} U \Rightarrow \downarrow_{[d,u]} U$. Then*

$$(I \models_{@[d,u]} U) \Leftrightarrow (I \models_{\downarrow[d,u]} U) \Leftrightarrow (I \models U).$$

Proof There is the following chain of equivalences:

$$@\exists b.\, Ib \wedge Ub \;\Leftrightarrow\; \exists b.\, @Ib \wedge @Ub \;\Leftrightarrow\; \exists b.\, Ib \wedge \downarrow Ub \;\Leftrightarrow\; \downarrow\exists b.\, Ib \wedge Ub \;\Leftrightarrow\; \exists b.\, Ib \wedge Ub,$$

where the first is Proposition 5.62, the second is by assumption (and Proposition 5.39), the third is Lemma 5.34, and the fourth is Lemma A.54. By Definition A.44, the left-hand side is $I \models_{@[d,u]} U$, the right-hand side is $I \models U$, and in the fourth position is $I \models_{\downarrow[d,u]} U$. □

Proposition 7.16 *The inclusions* $\underline{\bar{\mathbb{R}}}_{@[d,u]} \subseteq \underline{\bar{\mathbb{R}}}_{\downarrow[d,u]}$ *and* $\underline{\bar{\mathbb{R}}}_{@[d,u]} \subseteq \underline{\bar{\mathbb{R}}}$ *preserve inequalities. The same holds for any of the two-sided Dedekind numeric types.*

Proof Define $U_< : \Omega(\underline{\bar{\mathbb{R}}}_{\mathrm{pre}} \times \underline{\bar{\mathbb{R}}}_{\mathrm{pre}})$ by $U_<(d,u,d',u') := (u < d')$; it is open as seen in Proposition 7.10, and it is decidable, hence $@_{[d,u]}U \Rightarrow \downarrow_{[d,u]}U$ by Lemma 5.52. Thus by Lemma 7.15, for any $(x,y) \in (\underline{\bar{\mathbb{R}}}_{@[d,u]})^2 \cong \mathrm{RId}_{@[d,u]}(\underline{\bar{\mathbb{R}}}_{\mathrm{pre}} \times \underline{\bar{\mathbb{R}}}_{\mathrm{pre}})$, we have

$$(x,y) \models_{@[d,u]} U_< \quad\Leftrightarrow\quad (x,y) \models_{\downarrow[d,u]} U_< \quad\Leftrightarrow\quad (x,y) \models U_<$$

i.e., $(x <_@ y) \Leftrightarrow (x <_\downarrow y) \Leftrightarrow (x < y)$. □

Proposition 7.17 *We have the following inclusions, all of which preserve domain order, constants, arithmetic, and inequalities:*

$$\underline{\mathbb{R}}_{@0} \subseteq \underline{\mathbb{R}}_{\pi}, \qquad \bar{\mathbb{R}}_{@0} \subseteq \bar{\mathbb{R}}_{\pi}, \qquad \underline{\bar{\mathbb{R}}}_{@0} \subseteq \underline{\bar{\mathbb{R}}}_{\pi},$$
$$\underline{\mathbb{R}}^{\infty}_{@0} \subseteq \underline{\mathbb{R}}^{\infty}_{\pi}, \qquad \bar{\mathbb{R}}^{\infty}_{@0} \subseteq \bar{\mathbb{R}}^{\infty}_{\pi}, \qquad \underline{\bar{\mathbb{R}}}^{\infty}_{@0} \subseteq \underline{\bar{\mathbb{R}}}^{\infty}_{\pi}.$$

Proof We prove the first, the rest being similar. If $\delta : \mathbb{Q} \to \mathtt{Prop}$ is $@_0$-closed, it is π-closed. The result then follows from Propositions 7.12 and 7.16. □

7.1.1 Section Summary

We now attempt to summarize the results of this section for future reference. The j-numeric types were defined for a generic modality in Definition 4.21, and summarized in Table 4.1. For ease of reference, we repeat the list of axioms:

$0\mathrm{a}_j$.	$\forall (q : \mathbb{Q}).\, j(\delta q) \Rightarrow \delta q$	$0\mathrm{b}_j$.	$\forall (q : \mathbb{Q}).\, j(\upsilon q) \Rightarrow \upsilon q$
$1\mathrm{a}_j$.	$\forall q_1\, q_2.\, (q_1 < q_2) \Rightarrow \delta q_2 \Rightarrow \delta q_1$	$1\mathrm{b}_j$.	$\forall q_1\, q_2.\, (q_1 < q_2) \Rightarrow \upsilon q_1 \Rightarrow \upsilon q_2$
$2\mathrm{a}_j$.	$\forall q_1.\, \delta q_1 \Rightarrow j\exists q_2.\, (q_1 < q_2) \wedge \delta q_2$	$2\mathrm{b}_j$.	$\forall q_2.\, \upsilon q_2 \Rightarrow j\exists q_1.\, (q_1 < q_2) \wedge \upsilon q_1$
$3\mathrm{a}_j$.	$j\exists q.\, \delta q$	$3\mathrm{b}_j$.	$j\exists q.\, \upsilon q$
4_j.	$\forall q.\, (\delta q \wedge \upsilon q) \Rightarrow j\bot$		
5_j.	$\forall q_1\, q_2.\, (q_1 < q_2) \Rightarrow j(\delta q_1 \vee \upsilon q_2)$		

where we now annotate each with the modality by which it is modified. As usual, un-annotated axioms, such as 1a or 4, refer to the identity modality $j = \mathrm{id}$.

Table 7.1 updates Table 4.1 with the results of this section. The characterizations given follow from Propositions 7.5, 7.8, 7.10, 7.12 (recall in particular from Proposition 7.5 that $4_\downarrow$ and $4_@$ are equivalent). The unbounded versions simply remove Axiom 3; for example, $\mathbb{IR}_@^\infty = \{(\delta, \upsilon) \mid 0_@, 1, 2, 4_\downarrow\}$. This table clearly shows all subtype relationships. For instance, $\mathbb{IR}_\pi \subseteq \mathbb{IR}$, and in fact $\mathbb{IR}_\pi = \mathbb{IR} \cap \bar{\underline{\mathbb{R}}}_\pi$.

We also saw in this section that the various inclusions preserve much of the structure present, as summarized in the following corollary.

Corollary 7.18 *Let j be one of the following modalities: @, $\downarrow$, or π. Then we have the following subtype relationships, all of which preserve domain order and constants:*

$$\begin{array}{ccc} \underline{\mathbb{R}}_j & \subseteq & \underline{\mathbb{R}} \\ \cap\!\!| & & \cap\!\!| \\ \underline{\mathbb{R}}_j^\infty & \subseteq & \underline{\mathbb{R}}^\infty \end{array} \qquad \begin{array}{ccc} \bar{\mathbb{R}}_j & \subseteq & \bar{\mathbb{R}} \\ \cap\!\!| & & \cap\!\!| \\ \bar{\mathbb{R}}_j^\infty & \subseteq & \bar{\mathbb{R}}^\infty \end{array} \qquad \begin{array}{ccc} \bar{\underline{\mathbb{R}}}_j & \subseteq & \bar{\underline{\mathbb{R}}} \\ \cap\!\!| & & \cap\!\!| \\ \bar{\underline{\mathbb{R}}}_j^\infty & \subseteq & \bar{\underline{\mathbb{R}}}^\infty \end{array}$$

Moreover if j is @ or $\downarrow$, then all of the above inclusions also preserve arithmetic and inequalities.

Table 7.1 The local Dedekind numeric types; the characterization column refers to the axioms on page 141

Name of type in $\mathcal{E}$	Notation	Characterization
Lower reals	$\underline{\mathbb{R}}$	$\{\delta \mid 1a, 2a, 3a\}$
Upper reals	$\bar{\mathbb{R}}$	$\{\upsilon \mid 1b, 2b, 3b\}$
Improper intervals	$\bar{\underline{\mathbb{R}}}$	$\{(\delta, \upsilon) \mid 1, 2, 3\}$
(Proper) intervals	$\mathbb{IR}$	$\{(\delta, \upsilon) \mid 1, 2, 3, 4\}$
(Constant) real numbers	$\mathbb{R}$	$\{(\delta, \upsilon) \mid 1, 2, 3, 4, 5\}$
π-Local lower reals	$\underline{\mathbb{R}}_\pi$	$\{\delta \mid 0a_\pi, 1a, 2a, 3a\}$
π-Local upper reals	$\bar{\mathbb{R}}_\pi$	$\{\upsilon \mid 0b_\pi, 1b, 2b, 3b\}$
π-Local improper intervals	$\bar{\underline{\mathbb{R}}}_\pi$	$\{(\delta, \upsilon) \mid 0_\pi, 1, 2, 3\}$
π-Local (proper) intervals	$\mathbb{IR}_\pi$	$\{(\delta, \upsilon) \mid 0_\pi, 1, 2, 3, 4\}$
π-Local real numbers	$\mathbb{R}_\pi$	$\{(\delta, \upsilon) \mid 0_\pi, 1, 2, 3, 4, 5_\pi\}$
$\downarrow$-Local lower reals	$\underline{\mathbb{R}}_\downarrow$	$\{\delta \mid 0a_\downarrow, 1a, 2a, 3a\}$
$\downarrow$-Local upper reals	$\bar{\mathbb{R}}_\downarrow$	$\{\upsilon \mid 0b_\downarrow, 1b, 2b, 3b\}$
$\downarrow$-Local improper intervals	$\bar{\underline{\mathbb{R}}}_\downarrow$	$\{(\delta, \upsilon) \mid 0_\downarrow, 1, 2, 3\}$
$\downarrow$-Local (proper) intervals	$\mathbb{IR}_\downarrow$	$\{(\delta, \upsilon) \mid 0_\downarrow, 1, 2, 3, 4_\downarrow\}$
$\downarrow$-Local real numbers	$\mathbb{R}_\downarrow$	$\{(\delta, \upsilon) \mid 0_\downarrow, 1, 2, 3, 4_\downarrow, 5\}$
@-Local lower reals	$\underline{\mathbb{R}}_@$	$\{\delta \mid 0a_@, 1a, 2a, 3a\}$
@-Local upper reals	$\bar{\mathbb{R}}_@$	$\{\upsilon \mid 0b_@, 1b, 2b, 3b\}$
@-Local improper intervals	$\bar{\underline{\mathbb{R}}}_@$	$\{(\delta, \upsilon) \mid 0_@, 1, 2, 3\}$
@-Local (proper) intervals	$\mathbb{IR}_@$	$\{(\delta, \upsilon) \mid 0_@, 1, 2, 3, 4_\downarrow\}$
@-Local real numbers	$\mathbb{R}_@$	$\{(\delta, \upsilon) \mid 0_@, 1, 2, 3, 4_\downarrow, 5\}$

Proof In all three diagrams, the inclusion of top into bottom is obvious: in each case the subtype is cut out by some axiom. In the case of π, the horizontal inclusions were shown in Proposition 7.12. For $\downarrow$, Proposition 7.10 shows that the horizontal inclusions exist and preserve arithmetic and inequalities. Finally, for @, Proposition 7.8 shows that the horizontal inclusions exist and preserve arithmetic (by composing with the inclusions for $\downarrow$), and Proposition 7.16 shows that these inclusions also preserve inequalities. □

7.2 Semantics of Numeric Types in Various Modalities

Recall from Proposition 2.29 that there is a poset isomorphism $h\colon \mathbb{IR} \cong H$ between the interval domain and the upper half-plane $H = \{(x, y) \in \mathbb{R}^2 \mid y \geq 0\}$. It is given by $h([d, u]) := (m, r)$, where $m := (u + d)/2$ and $r := (u - d)/2$ are the midpoint and radius. We call $h([d, u])$ the *half-plane representation* of $[d, u]$. Recall from Eq. (2.4) that the sets $\uparrow(m, r)$ and $\Uparrow(m, r)$ can be visualized as the cones:

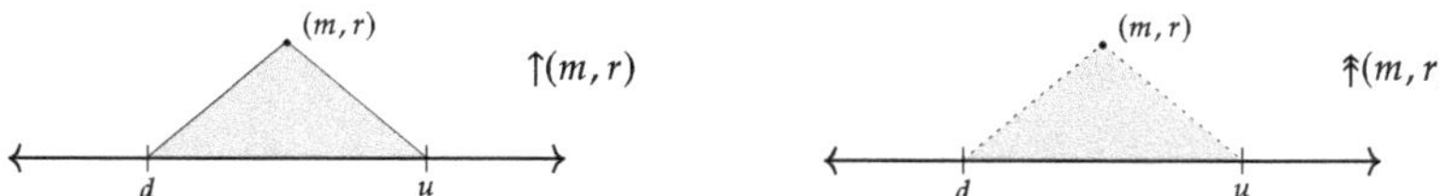

The horizontal axis in H represents the real numbers $\mathbb{R} \subseteq \mathbb{IR}$, so we call it the *real axis*. In particular the line segment from d to u is $\uparrow(m, r) \cap h(\mathbb{R})$, but we denote it $\uparrow(m, r) \cap \mathbb{R}$ for typographical simplicity. We will see that the π-numeric types are determined by their values on the real axis.

For visualizing the $\downarrow_{[d,u]}$- and $@_{[d,u]}$-numeric types, it will also be useful to have a picture of $\downarrow(m, r) = \{(m', r') \in \mathbb{R}^2 \mid |m' - m| \leq r' - r\}$, namely the gray area shown below:

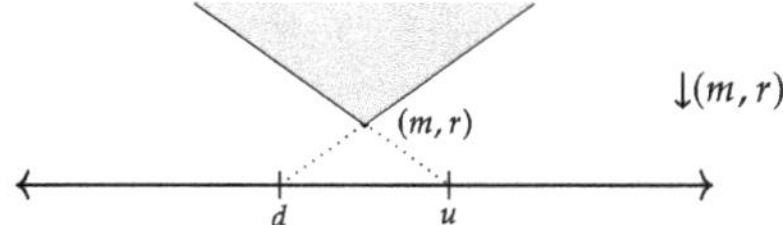

As discussed in Sect. 2.4, we consider H as having the usual Euclidean topology.

The following omnibus theorem collects characterizations of the semantics of all the numeric types we consider. The new characterizations follow easily from Corollary 6.7 and Proposition 6.15, together with the results of Sect. 7.1 as summarized in Table 7.1.

Theorem 7.19 *Suppose we are in a context with a variable* t : `Time`*, and we fix an element* $(d_t, u_t) \in [\![\texttt{Time}]\!](\ell)$*. Let* $(m_t, r_t) := h([d_t, u_t])$ *be its half-plane representation.*

1. $[\![\underline{\mathbb{R}}]\!](\ell) \cong \{\, f : \Uparrow(m_t, r_t) \to \mathbb{R} \cup \{\infty\} \mid$ *f is lower semi-continuous and order-preserving* $\}$
2. $[\![\bar{\mathbb{R}}]\!](\ell) \cong \{\, f : \Uparrow(m_t, r_t) \to \mathbb{R} \cup \{-\infty\} \mid$ *f is upper semi-continuous and order-reversing* $\}$
3. $[\![\bar{\underline{\mathbb{R}}}]\!](\ell) \cong [\![\underline{\mathbb{R}}]\!](\ell) \times [\![\bar{\mathbb{R}}]\!](\ell)$
4. $[\![\mathbb{IR}]\!](\ell) \cong \{\, (\underline{f}, \bar{f}) \in [\![\bar{\underline{\mathbb{R}}}]\!](\ell) \mid \forall (p \in \Uparrow(m_t, r_t)).\, \underline{f}(p) \leq \bar{f}(p) \,\}$
5. $[\![\mathbb{R}]\!](\ell) \cong \{\, (\underline{f}, \bar{f}) \in [\![\bar{\underline{\mathbb{R}}}]\!](\ell) \mid \forall (p \in \Uparrow(m_t, r_t)).\, \underline{f}(p) = \bar{f}(p) \,\} \cong \mathbb{R}$

In the π*-modality, the* f*'s are determined by their values on the real axis:*

6. $[\![\underline{\mathbb{R}}_\pi]\!](\ell) \cong \{\, f \in [\![\underline{\mathbb{R}}]\!](\ell) \mid \forall (p \in \Uparrow(m_t, r_t)).\, f(p) = \inf f(\uparrow p \cap \mathbb{R}) \,\}$
7. $[\![\bar{\mathbb{R}}_\pi]\!](\ell) \cong \{\, f \in [\![\bar{\mathbb{R}}]\!](\ell) \mid \forall (p \in \Uparrow(m_t, r_t)).\, f(p) = \sup f(\uparrow p \cap \mathbb{R}) \,\}$
8. $[\![\bar{\underline{\mathbb{R}}}_\pi]\!](\ell) \cong [\![\underline{\mathbb{R}}_\pi]\!](\ell) \times [\![\bar{\mathbb{R}}_\pi]\!](\ell)$
9. $[\![\mathbb{IR}_\pi]\!](\ell) \cong \{\, (\underline{f}, \bar{f}) \in [\![\bar{\underline{\mathbb{R}}}_\pi]\!](\ell) \mid \forall (p \in \Uparrow(m_t, r_t)).\, \underline{f}(p) \leq \bar{f}(p) \,\}$
10. $[\![\mathbb{R}_\pi]\!](\ell) \cong \{\, (\underline{f}, \bar{f}) \in [\![\bar{\underline{\mathbb{R}}}_\pi]\!](\ell) \mid \forall (p \in \Uparrow(m_t, r_t) \cap \mathbb{R}).\, \underline{f}(p) = \bar{f}(p) \,\}$

Let d, u : $\mathbb{R}$*, and let* $(m, r) = h([d, u])$ *be its half-plane representation. In the* $\downarrow$ *modality, the* f*'s are determined by their values in* $\downarrow(m, r)$*:*

11. $[\![\underline{\mathbb{R}}_{\downarrow[d,u]}]\!](\ell) \cong \{\, f \in [\![\underline{\mathbb{R}}]\!](\ell) \mid \forall (p \in \Uparrow(m_t, r_t)).\, p \in \downarrow(m, r) \vee f(p) = \infty \,\}$
12. $[\![\bar{\mathbb{R}}_{\downarrow[d,u]}]\!](\ell) \cong \{\, f \in [\![\bar{\mathbb{R}}]\!](\ell) \mid \forall (p \in \Uparrow(m_t, r_t)).\, p \in \downarrow(m, r) \vee f(p) = -\infty \,\}$
13. $[\![\bar{\underline{\mathbb{R}}}_{\downarrow[d,u]}]\!](\ell) \cong [\![\underline{\mathbb{R}}_{\downarrow[d,u]}]\!](\ell) \times [\![\bar{\mathbb{R}}_{\downarrow[d,u]}]\!](\ell)$
14. $[\![\mathbb{IR}_{\downarrow[d,u]}]\!](\ell) \cong \{\, (\underline{f}, \bar{f}) \in [\![\bar{\underline{\mathbb{R}}}_{\downarrow[d,u]}]\!](\ell) \mid \forall (p \in \Uparrow(m_t, r_t)).\, p \in \downarrow(m, r) \Rightarrow \underline{f}(p) \leq \bar{f}(p) \,\}$
15. $[\![\mathbb{R}_{\downarrow[d,u]}]\!](\ell) \cong \{\, (\underline{f}, \bar{f}) \in [\![\bar{\underline{\mathbb{R}}}_{\downarrow[d,u]}]\!](\ell) \mid \forall p \in \Uparrow(m_t, r_t).\, p \in \downarrow(m, r) \Rightarrow \underline{f}(p) = \bar{f}(p) \,\}$

Let (m, r) *be as above. In the @ modality, the* f*'s are constant in* $\downarrow(m, r)$*:*

16. $[\![\underline{\mathbb{R}}_{@[d,u]}]\!](\ell) \cong \{\, f \in [\![\underline{\mathbb{R}}_{\downarrow[d,u]}]\!](\ell) \mid \forall (p \in \Uparrow(m_t, r_t)).\, p \in \downarrow(m, r) \Rightarrow f(p) = f((m, r)) \,\}$
17. $[\![\bar{\mathbb{R}}_{@[d,u]}]\!](\ell) \cong \{\, f \in [\![\bar{\mathbb{R}}_{\downarrow[d,u]}]\!](\ell) \mid \forall (p \in \Uparrow(m_t, r_t)).\, p \in \downarrow(m, r) \Rightarrow f(p) = f((m, r)) \,\}$
18. $[\![\bar{\underline{\mathbb{R}}}_{@[d,u]}]\!](\ell) \cong [\![\underline{\mathbb{R}}_{@[d,u]}]\!](\ell) \times [\![\bar{\mathbb{R}}_{@[d,u]}]\!](\ell)$
19. $[\![\mathbb{IR}_{@[d,u]}]\!](\ell) \cong \{\, (\underline{f}, \bar{f}) \in [\![\bar{\underline{\mathbb{R}}}_{@[d,u]}]\!](\ell) \mid (m, r) \in \Uparrow(m_t, r_t) \Rightarrow \underline{f}((m, r)) \leq \bar{f}((m, r)) \,\}$
20. $[\![\mathbb{R}_{@[d,u]}]\!](\ell) \cong \{\, (\underline{f}, \bar{f}) \in [\![\bar{\underline{\mathbb{R}}}_{@[d,u]}]\!](\ell) \mid (m, r) \in \Uparrow(m_t, r_t) \Rightarrow \underline{f}((m, r)) = \bar{f}((m, r)) \,\}$

Remark 7.20 Up to isomorphism, the sets given in Theorem 7.19 numbers 1–10 are independent of the choice of t : `Time` and (d_t, u_t), and hence (m_t, r_t). The reason they are written in those terms is in order to facilitate the comparison between numbers 1–10 and numbers 11–20.

Example 7.21 The internal fact that $\mathbb{R}_{@[d,u]} = \mathbb{R}_{\downarrow[d,u]}$ proven in Proposition 7.9 is reflected semantically in the fact that the sets described in Theorem 7.19 numbers 15 and 20 are the same.

One might imagine an element of $[\![\mathbb{R}_{@[d,u]}]\!](\ell)$ like this[2]:

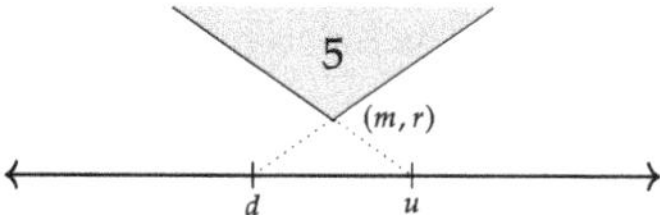

where $(m, r) = h([d, u])$ is the half-plane representation. Roughly speaking, the fact that this function has constant value (say 5) everywhere above (m, r) is a reflection of the fact that we are looking at a stalk there, whereas the fact that this function is ∞ everywhere else is a reflection of the added @-closure property $0\mathrm{a}_j$; see Table 7.1.

Corollary 7.22 *Let* $(0, \ell) \subseteq \mathbb{R}$ *denote the open interval of length* ℓ. *There is an isomorphism*

$$[\![\mathbb{R}_\pi]\!](\ell) \cong \{ f : (0, \ell) \to \mathbb{R} \mid f \text{ is continuous} \}.$$

Proof For $f : (0, \ell) \to \mathbb{R}$, define $\underline{f} : \Uparrow[0, \ell] \to \mathbb{R}$ and $\bar{f} : \Uparrow[0, \ell] \to \mathbb{R}$ by

$$\underline{f}(p) := \inf f(\Uparrow p \cap \mathbb{R}) \qquad \text{and} \qquad \bar{f}(p) := \sup f(\Uparrow p \cap \mathbb{R})$$

It is easy to see that $\underline{f}$ is automatically order-preserving and $\bar{f}$ is automatically order-reversing. Thus by Theorem 7.19 and Remark 7.20 we have an isomorphism:

$$[\![\mathbb{R}_\pi]\!] \cong \{ f : (0, \ell) \to \mathbb{R} \mid \underline{f} \text{ is lower semi-continuous and } \bar{f} \text{ is upper semi-continuous}\}.$$

If f is continuous, then $\underline{f}$ is lower semi-continuous and $\bar{f}$ is upper semi-continuous; it remains to show the converse. For any $0 < x < \ell$ and $\epsilon > 0$, the neighborhood $U := (f(x) - \epsilon, f(x) + \epsilon)$ is the intersection of $U_1 := \{y \in \mathbb{R} \mid f(x) - \epsilon < y\}$ and $U_2 := \{y \in \mathbb{R} \mid y < f(x) + \epsilon)$. By definition of lower and upper semi-continuity, the pullbacks $\underline{f}^{-1}(U_1)$ and $\bar{f}^{-1}(U_2)$ are open neighborhoods of $x \in \Uparrow[0, \ell]$, so $f^{-1} = \underline{f}^{-1}(U_1) \cap \bar{f}^{-1}(U_2) \cap \mathbb{R}$ is open as well. □

Remark 7.23 Our original *external* definition of `Time` was Eq. (3.2), $\texttt{Time}(\ell) = \{ (d, u) \in \mathbb{R}^2 \mid u - d = \ell \}$. Our original *internal* definition of `Time` was as a subtype of $\underline{\bar{\mathbb{R}}}$; see Eq. (5.2). In Lemma 6.10 and throughout Sect. 6.5 we showed that

[2]We are suppressing the time context for readability. To be explicit, this example takes place in the context of a time interval $(d_t, u_t) \in [\![\texttt{Time}]\!](\ell)$, where $d_t \leq d \leq u \leq u_t$.

the former is a sound semantics for the latter. With Theorem 7.19 and Corollary 7.22 in front of us, it may be worthwhile to summarize how `Time` fits in with $\bar{\underline{\mathbb{R}}}$, $\mathbb{IR}$, and $\mathbb{R}_\pi$.

The elements of $[\![\bar{\underline{\mathbb{R}}}]\!]$ are pairs of lower/upper semi-continuous order-preserving/reversing functions. As a subsheaf, $[\![\texttt{Time}]\!] \subseteq [\![\bar{\underline{\mathbb{R}}}]\!]$ is the set of pairs $(\underline{t}, \bar{t})\colon \uparrow[0, \ell] \to \mathbb{R}$, for which there exists $r \in \mathbb{R}$ with $\underline{t}([d, u]) := d + r$ and $\bar{t}([d, u]) := u + r$. Clearly $\underline{t} \leq \bar{t}$, so we also have $[\![\texttt{Time}]\!] \subseteq [\![\mathbb{IR}]\!]$.

In Remark 5.56 we identified `Time` internally as a subtype of $\mathbb{R}_\pi$. Under this identification, we could write

$$[\![\texttt{Time}]\!](\ell) \cong \big\{ (x \mapsto r + x)\colon (0, \ell) \to \mathbb{R} \mid r \in \mathbb{R} \big\}$$

i.e., the set of real-valued functions on the open interval $(0, \ell) \subseteq \mathbb{R}$ that have unit slope. In Remark 3.13 we called these "clock behaviors."

Remark 7.24 Semantically, $\mathbb{R}_\pi$ denotes the sheaf of continuous real-valued functions—which is defined in the usual $\forall(\epsilon > 0). \exists(\delta > 0) \ldots$ style—whereas internally $\mathbb{R}_\pi$ is just the type of π-local Dedekind cuts. One may be curious whether we can prove some sort of ϵ, δ continuity property internally. The answer is yes, as we now explain, though note that we have not seen any use for this rather unwieldy perspective; the only reason to include it is to give evidence that our axioms are sufficiently powerful to reproduce external facts internally.

To obtain an ϵ, δ notion of continuity, we need to define yet another numeric object $\mathbb{R}^?$, which is very similar to the constant reals $\mathbb{R} = \{(\delta, \upsilon) : \bar{\underline{\mathbb{R}}} \mid$ located and disjoint$\}$, except with a slightly weakened disjointness axiom:

weakly disjoint: $(\exists(q : \mathbb{Q}). \delta q \wedge \upsilon q) \Rightarrow \forall(q : \mathbb{Q}). \delta q \wedge \upsilon q.$

Define $\mathbb{R}^? = \{(\delta, \upsilon) : \bar{\underline{\mathbb{R}}} \mid$ located and weakly disjoint$\}$. If $x = (\delta, \upsilon) : \mathbb{R}^?$ satisfies $\forall(q : \mathbb{Q}). \delta q \wedge \upsilon q$, we write $x = \texttt{undef}$.

We can now internally define a function $\mathbb{R}_\pi \to \texttt{Time} \to \mathbb{R}^?$, sending $x = (\delta, \upsilon) : \mathbb{R}_\pi$ to the function $f_x : \texttt{Time} \to \mathbb{R}^?$ given by $f_x(t) := @_0^t x$, or more precisely $f_x(t) = (\delta_{x,t}, \upsilon_{x,t})$ where $\delta_{x,t} q \Leftrightarrow @_0^t \delta q$ and $\upsilon_{x,t} q \Leftrightarrow @_0^t \upsilon q$.

Soon we will prove that f_x is continuous, but first we want to check that $f_x(t) \in \mathbb{R}^?$ for any given x, t. It is easy to see that $f_x(t)$ is down/up-closed and bounded. It is rounded by Proposition 5.62. We check that it is located and weakly disjoint. Given $q_1 < q_2$ we have $\pi(\delta q_1 \vee \upsilon q_2)$, which by definition is $\forall(t : \texttt{Time}). @_0^t(\delta q_1 \vee \upsilon q_2)$, and we obtain locatedness by Lemma 5.32. If $@_0^t(\delta q \wedge \upsilon q)$, then we have $@_0^t(\pi \bot)$ by π-disjointness, and hence $@_0^t \bot$ by Corollary 5.53. Thus we obtain $\forall(q : \mathbb{Q}). @_0^t(\delta q \wedge \upsilon q)$, proving $f_x(t)$ is weakly disjoint.

Finally, we want to show that $f_x : \texttt{Time} \to \mathbb{R}^?$ is continuous in the sense that the following holds:

$$\forall(\epsilon > 0)(t : \texttt{Time}). (f_x(t) = \texttt{undef}) \vee \\ \exists(\delta > 0). \forall(r : \mathbb{R}). (-\delta < t + r < \delta) \Rightarrow |f_x(t) - f_x(t + r)| < \epsilon.$$

By arithmetic locatedness (see Proposition 4.26, which also holds for $\mathbb{R}^?$) and Lemma 5.42, it suffices to show

$$\forall(q_1, q_2 : \mathbb{Q})(t : \texttt{Time}).\,(q_1 < f_x(t) < q_2) \Rightarrow (f_x(t) = \texttt{undef}) \vee \\ \exists(d, u : \mathbb{Q}).\,(d < u) \wedge \uparrow^t_{[d,u]}(q_1 < f_x < q_2).$$

and this follows from Axiom 9.

7.3 Derivatives of Interval-Valued Functions

In this section we will define what it means for one interval-valued function to be the derivative of another. We use Newton's notation $\dot{x}$ for the derivative because its semantics will be differentiation with respect to $t : \texttt{Time}$. If x has a derivative $\dot{x}$ and both $x, \dot{x} : \mathbb{R}_\pi$ are variable reals, we say that x is differentiable. For example, we will see that $x(t) = t^2$ is differentiable and its derivative is $2 * t$. We will show that differentiation is linear and satisfies the Leibniz rule in Theorem 7.49.

The derivative can be defined internally for any $x : \bar{\underline{\mathbb{R}}}^\infty_\pi$. Because $\bar{\underline{\mathbb{R}}}^\infty_\pi$ is a domain (see Propositions 4.30 and 4.31), we can define $\dot{x}$ by giving a directed family of $y \in \bar{\underline{\mathbb{R}}}^\infty_\pi$ which approximate $\dot{x}$, and take $\dot{x}$ to be the directed join of the approximating family; see Definition 7.31. The semantics of derivatives will be given in Sect. 7.3.2.

7.3.1 Internal Definition of the Derivative

Definition 7.25 (Approximates the Derivative) Let $x, y : \bar{\underline{\mathbb{R}}}^\infty_\pi$ be π-local improper intervals. We say that y *approximates the derivative of* x, if for any $t : \texttt{Time}$ it satisfies the following:

$$\forall(r_1, r_2 : \mathbb{R}).\,r_1 < r_2 \Rightarrow y \sqsubseteq \frac{x^@(r_2) - x^@(r_1)}{r_2 - r_1}. \tag{7.3}$$

For now we write $\mathrm{AD}(y, x)$ iff y satisfies Eq. (7.3).

By Axiom 4, $\mathrm{AD}(y, x)$ is independent of $t : \texttt{Time}$, i.e., it is true for one choice iff it is true for any another. Thus in the results below, a choice of $t : \texttt{Time}$ is implied, but is left out of the statements.

Lemma 7.26 *For all* $x, y : \bar{\underline{\mathbb{R}}}^\infty_\pi$ *and* $r_1 < r_2 : \mathbb{R}$, *the fraction* $\frac{x^@(r_2)-x^@(r_1)}{r_2-r_1}$ *is in* $\bar{\underline{\mathbb{R}}}_{@[r_1,r_2]}$, *and the proposition* $y \sqsubseteq \frac{x^@(r_2)-x^@(r_1)}{r_2-r_1}$ *is* $@_{[r_1,r_2]}$*-closed.*

Proof By Proposition 7.8 and Remark 7.11, the difference $x^@(r_2) - x^@(r_1)$ is a constant in $\bar{\mathbb{R}}^\infty_{\downarrow[r_1,r_2]}$, and thus it is in $\bar{\mathbb{R}}^\infty_{@[r_1,r_2]}$ by Proposition 7.8. The reciprocal $1/(r_2 - r_1)$ is a (constant) real by Proposition 4.52, so the product $(\delta, \upsilon) := \frac{x^@(r_2)-x^@(r_1)}{r_2-r_1}$ is in $\bar{\mathbb{R}}^\infty_{@[r_1,r_2]}$. By Corollary 7.18, we can consider both sides of the expression $y \sqsubseteq (\delta, \upsilon)$ as terms in $\bar{\mathbb{R}}$, and $\sqsubseteq$ is simply implication: $q < y \Rightarrow \delta q$ and $y < q \Rightarrow \upsilon q$. Since δ and υ are $@_{[r_1,r_2]}$-closed, so are these implications. □

Lemma 7.27 *Choose* $x, y : \bar{\mathbb{R}}^\infty_\pi$ *and* $d, u : \mathbb{R}^\infty$. *Then* $\Uparrow_{[d,u]}\mathrm{AD}(y, x)$ *is equivalent to*

$$\forall (r_1, r_2 : \mathbb{R}).\, d < r_1 < r_2 < u \Rightarrow y \sqsubseteq \frac{x^@(r_2) - x^@(r_1)}{r_2 - r_1}.$$

Proof By Eq. (5.8), $\Uparrow_{[d,u]}\mathrm{AD}(y, x)$ is equivalent to

$$\forall (r_1, r_2 : \mathbb{R}).\, r_1 < r_2 \Rightarrow \Uparrow_{[d,u]}\left(y \sqsubseteq \frac{x^@(r_2) - x^@(r_1)}{r_2 - r_1}\right).$$

By Lemma 7.26 and Proposition 5.44, this is equivalent to

$$\forall (r_1, r_2 : \mathbb{R}).\, r_1 < r_2 \Rightarrow ((d < r_1) \wedge (r_2 < u)) \Rightarrow \left(y \sqsubseteq \frac{x^@(r_2) - x^@(r_1)}{r_2 - r_1}\right),$$

which is clearly equivalent to the desired conclusion. □

Proposition 7.28 *For any* $x, y : \bar{\mathbb{R}}^\infty_\pi$, *the proposition* $\mathrm{AD}(y, x)$ *is* π*-closed.*

Proof We prove this using Axiom 10. Let $q_1 < q_2$ be rationals, and suppose $t \,\#\, [q_2, q_1] \Rightarrow \mathrm{AD}(y, x)$. This is equivalent to $\Uparrow_{[-\infty,q_2]}\mathrm{AD}(y, x) \wedge \Uparrow_{[q_1,\infty]}\mathrm{AD}(y, x)$. We want to show $\mathrm{AD}(y, x)$. By Lemma 7.27, we have

$$\forall (r_1, r_2 : \mathbb{R}).\, r_1 < r_2 < q_2 \Rightarrow y \sqsubseteq \frac{x^@(r_2) - x^@(r_1)}{r_2 - r_1} \tag{7.4}$$

$$\forall (r_1, r_2 : \mathbb{R}).\, q_1 < r_1 < r_2 \Rightarrow y \sqsubseteq \frac{x^@(r_2) - x^@(r_1)}{r_2 - r_1}, \tag{7.5}$$

and we want to show

$$\forall (r_1, r_2 : \mathbb{R}).\, r_1 < r_2 \Rightarrow y \sqsubseteq \frac{x^@(r_2) - x^@(r_1)}{r_2 - r_1}.$$

Take any $r_1 < r_2 : \mathbb{R}$. By trichotomy (5.4) $(q_1 < r_1) \vee (r_1 \leq q_1)$ and $(r_2 < q_2) \vee (q_2 \leq r_2)$. If $q_1 < r_1$, then we are done by (7.5), and if $r_2 < q_2$, we are done by (7.4).

Now assume $r_1 \leq q_1$ and $q_2 \leq r_2$, and choose any r_m with $q_1 < r_m < q_2$. Then (7.4) and (7.5) give us

$$y \sqsubseteq \frac{x^@(r_m) - x^@(r_1)}{r_m - r_1} \qquad \text{and} \qquad y \sqsubseteq \frac{x^@(r_2) - x^@(r_m)}{r_2 - r_m}.$$

By continuity of multiplication (Theorem 4.50), we get $yr_m - yr_1 \sqsubseteq x^@(r_m) - x^@(r_1)$ and $yr_2 - yr_m \sqsubseteq x^@(r_2) - x^@(r_m)$, and by continuity of addition (Theorem 4.46), we can add these together to get

$$y(r_2 - r_1) = yr_2 - yr_1 \sqsubseteq x^@(r_2) - x^@(r_1)$$

as desired. □

Recall from Notation 4.41 the notation for constant intervals $[d, u] : \mathrm{c}\bar{\underline{\mathbb{R}}}_j^\infty$, where $d, u : \mathrm{c}\underline{\mathbb{R}}_j^\infty$, or in particular where $d, u : \mathbb{Q}$.

Proposition 7.29 *For a fixed* $x : \bar{\underline{\mathbb{R}}}_\pi^\infty$, *the family* $\{\, y : \bar{\underline{\mathbb{R}}}_\pi^\infty \mid \mathrm{AD}(y, x) \,\}$ *is directed and down-closed.*

Proof Clearly $\mathrm{AD}([-\infty, \infty], x)$ since $[-\infty, \infty] \sqsubseteq z$ for any $z : \bar{\underline{\mathbb{R}}}^\infty$. If $\mathrm{AD}(y_1, x)$ and $\mathrm{AD}(y_2, x)$, it is equally clear that $\mathrm{AD}(y_1 \sqcup y_2, x)$. Down-closedness, that if $y_1 \sqsubseteq y_2$ and $\mathrm{AD}(y_2, x)$ then $\mathrm{AD}(y_1, x)$, is obvious from Definition 7.25. □

Proposition 7.30 *Let* $x : \bar{\underline{\mathbb{R}}}_\pi^\infty$, *and let* $(\dot{\delta}, \dot{\upsilon}) : \bar{\underline{\mathbb{R}}}_\pi^\infty$ *denote the join of the directed family* $\{\, y : \bar{\underline{\mathbb{R}}}_\pi^\infty \mid \mathrm{AD}(y, x) \,\}$. *Then for any* $q : \mathbb{Q}$,

$$\begin{aligned} \dot{\delta} d &:= \exists (d' : \mathbb{Q}).\, (d < d') \wedge \forall (r_1, r_2 : \mathbb{R}).\, r_1 < r_2 \Rightarrow d' < \frac{x^@(r_2) - x^@(r_1)}{r_2 - r_1} \\ \dot{\upsilon} u &:= \exists (u' : \mathbb{Q}).\, (u > u') \wedge \forall (r_1, r_2 : \mathbb{R}).\, r_1 < r_2 \Rightarrow u' > \frac{x^@(r_2) - x^@(r_1)}{r_2 - r_1} \end{aligned} \tag{7.6}$$

Proof Let $D : \bar{\underline{\mathbb{R}}}_\pi^\infty \to \mathtt{Prop}_\pi$ be given by $D(y) := \mathrm{AD}(y, x)$. This defines a subobject of $\bar{\underline{\mathbb{R}}}^\infty$, i.e., a family of elements of $\bar{\underline{\mathbb{R}}}^\infty$. Clearly the join of this family should be given by the supremum of all lower bounds together with the infimum of all upper bounds, taken in the π-modality. In other words, the join $(\dot{\delta}, \dot{\upsilon})$ of D satisfies the following formulas for any $(d, u) : \bar{\underline{\mathbb{R}}}_{\mathrm{pre}}^\infty$:

$$\dot{\delta} d := \pi \exists (y : \bar{\underline{\mathbb{R}}}^\infty).\, \mathrm{AD}(y, x) \wedge d < y, \qquad \dot{\upsilon} u := \pi \exists (y : \bar{\underline{\mathbb{R}}}^\infty).\, \mathrm{AD}(y, x) \wedge y < u.$$

It is simple to show that these are equivalent to

$$\begin{aligned} \dot{\delta} d &\Leftrightarrow \pi \exists (d' : \underline{\mathbb{R}}_{\mathrm{pre}}^\infty).\, \mathrm{AD}([d', +\infty], x) \wedge (d < d'), \\ \dot{\upsilon} u &\Leftrightarrow \pi \exists (u' : \underline{\mathbb{R}}_{\mathrm{pre}}^\infty).\, \mathrm{AD}([-\infty, u'], x) \wedge (u' < u). \end{aligned}$$

Indeed, that the latter implies the former is trivial. Conversely, suppose $\mathrm{AD}(y, x)$ and $d < y$. Then there exists a d' with $d < d' < y$, and $[d', +\infty] \sqsubseteq y$ implies $\mathrm{AD}([d', +\infty], x)$ by Proposition 7.29.

We next want to apply Proposition 5.72. The predomain $\bar{\mathbb{R}}^{\infty}_{\mathrm{pre}}$ is constant, and the corresponding poset $(\bar{\mathbb{R}}^{\infty}_{\mathrm{pre}}, \leqslant)^{\mathrm{op}}$ is directed (in fact, it is linear); see Examples A.5 and A.57 and Proposition 4.30. It follows that the subobject $\{d' : \bar{\mathbb{R}}^{\infty}_{\mathrm{pre}} \mid d < d'\}$ is also constant and directed (linear), and it follows from Proposition 7.29 that $d'' \leq d'$ implies $\mathrm{AD}([d', +\infty], x) \Rightarrow \mathrm{AD}([d'', +\infty])$. Hence the hypotheses of Lemma 5.72 are satisfied, which shows the first line of

$$\dot{\delta} d \Leftrightarrow \exists (d' : \mathbb{R}^{\infty}_{\mathrm{pre}}).\, \mathrm{AD}([d', +\infty], x) \wedge (d < d'),$$
$$\dot{\upsilon} u \Leftrightarrow \exists (u' : \mathbb{R}^{\infty}_{\mathrm{pre}}).\, \mathrm{AD}([-\infty, u'], r) \wedge (u' < u).$$

The second line follows similarly.

Finally, by restricting to the case where $d \in \mathbb{Q} \subseteq \mathbb{R}^{\infty}_{\mathrm{pre}}$ and $u \in \mathbb{Q} \subseteq \bar{\mathbb{R}}^{\infty}_{\mathrm{pre}}$, and unfolding the definition of AD, we have shown (7.6). □

Definition 7.31 (Derivative) Let $x : \bar{\mathbb{R}}^{\infty}_{\pi}$. We define the *derivative of* x, denoted $\dot{x} : \bar{\mathbb{R}}^{\infty}_{\pi}$, to be the join of the directed family $\{\, y : \bar{\mathbb{R}}^{\infty}_{\pi} \mid \mathrm{AD}(y, x) \,\}$. Its cuts $\dot{x} = (\dot{\delta}, \dot{\upsilon})$ are given by Eq. (7.6). We sometimes denote the derivative $\dot{x}$ by $\frac{d}{dt}(x)$.

Lemma 7.32 *For any* $x : \bar{\mathbb{R}}^{\infty}_{\pi}$, $\dot{x}$ *approximates the derivative of* x, *i.e.*, $\mathrm{AD}(\dot{x}, x)$.

Proof We need to show that for any reals $r_1 < r_2$, $\dot{x} \sqsubseteq \frac{x^{@}(r_2) - x^{@}(r_2)}{r_2 - r_1}$. This follows directly from the fact that $\dot{x}$ is the join of the family $\{y : \bar{\mathbb{R}}^{\infty}_{\pi} \mid \mathrm{AD}(y, x)\}$ and that by definition $y \sqsubseteq \frac{x^{@}(r_2) - x^{@}(r_2)}{r_2 - r_1}$ for any such y. □

Lemma 7.32 immediately implies:

Proposition 7.33 *For any* $x, y : \bar{\mathbb{R}}^{\infty}_{\pi}$, $\mathrm{AD}(y, x) \Leftrightarrow y \sqsubseteq \dot{x}$.

At this point we have internally defined the derivative of any π-local improper interval and justified the terminology "y approximates the derivative of x" from Definition 7.25. In the next section we will show that this internal definition has the correct external semantics. Note that it should also be possible to show that variable reals $x : \mathbb{R}_{\pi}$ have an antiderivative as well. We leave this as an open question for interested readers.

Recall the $\ll$ relation on $\mathbb{R}^{\infty}$ from Proposition 4.31; for rationals q_1, q_2 and $y = (\delta, \upsilon) \in \bar{\mathbb{R}}^{\infty}_{\pi}$, the notation $[q_1, q_2] \ll y$ means $(\delta q_1) \wedge (\upsilon q_2)$. Recall also that π-disjointness is equivalent to disjointness by Proposition 7.5.

Lemma 7.34 *Let* $x \in \bar{\mathbb{R}}^{\infty}_{\pi}$ *be disjoint, i.e.*, $x \in \mathbb{IR}^{\infty}_{\pi}$. *Then for any* $y \in \bar{\mathbb{R}}^{\infty}_{\pi}$, *if* $\mathrm{AD}(y, x)$, *then* y *is also disjoint.*

Proof Suppose $[q_1, q_2] \ll y$ for rationals $q_1, q_2 : \mathbb{Q}$. For any reals $r_1 < r_2$, $y \sqsubseteq \frac{x^{@}(r_2) - x^{@}(r_1)}{r_2 - r_1}$. But $\frac{x^{@}(r_2) - x^{@}(r_1)}{r_2 - r_1}$ is in $\mathbb{IR}^{\infty}_{@[r_1, r_2]}$ by Remark 7.11, hence is $@_{[r_1, r_2]}$-

disjoint, which implies $@_{[r_1,r_2]}(q_1 < q_2)$. So $\forall(r_1, r_2 : \mathbb{R}). r_1 < r_2 \Rightarrow @_{[r_1,r_2]}(q_1 < q_2)$, which by Axiom 8 implies $q_1 < q_2$. Therefore y is disjoint. □

Corollary 7.35 *Taking derivatives preserves disjointness: if $x \in \mathbb{IR}_\pi^\infty$, then $\dot{x} \in \mathbb{IR}_\pi^\infty$.*

Remark 7.36 Given a bounded improper interval $x : \bar{\underline{\mathbb{R}}}_\pi$, it may not be the case that $\dot{x}$ is π-bounded. For example, for any t : `Time` the function $x = t^{\frac{1}{3}}$, defined by $\delta q \Leftrightarrow q^3 < t$ and $\upsilon q \Leftrightarrow t < q^3$, has unbounded derivative at 0.

Proposition 7.37 *For any variable real $x : \mathbb{R}_\pi$ and real numbers $r, r_1, r_2 : \mathbb{R}$ with $r_1 < r_2$, we have the following:*

$$\forall(t : \texttt{Time}).\ \Uparrow_{[r_1,r_2]}(r \leq \dot{x}) \Rightarrow r \leq \frac{x^@(r_2) - x^@(r_1)}{r_2 - r_1},$$

$$\forall(t : \texttt{Time}).\ \Uparrow_{[r_1,r_2]}(\dot{x} \leq r) \Rightarrow \frac{x^@(r_2) - x^@(r_1)}{r_2 - r_1} \leq r.$$

Proof The two are similar, so we prove the first. Take any $r, r_1, r_2 : \mathbb{R}$ with $r_1 < r_2$ and t : `Time`, and let $y := \frac{x^@(r_2)-x^@(r_1)}{r_2-r_1}$. It is easy to prove that

$$(y < r) \Rightarrow [(r \leq \dot{x}) \Rightarrow \bot]$$

From this follows:

$$(y < r) \Rightarrow [\Uparrow_{[r_1,r_2]}(r \leq \dot{x}) \Rightarrow \Uparrow_{[r_1,r_2]}\bot]$$

$$(y < r) \Rightarrow [\Uparrow_{[r_1,r_2]}(r \leq \dot{x}) \Rightarrow @_{[r_1,r_2]}\bot]$$

$$\Uparrow_{[r_1,r_2]}(r \leq \dot{x}) \Rightarrow [(y < r) \Rightarrow @_{[r_1,r_2]}\bot]$$

$$\Uparrow_{[r_1,r_2]}(r \leq \dot{x}) \Rightarrow (r \leq y)$$

where in the first line we used the functoriality of the modality $\Uparrow_{[r_1,r_2]}$, in the second we used $\Uparrow_{[r_1,r_2]}\bot = t\#[r_2, r_1] = @_{[r_1,r_2]}\bot$, and in the last line we used Proposition 4.35 for the modality $@_{[r_1,r_2]}$, together with the fact that y is $@_{[r_1,r_2]}$-closed so $y < r$ is equivalent to $y < r^@([r_1, r_2])$. □

7.3.2 Semantics of Derivatives

Recall from Theorem 7.19 that $[\![\bar{\underline{\mathbb{R}}}_\pi^\infty]\!](\ell)$ is isomorphic to the set of pairs $(\underline{f}, \bar{f})$ of functions $\Uparrow[0, \ell] \to \underline{\mathbb{R}}^\infty$, such that $\underline{f}$ is lower semi-continuous, $\bar{f}$ is upper semi-continuous, and for all $x \in \Uparrow(0, \ell)$, $\underline{f}(x) = \inf \underline{f}(\uparrow x \cap \mathbb{R})$ and $\bar{f}(x) = \sup \bar{f}(\uparrow x \cap \mathbb{R})$. Equivalently, $[\![\bar{\underline{\mathbb{R}}}_\pi^\infty]\!](\ell)$ is isomorphic to the set of pairs $(\underline{f}, \bar{f})$ of functions $(0, \ell) \to \mathbb{R}^\infty$, where $(0, \ell) = \{r \in \mathbb{R} \mid 0 < r < \ell\}$ is the usual open interval, and where $\underline{f}$ is lower semi-continuous and $\bar{f}$ is upper semi-continuous.

Definition 7.38 Let $V \subseteq \mathbb{R}$ be an open subset of the real line, and let $(\underline{f}, \bar{f})$ be a pair of functions $V \to \mathbb{R}^\infty$ such that $\underline{f}$ is lower semi-continuous and $\bar{f}$ is upper semi-continuous. We define the *lower derivative* $\underline{f}'$ and the *upper derivative* $\bar{f}'$ to be the functions

$$\underline{f}'(x) := \sup_{x \in U \subseteq V} \inf_{d < u \in U} \frac{\underline{f}(u) - \bar{f}(d)}{u - d}$$

$$\bar{f}'(x) := \inf_{x \in U \subseteq V} \sup_{d < u \in U} \frac{\bar{f}(u) - \underline{f}(d)}{u - d}.$$

Remark 7.39 The $\sup_{x \in U \subseteq V}$ and $\inf_{x \in U \subseteq V}$ in Definition 7.38 can be replaced by lim, because, for example, the infimum $\inf_{d < u \in U} \cdots$ increases as U shrinks around x. Thus, for example, it is easy to see that if $\underline{f} \leq \bar{f}$ then $\underline{f}' \leq \bar{f}'$. This is the external version of Corollary 7.35.

The following proposition shows that the above notions of lower derivative and upper derivative are reasonable.

Proposition 7.40 *Suppose $V \subseteq \mathbb{R}$ is open, $x \in V$ is a point, $f : V \to \mathbb{R}$ is continuous, let $\underline{f} := f$ and $\bar{f} := f$, and let $\underline{f}'$ and $\bar{f}'$ be the lower and upper derivatives. They agree at x, i.e., $\underline{f}'(x) = \bar{f}'(x)$, iff f is continuously differentiable at x, and in this case all three coincide $\underline{f}'(x) = f'(x) = \bar{f}'(x)$.*

Proof The condition $\underline{f}'(x) = \bar{f}'(x)$ is easily shown equivalent to the statement that there exists $y \in \mathbb{R}$ such that for all $\epsilon > 0$ there exists $\delta > 0$ such that for all $d < u$ in the ball $(x - \delta, x + \delta)$,

$$\frac{f(u) - f(d)}{u - d} \in (y - \epsilon, y + \epsilon) \tag{7.7}$$

This clearly implies that for all $\epsilon > 0$ there exists $\delta > 0$ such that for all $0 < h < \delta$,

$$\frac{f(x + h) - f(x - h)}{2h} \in (y - \epsilon, y + \epsilon) \tag{7.8}$$

which is the symmetric definition of $f'(x) = y$.

For the other direction, choose $\epsilon > 0$; since f' is continuous, there is some $\delta_1 > 0$ such that $f'(x) - f'(x_1) \in (y - \epsilon/2, y + \epsilon/2)$ for all $x_1 \in (x - \delta_1, x + \delta_1)$. Again using $\epsilon/2$, we obtain $\delta_2 > 0$ satisfying Eq. (7.8) for all $0 < h < \delta_2$. Let $\delta := \min(\delta_1, \delta_2)$ and for any $d < u$ in $(x - \delta, x + \delta)$, let $x_0 := (u + d)/2$ be the midpoint and let $h_0 := (u - d)/2$ the radius, and we obtain Eq. (7.7). □

Lemma 7.41 *Let* $f = (\underline{f}, \bar{f})$ *and* $g = (\underline{g}, \bar{g})$ *be elements of* $[\![\bar{\underline{\mathbb{R}}}_\pi^\infty]\!](\ell)$. *Then* $\ell \Vdash \mathrm{AD}(f, g)$ *if and only if for all* $[d, u] \in \Uparrow[0, \ell]$ *and all* $r_1 < r_2$ *with* $d \leq r_1 < r_2 \leq u$,

$$\underline{f}([d, u]) \leq \frac{\underline{g}([r_2, r_2]) - \bar{g}([r_1, r_1])}{r_2 - r_1} \quad \text{and} \quad \bar{f}([d, u]) \geq \frac{\bar{g}([r_2, r_2]) - \underline{g}([r_1, r_1])}{r_2 - r_1}.$$

Proof Choose $(d_t, u_t) \in [\![\mathtt{Time}]\!](\ell)$ and for $i = 1, 2$ let $(\underline{g_i}, \bar{g}_i) := g^{@}(r_i)$.[3] Since it is $@_{[r,r]}$-local, Theorem 7.19 (18) tells us that for any point $[d, u] \in \Uparrow[d_t, u_t]$, we have

$$\underline{g}_i([d, u]) = \begin{cases} \underline{g}_i([r_i, r_i]) & \text{if } d \leq r_i \leq u \\ -\infty & \text{otherwise} \end{cases} \quad \text{and}$$

$$\bar{g}_i([d, u]) = \begin{cases} \bar{g}_i([r_i, r_i]) & \text{if } d \leq r_i \leq u \\ \infty & \text{otherwise} \end{cases}$$

Letting $(\underline{y}, \bar{y}) := \frac{g^{@}(r_2) - g^{@}(r_1)}{r_2 - r_1}$ and tracing through the Joyal–Kripke semantics for $\ell \Vdash \mathrm{AD}(f, g)$ (see Definition 7.25), we find that it holds iff for all real numbers $r_1, r_2 : \mathbb{R}$ with $r_1 < r_2$ and $q \in \mathbb{Q}$, if $\ell \Vdash \underline{f} q$, then $\ell \Vdash \underline{y} q$ and if $\ell \Vdash \bar{f} q$, then $\ell \Vdash \bar{y} q$. This holds iff $\underline{f} \leq \underline{y}$ and $\bar{f} \geq \bar{y}$, as desired. □

Proposition 7.42 *For any* $f, g \in [\![\bar{\underline{\mathbb{R}}}_\pi^\infty]\!](\ell)$, *if* $\mathrm{AD}(f, g)$, *then for all* r, *if* $0 < r < \ell$, *then* $\underline{f}([r, r]) \leq \underline{g}'(r)$ *and* $\bar{f}([r, r]) \geq \bar{g}'(r)$.

Proof We show that $\underline{f}([r, r]) \leq \underline{g}'(r)$; the second part is analogous. We want to show that for any $\epsilon > 0$, there exists an open neighborhood $r \in U \subseteq (0, \ell)$ such that for all $d < u \in U$, $\underline{f}([r, r]) - \epsilon < \frac{\underline{g}(u) - \bar{g}(d)}{u - d}$. Using that $\underline{f}([r, r]) = \sup_{p \in \Downarrow[r,r]} \underline{f}(p)$, there exists a point $p \ll [r, r]$ such that $\underline{f}([r, r]) - \epsilon < \underline{f}(p)$. By the assumption that $\mathrm{AD}(f, g)$ and Lemma 7.41, letting $U = \Uparrow p$, we have that for any $d < u \in U$, $\underline{f}([r, r]) - \epsilon < \underline{f}(p) \leq \frac{\underline{g}([u,u]) - \bar{g}([d,d])}{u - d}$, as desired. □

The next lemma is a sort of mean-value theorem for upper/lower derivatives.

Lemma 7.43 *Let* $f \in [\![\mathbb{IR}_\pi^\infty]\!](\ell)$, *i.e.*, $(\underline{f}, \bar{f}) \in [\![\bar{\underline{\mathbb{R}}}_\pi^\infty]\!](\ell)$ *with* $\underline{f} \leq \bar{f}$. *Then for any* a, b *with* $0 < a < b < \ell$, *we have*

$$\inf_{r \in [a,b]} \underline{f}'(r) \leq \frac{\underline{f}(b) - \bar{f}(a)}{b - a} \quad \text{and} \quad \sup_{r \in [a,b]} \bar{f}'(r) \geq \frac{\bar{f}(b) - \underline{f}(a)}{b - a}.$$

[3] Technically, the semantics of the term-in-context $x : \bar{\underline{\mathbb{R}}}_\pi^\infty, y : \mathbb{R} \vdash x^{@}(y) : \bar{\underline{\mathbb{R}}}_{@[r,r]}^\infty$ is a sheaf homomorphism $[\![\bar{\underline{\mathbb{R}}}_\pi^\infty \times \mathbb{R}]\!] \to [\![\bar{\underline{\mathbb{R}}}_{@[r,r]}^\infty]\!]$, and we are denoting the image of (g, r) under the ℓ-component of that map.

Proof Let $s = \inf_{r\in[a,b]} \underline{f}'(r)$. Then for any $r \in [a, b]$, $s \le \underline{f}'(r)$. Fix an $\epsilon > 0$. Then for any $r \in [a, b]$, there exists a neighborhood $r \in U \subseteq (0, \ell)$ such that for all $d < u \in U$, $s - \epsilon < \frac{\underline{f}(u)-\bar{f}(d)}{u-d}$. These opens $\{U_j\}$ thus cover $[a, b]$, so by compactness there must exist a sequence $a = r_0 < r_1 < \cdots < r_{n+1} = b$ such that for all i, $[r_i, r_{i+1}] \subseteq U_j$ for some j, and in particular $s - \epsilon < \frac{\underline{f}(r_{i+1})-\bar{f}(r_i)}{r_{i+1}-r_i}$. Then we have that

$$\frac{\underline{f}(b) - \bar{f}(a)}{b-a} \ge \sum_{i=0}^{n} \frac{r_{i+1}-r_i}{b-a} \frac{\underline{f}(r_{i+1}) - \bar{f}(r_i)}{r_{i+1}-r_i} > \sum_{i=0}^{n} \frac{r_{i+1}-r_i}{b-a}(s-\epsilon) = s - \epsilon,$$

where the first inequality makes use of $\underline{f}(r_i) \le \bar{f}(r_i)$ for $1 \le i \le n$. As ϵ was arbitrary, we have $s \le \frac{\underline{f}(b)-\bar{f}(a)}{b-a}$.

The proof of the second part is similar. □

Proposition 7.44 *For any* $f, g \in [\![\mathbb{IR}_\pi^\infty]\!](\ell)$, $\mathrm{AD}(f, g)$ *if and only if* $\underline{f}([r, r]) \le \underline{g}'(r)$ *and* $\bar{f}([r, r]) \ge \bar{g}'(r)$ *for all* $r \in (0, \ell)$.

Proof The forward direction is Proposition 7.44. For the converse, assume that for all $r \in (0, \ell)$, $\underline{f}([r, r]) \le \underline{g}'(r)$ and $\bar{f}([r, r]) \ge \bar{g}'(r)$. By Lemma 7.41, it suffices to show that for any $0 < d \le r_1 < r_2 \le u < \ell$, $\underline{f}([d, u]) \le \frac{\underline{g}([r_2,r_2])-\bar{g}([r_1,r_1])}{r_2-r_1}$ and $\bar{f}([d, u]) \ge \frac{\bar{g}([r_2,r_2])-\underline{g}([r_1,r_1])}{r_2-r_1}$.

For the first, we have

$$\underline{f}([d, u]) = \inf_{r\in[d,u]} \underline{f}([r, r]) \le \inf_{r\in[d,u]} \underline{g}'(r) \le \frac{\underline{f}(u) - \bar{f}(d)}{u-d},$$

where we have used that f is π-closed for the first equality, the first inequality is by assumption, and the second is Lemma 7.43. □

In the following corollary, we use Corollary 7.22 to identify $[\![\mathbb{R}_\pi]\!](\ell)$ with the set of continuous functions $(0, \ell) \to \mathbb{R}$.

Corollary 7.45 *Suppose that* $f \in [\![\mathbb{R}_\pi]\!](\ell)$ *is internally continuously differentiable in the sense that its internal derivative (Definition 7.31) is also continuous* $\dot{f} : [\![\mathbb{R}_\pi]\!](\ell)$. *Then* $\dot{f}$ *is externally the derivative of* f.

Proof Let $g = \dot{f}$ as in Definition 7.31. Since $f, g \in [\![\mathbb{R}_\pi]\!](\ell)$, we have $\underline{f} = \bar{f}$ and $\underline{g} = \bar{g}$. By Remark 7.39 and Proposition 7.44, we have $\underline{g}' = f = \bar{g}'$. The result then follows from Propositions 7.40. □

7.3.3 Differentiability

We know by Corollary 7.45 that the semantics of our notion of derivative matches the usual definition. In this section we define what it means for a variable real to be differentiable and we give internal proofs of certain well-known formulas: that time has unit derivative, $\dot{t} = \frac{dt}{dt} = 1$, that the Leibniz rule—often called the product rule—holds, and that differentiation is linear.

Definition 7.46 (Differentiable) Suppose that $x : \mathbb{R}_\pi$ is a variable real, so its derivative $\dot{x}$ is a π-local extended interval; see Definition 7.25. We say x is *differentiable* if it is in fact a variable real $\dot{x} \in \mathbb{R}_\pi$, i.e., if $\dot{x}$ is π-bounded and π-located.

Proposition 7.47 *Any* t : `Time` *is differentiable, and its derivative is* $\dot{t} = 1$.

Proof By Axiom 3b, Corollary 7.35, and Lemma 4.33, if $1 \sqsubseteq \dot{t}$, then $1 = \dot{t}$. Thus, by Lemma 7.26, it suffices to show

$$q < 1 \Rightarrow \downarrow_{[r_1,r_2]} \left(q < \frac{t^@(r_2) - t^@(r_1)}{r_2 - r_1} \right)$$

the upper cut being similar. Assuming $q < 1$, it suffices to show

$$\downarrow_{[r_1,r_2]} \exists (q_1, q_2 : \mathbb{Q}).\, (q * (r_2 - r_1) < q_2 - q_1) \wedge (q_2 < t^@(r_2)) \wedge (t^@(r_1) < q_1).$$

It is easy to check that $q * (r_2 - r_1) < r_2 - r_1$, e.g., using Proposition 4.26. Thus by definition we have $\exists (q_1, q_2).\, (q * (r_2 - r_1) < q_2 - q_1) \wedge (q_2 < r_2) \wedge (r_1 < q_1)$, and the result follows by Example 7.3. □

Proposition 7.48 *Suppose* $x : \mathbb{R}_\pi$ *is differentiable. Then for any* $y : \mathbb{R}_\pi$, *we have* $y = \dot{x}$ *iff* AD(y, x), *i.e.,*

$$\forall (r_1, r_2 : \mathbb{R}).\, r_1 < r_2 \Rightarrow y \sqsubseteq \frac{x^@(r_2) - x^@(r_1)}{r_2 - r_1}$$

Proof By Proposition 7.33, we have AD(y, x) iff $y \sqsubseteq \dot{x}$. Because y is π-located and $\dot{x}$ is π-disjoint, we conclude by Lemma 4.33. □

In the following theorem, all arithmetic operations take place in the π modality, i.e., $x + y$ and $x * y$ mean $x +_\pi y$ and $x *_\pi y$, in the sense of Sect. 4.3.7.

Theorem 7.49 *Let* $x, y : \mathbb{R}_\pi$. *If* x *and* y *are differentiable, then so is* $x * y$, *and the Leibniz rule holds:*

$$\frac{d}{dt}(x * y) = x * \dot{y} + \dot{x} * y.$$

Moreover, the derivative is $\mathbb{R}$*-linear,* $\frac{d}{dt}(r * x + y) = r * \dot{x} + \dot{y}$ *for any* $r : \mathbb{R}$.

Proof As mentioned above the theorem, the arithmetic operations are taking place in the π-modality. So assume x and y are differentiable; we really need to prove that $x *_\pi y$ is differentiable, that $\frac{d}{dt}(x *_\pi y) = x *_\pi \dot{y} +_\pi \dot{x} *_\pi y$, and that $\frac{d}{dt}(r *_\pi x +_\pi y) = r *_\pi \dot{x} +_\pi \dot{y}$. We continue to leave the π-subscript off of arithmetic operations, for readability purposes.

Since $\dot{x}$ and $\dot{y}$ are π-bounded and π-located, so are $x * \dot{y} + \dot{x} * y$ and $r * \dot{x} + \dot{y}$. We begin by proving that the Leibniz rule holds. Since $\frac{d}{dt}(x * y)$ is π-disjoint by Corollary 7.35, it suffices to show that $(x * \dot{y}) + (\dot{x} * y) \sqsubseteq \frac{d}{dt}(x * y)$, by Lemma 4.33.

Because $\mathrm{c}\bar{\underline{\mathbb{R}}}_\pi \subseteq \bar{\underline{\mathbb{R}}}_\pi$ is a basis (see Remark 4.38), it suffices to show that for any $c : \mathrm{c}\bar{\underline{\mathbb{R}}}$, if $c \ll (x * \dot{y}) + (\dot{x} * y)$, then $c \sqsubseteq \frac{d}{dt}(x * y)$. Let $c : \mathrm{c}\bar{\underline{\mathbb{R}}}$, and suppose $c \ll (x * \dot{y}) + (\dot{x} * y)$. Then

$$\begin{aligned} &\pi\exists(c_1, c_2, c_3, c_4 : \mathrm{c}\bar{\underline{\mathbb{R}}}).\, (c_1 \sqsubseteq x) \wedge (c_2 \sqsubseteq \dot{y}) \wedge (c_3 \sqsubseteq \dot{x}) \\ &\quad \wedge (c_4 \sqsubseteq y) \wedge (c \ll c_1 * c_2 + c_3 * c_4). \end{aligned} \tag{7.9}$$

Ignoring the preceding π for now, suppose we have $c_1, c_2, c_3, c_4 : \mathrm{c}\bar{\underline{\mathbb{R}}}$ such that $c_1 \sqsubseteq x$, $c_2 \sqsubseteq \dot{y}$, $c_3 \sqsubseteq \dot{x}$, $c_4 \sqsubseteq y$, and $c \ll c_1 * c_2 + c_3 * c_4$. We want to show that $c \sqsubseteq \frac{d}{dt}(x * y)$, or equivalently $\mathrm{AD}(c, x * y)$. So letting $r_1 < r_2$ be arbitrary reals, we want to show that $c \sqsubseteq \frac{x^@(r_2)*y^@(r_2) - x^@(r_1)*y^@(r_1)}{r_2 - r_1}$, where we have used that the maps $@_{r_1} : \bar{\underline{\mathbb{R}}}_\pi \to \bar{\underline{\mathbb{R}}}_{@r_1}$ and $@_{r_2} : \bar{\underline{\mathbb{R}}}_\pi \to \bar{\underline{\mathbb{R}}}_{@r_2}$ preserve multiplication; see Proposition 4.54.

From $c_2 \sqsubseteq \dot{y}$, we have $c_2 \sqsubseteq \frac{y^@(r_2) - y^@(r_1)}{r_2 - r_1}$ hence $c_1 * c_2 \sqsubseteq c_1 * \frac{y^@(r_2) - y^@(r_1)}{r_2 - r_1} \sqsubseteq x^@(r_1)\frac{y^@(r_2) - y^@(r_1)}{r_2 - r_1}$. Similarly, $c_3 * c_4 \sqsubseteq \frac{x^@(r_2) - x^@(r_1)}{r_2 - r_1} y^@(r_2)$. Adding these together and simplifying yields $c \ll c_1 * c_2 + c_3 * c_4 \sqsubseteq \frac{x^@(r_2)*y^@(r_2) - x^@(r_1)*y^@(r_1)}{r_2 - r_1}$, as desired.

Thus we have shown, assuming the existence of $c_1, \ldots, c_4$, that $\mathrm{AD}(c, x * y)$ holds. Hence (7.9) implies $\pi\mathrm{AD}(c, x * y)$, which by Proposition 7.28 implies $\mathrm{AD}(c, x * y)$. We therefore have shown that $c \ll x * \dot{y} + \dot{x} * y$ implies $\mathrm{AD}(c, x * y)$, and hence $c \sqsubseteq \frac{d}{dt}(x * y)$, as desired.

Showing linearity is similar: $\frac{d}{dt}(r * x + y)$ is again π-disjoint, so it suffices to show that $r * \dot{x} + \dot{y} \sqsubseteq \frac{d}{dt}(r * x + y)$. This follows as above, using the fact that localizing is linear, i.e., $(r * x)^@(r') = r * x^@(r')$; see Proposition 4.54. □

Chapter 8
Applications

As mentioned in the introduction, we believe that our temporal type theory can serve as a "big tent" into which to embed many disparate formalisms for proving property of behaviors. In this chapter we discuss a few of these, including hybrid dynamical systems in Sect. 8.1, delays in Sect. 8.2, differential equations in Sect. 8.3, and labeled transition systems in Sect. 8.4.4. This last occurs in Sect. 8.4 where we briefly discuss a general framework on machines, systems, and behavior contracts. Next in Sect. 8.5 we give a toy example—an extreme simplification of the National Airspace System—and prove a safety property. The idea is to show that we really can mix continuous, discrete, and delay properties without ever leaving the temporal type theory. We conclude this chapter by discussing how some temporal logics, e.g., metric temporal logic, embeds into the temporal type theory.

8.1 Hybrid Sheaves

The NAS is a hybrid system: it includes both continuous aspects, like the differential equations governing airplane motion, and discrete aspects, like the signals and messages that alert pilots to possible danger. We begin by explaining how one might integrate continuous and discrete behavior into a single sheaf, which we call a "hybrid sheaf" or "hybrid type," depending on if we are working externally or internally.

Our method of specifying (and constructing) a hybrid type takes an arbitrary behavior type—a sheaf C—which we think of as a type of *continuous* behaviors, and produces a new behavior type which extends C with certain allowed discontinuities or discrete transitions, coming from a type D.

P. Schultz, D. I. Spivak, *Temporal Type Theory*, Progress in Computer Science and Applied Logic 29, https://doi.org/10.1007/978-3-030-00704-1_8

8.1.1 Constructing Hybrid Sheaves

Recall the pointwise modality $\pi : \mathtt{Prop} \to \mathtt{Prop}$ from Notation 5.28 and the π-sheafification functor sh_π from Definition 4.11. Recall also the notion of sum types and quotient types from Sect. 4.1.

Definition 8.1 (Hybrid Type) A *hybrid datum* is a tuple (C, D, src, tgt, τ), where C and D are arbitrary types, and where $src, tgt : D \to C$ and $\tau : D \to \mathtt{Time}$ are functions. Consider the quotient type $\mathrm{Hyb}_1(C, D) := (C + D)/\sim$, where for $d : D$ we put

$$d \sim src(d) \text{ if } \tau(d) < 0, \qquad \text{and} \qquad d \sim tgt(d) \text{ if } \tau(d) > 0. \tag{8.1}$$

The *hybrid behavior type* on the hybrid datum (C, D, src, tgt, τ) is defined to be the π-sheafification of this quotient, which we denote $\mathrm{Hyb}(C, D) := \mathsf{sh}_\pi \mathrm{Hyb}_1(C, D)$. There are natural maps $C \to \mathrm{Hyb}(C, D)$ and $D \to \mathrm{Hyb}(C, D)$; we refer to the image of C as the *continuous part*.

Intuitively, a section d of the sheaf D represents a behavior which has a discontinuity at the point $\tau(d) = 0$. We want to adjoin it to C, by connecting the continuous behavior $src(d)$ on the left of the discontinuity to the continuous behavior $tgt(d)$ on the right. When we sheafify, we allow continuous behaviors that have been interrupted by finitely many of these discontinuity events.

The following is straightforward.

Lemma 8.2 *In the notation of Definition 8.1, the equivalence relation on $C + D$ generated by* (8.1) *is as follows for $c, c' : C$ and $d, d' : D$,*

$$\begin{aligned}
c \sim c' &\Leftrightarrow c = c' \\
c \sim d &\Leftrightarrow (\tau(d) < 0 \wedge c = src(d)) \vee (\tau(d) > 0 \wedge c = tgt(d)) \\
d \sim c &\Leftrightarrow (\tau(d) < 0 \wedge c = src(d)) \vee (\tau(d) > 0 \wedge c = tgt(d)) \\
d \sim d' &\Leftrightarrow \big(d = d' \\
&\qquad \vee (src(d) = src(d') \wedge (\tau(d) < 0) \wedge (\tau(d') < 0)) \\
&\qquad \vee (tgt(d) = tgt(d') \wedge (\tau(d) > 0) \wedge (\tau(d') > 0)) \\
&\qquad \vee (tgt(d) = src(d') \wedge \tau(d) < 0 < \tau(d')) \\
&\qquad \vee (src(d) = tgt(d') \wedge \tau(d') < 0 < \tau(d))\big)
\end{aligned}$$

Proposition 8.3 *Let (C, D, src, tgt, τ) be a hybrid sheaf datum, and suppose that C and D have decidable equality. Then the type* $\mathrm{Hyb}_1(C, D)$ *from Definition 8.1 is π-separated.*

Proof Let $x, x' : C + D$; we need to show that $\pi(x \sim x')$ implies $x \sim x'$. We will work by cases on the components of x and x', using Lemma 8.2. If $x, x' \in C$, then $\pi(x = x')$, so $x = x'$ by Corollary 5.54. For the remaining cases we will apply Lemma 5.71. For example, it directly implies that if $x \in C$ and $x' \in D$ (or vice versa), then $x \sim x'$ is π-closed. Finally, in the case $x, x' \in D$, one uses Proposition 4.10 and then repeats the above argument twice more:

$$\begin{aligned}
&\pi(x \sim x')\\
\Rightarrow&\pi[(x = x') \vee (P \wedge t < 0 \wedge t' < 0) \vee (Q \wedge 0 < t \wedge 0 < t'\}) \vee (R \wedge t < 0 < t') \vee\\
&\quad (S \wedge t' < 0 < t)]\\
\Rightarrow&(x = x') \vee \pi[(t < 0 \wedge \pi((P \wedge t' < 0) \vee (R \wedge 0 < t'))) \vee (0 < t \wedge \pi((Q \wedge 0 < t') \vee\\
&\quad (S \wedge t' < 0)))]\\
\Rightarrow&(x = x') \vee (t < 0 \wedge \pi((P \wedge t' < 0) \vee (R \wedge 0 < t'))) \vee (0 < t \wedge \pi((Q \wedge 0 < t') \vee\\
&\quad (S \wedge t' < 0)))\\
\Rightarrow&(x = x') \vee (t < 0 \wedge (P \wedge t' < 0) \vee (R \wedge 0 < t')) \vee (0 < t \wedge (Q \wedge 0 < t') \vee (S \wedge t' < 0))\\
&\quad \Leftrightarrow (x \sim x'),
\end{aligned}$$

where $t := \tau(x)$, $t' := \tau(x')$, and where $P := (src(x) = src(x'))$, $Q := (tgt(x) = tgt(x'))$, $R := (tgt(x) = src(x'))$, and $S := (src(x) = tgt(x'))$, all of which are π-closed. □

Remark 8.4 The only information in a section d of D which is relevant to this construction is the restriction of d to arbitrarily small neighborhoods of the point $\tau(d) = 0$. Externally, we can make this precise as follows: consider the stalk functor $\mathsf{stk}_0 \colon \mathcal{B} \to \mathbf{Set}$ sending a sheaf to the set of zero-length germs. It can be obtained by composing the geometric morphism $\mathsf{stk}_{@0} \colon \mathbf{Set} \to \mathsf{Shv}(S_{\mathbb{IR}}) \cong \mathcal{B}/\mathtt{Time}$, corresponding to the point $[0, 0] \in \mathbb{IR}$, with the geometric morphism $\mathcal{B}/\mathtt{Time} \to \mathcal{B}$; see also Proposition 3.7. Given a sheaf over $\mathtt{Time}$, $\tau \colon D \to \mathtt{Time}$, the following square is a pullback:

$$\begin{array}{ccc}
\mathsf{stk}_{@0}(D) & \longrightarrow & 1\\
\downarrow & \lrcorner & \downarrow{\scriptstyle 0}\\
\mathsf{stk}_0(D) & \xrightarrow[\mathsf{stk}_0(\tau)]{} & \mathsf{stk}_0(\mathtt{Time})
\end{array}$$

Then any hybrid datum (C, D, src, tgt, τ) determines a pair of functions $\mathsf{stk}_{@0}(D) \rightrightarrows \mathsf{stk}_0(C)$. It also determines a function $\mathsf{stk}_{@0}(D) \to \mathsf{stk}_0 D \to \mathsf{stk}_0\mathrm{Hyb}(C, D)$, and the image of this function might be called the set of *discontinuities* in the hybrid type.

The hybrid behavior type construction really only depends on the sheaf C, the set $S = \mathsf{stk}_{@0}(D)$, and the source and target functions $S \rightrightarrows \mathsf{stk}_0(C)$. That is, if $(C, D', src', tgt', \tau')$ is a different hybrid datum on the same continuous part C, then to every bijection $\mathsf{stk}_{@0}(D) \cong \mathsf{stk}_{@0}(D')$ commuting with the source and target functions, there is a canonical isomorphism $\mathrm{Hyb}(C, D) \cong \mathrm{Hyb}(C, D')$. Conversely, given any set S and pair of functions $S \to \mathsf{stk}_0(C)$, there is a hybrid datum (C, D, src, tgt, τ) realizing it.

In particular, if $f: D \to D'$ is a morphism commuting with *src*, *src'*, *tgt*, *tgt'*, τ, and τ', then f induces an isomorphism $\mathrm{Hyb}(C, D) \cong \mathrm{Hyb}(C, D')$ if and only if $\mathsf{stk}_{@0}(f)$ is a bijection.

The case when f is a monomorphism is relatively easy to capture internally. Let $\phi: D \to \mathtt{Prop}$ be the predicate characterizing $f: D' \rightarrowtail D$. Then f induces a bijection on stalks if and only if the proposition

$$\forall (d : D).\, @_0^{\tau(d)} \phi(d)$$

holds internally. This simply says that for all sections $d \in D(\ell)$, there is a neighborhood of $\tau(d) = 0$ on which the restriction of d is in D', i.e., the stalk of d at $\tau(d) = 0$ is in the image of $\mathsf{stk}_0(f)$. Then it is easy to see that the induced map $(C + D')/\sim\, \to (C + D)/\sim$ will again be mono, characterized by the predicate $\phi' : (C + D)/\sim\, \to \mathtt{Prop}$ which is $\top$ on C and $\downarrow_0^{\tau(d)} \phi(d)$ on $d \in D$. This predicate ϕ' is π*-dense*, meaning $\pi \circ \phi' = \top$, since $\pi \top \Leftrightarrow \top$ and, by Proposition 5.44, $\pi(\downarrow_0^{\tau(d)} \phi(d)) \Leftrightarrow @_0^{\tau(d)} \phi(d) \Leftrightarrow \top$ for all $d \in D$. Therefore the mono $(C + D')/\sim\, \rightarrowtail (C + D)/\sim$ characterized by ϕ' induces an isomorphism between the sheafifications $\mathrm{Hyb}(C, D') \cong \mathrm{Hyb}(C, D)$.

Later, we will need to know that a section of a hybrid sheaf is almost always in the continuous part, C. Let $\mathtt{is\_cts} : \mathrm{Hyb}(C, D) \to \mathtt{Prop}$ classify the image of the map $C \to \mathrm{Hyb}(C, D)$.

Proposition 8.5 *Let* $\mathrm{Hyb}(C, D)$ *be the hybrid type on the data* (C, D, src, tgt, τ)*. A section of* $\mathrm{Hyb}(C, D)$ *is almost-always in the continuous part, i.e.,*

$$\forall (h : \mathrm{Hyb}(C, D)).\, \neg\neg \mathtt{is\_cts}(h).$$

Proof Choose $h : \mathrm{Hyb}(C, D)$; we first spell out its definition and that of $\mathtt{is\_cts}(h)$. Let $S = C + D/\sim$. By Definition 4.11, $h : S \to \mathtt{Prop}$ is a predicate satisfying

$$\forall (s : S).\, \pi h(s) \Rightarrow h(s) \qquad \text{and} \qquad \pi \exists (s_0 : S).\, \forall (s : S).\, h(s) \Leftrightarrow \pi(s_0 = s).$$

Unwinding definitions and using Eq. (4.8), we see that $\mathtt{is\_cts}(h)$ is equivalent to the proposition

$$\exists (c : C).\, \forall (s : S).\, h(s) \Leftrightarrow \pi(s = c).$$

Assume $\neg\mathtt{is\_cts}(h)$; we will show $\pi\bot$, which is enough by Corollary 5.53. By Remark 4.9 we may choose $s_0 : S$ such that $\forall (s : S).\, h(s) \Leftrightarrow \pi(s_0 = s)$. If $s_0 \sim c$ for some $c : C$, we obtain the desired contradiction, so we may assume $s_0 : D$. By Lemma 8.2, we obtain a contradiction if $\tau(d) < 0$ or $\tau(d) > 0$, so we have $\neg(\tau(d) < 0) \wedge \neg(\tau(d) > 0)$. But this is a contradiction by Axiom 3b. □

Definition 8.6 Let (C, D, src, tgt, τ) and $(C', D', src', tgt', \tau')$ be hybrid data as in Definition 8.1. A *morphism* between them consists of a function $f: C \to C'$ and a function $g: D \to D'$, commuting with the rest of the structure, i.e., $src' \circ g = f \circ src$ and $tgt' \circ g = f \circ tgt$ and $\tau' \circ g = \tau$. This defines a category of hybrid data.

Proposition 8.7 *A morphism of hybrid data induces a morphism of behavior types* $\mathrm{Hyb}(C, D) \to \mathrm{Hyb}(C', D')$. *In fact,* Hyb *is a functor.*

Proof Suppose given $f: C \to C'$ and $g: D \to D'$, such that $src \circ g = f \circ src'$ and $tgt \circ g = f \circ tgt'$ and $\tau' \circ g = \tau$. There is an induced map $C + D \to C' + D'$, and this map respects the equivalence relation in Eq. (8.1) because if $\tau(d) < 0$, then $\tau'(g(d)) < 0$, in which case $d \sim src(d)$ implies $g(d) \sim src'(g(d))$. Thus we obtain a map between their quotients $S \to S'$, and hence between their π-sheafifications $\mathsf{sh}_\pi S \to \mathsf{sh}_\pi S'$. All of the induced maps along the way have been functorial. □

8.1.2 Special Case: Walks Through a Graph

Suppose given a graph G (in **Set**)

$$E \underset{c}{\overset{d}{\rightrightarrows}} V \tag{8.2}$$

where d stands for "domain" and c stands for "codomain." Define a *walk of length ℓ through G* to be a finite path in G, together with a positive duration of time assigned to each vertex in the path, such that the total of these durations is ℓ. In other words, we imagine that each edge is traversed instantaneously.

Considering E and V as the corresponding constant types (see Sect. 6.2.1), we can build the sheaf of walks W_G in G as a hybrid type using the following as a hybrid sheaf datum (in the sense of Definition 8.1):

$$\mathtt{Time} \xleftarrow{p_{\mathtt{Time}}} E \times \mathtt{Time} \underset{c \circ p_E}{\overset{d \circ p_E}{\rightrightarrows}} V. \tag{8.3}$$

Here $\mathtt{Time} \xleftarrow{p_{\mathtt{Time}}} E \times \mathtt{Time} \xrightarrow{p_E} E$ are the projections. Unraveling Definition 8.1 in this case, we find that the hybrid behavior type $\mathrm{Hyb}(V, E \times \mathtt{Time})$ is the π-sheafification of the quotient $W'_G := \mathrm{Hyb}_1(V, E \times \mathtt{Time}) = (V + (E \times \mathtt{Time}))/\sim$, where

$$(e, t) \sim d(e) \text{ if } t < 0 \qquad \text{and} \qquad (e, t) \sim c(e) \text{ if } t > 0 \tag{8.4}$$

Semantically, the sections of W'_G are walks with at most one transition.

Corollary 8.8 *The type W'_G is π-separated.*

Proof This follows from Proposition 8.3 because constant types and $\mathtt{Time}$—and hence V and $E \times \mathtt{Time}$—have decidable equality; see Proposition 5.19 and Corollary 5.8. □

We can now further unravel the definition of hybrid sheaf in this case, using Definition 4.11 and Corollary 8.8. The π-sheafification $W_G := \mathsf{sh}_\pi(W'_G) = \mathrm{Hyb}(V, E \times \mathtt{Time})$ is the type of functions $\phi : (V + E \times \mathtt{Time}) \to \mathtt{Prop}$ satisfying

1. $\forall(t : \mathtt{Time})(e : E).\, t < 0 \Rightarrow (\phi(e, t) \Leftrightarrow \phi(d(e)))$,
2. $\forall(t : \mathtt{Time})(e : E).\, 0 < t \Rightarrow (\phi(e, t) \Leftrightarrow \phi(c(e)))$,
3. $\forall(g : V + E \times \mathtt{Time}).\, \pi(\phi g) \Rightarrow \phi g$, and
4. $\pi\exists(g : V + E \times \mathtt{Time}).\, \phi = \{g\}$

where, as usual, $\{g\} : (V + E \times \mathtt{Time}) \to \mathtt{Prop}$ is the predicate sending g' to $(g = g') : \mathtt{Prop}$. Semantically, W_G is the sheaf of walks through G with finitely many transitions over any (finite-length) interval ℓ.

Example 8.9 It is often useful to know which edge transitions occur in a walk through a graph $G = (E \rightrightarrows V)$.[1] Given a section $\phi : W_G$ and a time $t : \mathtt{Time}$, consider the predicate

$$\mathtt{trav}_\phi : (\mathbb{R} \times E) \to \mathtt{Prop}, \qquad \mathtt{trav}_\phi(r, e) := @^t_{[r,r]}\big(\phi = \{(e, t - r)\}\big).$$

Semantically, this proposition is true on an interval $[d, u]$ if either r is outside the interval, $(r < d) \vee (u < r)$, or if at time r the walk ϕ is traversing edge e.

Example 8.10 Consider the terminal graph $T :=$. The corresponding hybrid datum $\mathtt{Time} = \mathtt{Time} \rightrightarrows 1$ is also the terminal object in the category of hybrid data (see Definition 8.6).

Walks through the terminal graph W_T gives a model for simple pulsing. The only information in a section of W_T is the finite set of instants at which pulses occur, i.e., at which the unique edge is traversed. This could be used to model musical or neuron-spiking behavior.

Proposition 8.11 *The above construction, taking a graph G and returning a behavior type W_G, is functorial.*

Proof We need to show that given a morphism of graphs, one obtains a morphism of behavior types, respecting identities and composition. We know by Proposition 8.7 that Hyb is functorial, so it suffices to see that our construction of the hybrid sheaf datum in Eq. (8.3) was functorial. But this is obvious: given $f : E \to E'$ and $g : V \to V'$, it is straightforward to check that the following diagram commutes:

[1] The construction of trav described here extends to hybrid sheaves in general; it is only a bit more complicated in the general case because we cannot assume $\mathrm{Hyb}_1(C, D)$ is π-separated.

$$\begin{array}{ccccc} \texttt{Time} & \xleftarrow{p_{\texttt{Time}}} & E \times \texttt{Time} & \overset{d \circ p_E}{\underset{c \circ p_E}{\rightrightarrows}} & V \\ \| & & \downarrow{\scriptstyle f \times \texttt{Time}} & & \downarrow{\scriptstyle g} \\ \texttt{Time} & \xleftarrow[p_{\texttt{Time}}]{} & E' \times \texttt{Time} & \overset{d \circ p_E}{\underset{c \circ p_E}{\rightrightarrows}} & V' \end{array}$$

□

Timed Random Walks

Fix $0 \leq a \leq b$; suppose we want random walks through G whose time between transitions is required to be between a and b. We can also construct such a sheaf internally using the hybrid type construction. Let G be as in Eq. (8.2), and use the following hybrid datum:

$$\begin{array}{ccc} E \times \{(r,t) : \mathbb{R} \times \texttt{Time} \mid (a < r < b) \wedge (-r < t < b - r)\} & \overset{src}{\underset{tgt}{\rightrightarrows}} & V \times \{t : \texttt{Time} \mid 0 < t < b\} \\ \downarrow{\scriptstyle \tau} & & \\ \texttt{Time} & & \end{array}$$

where $\tau(e, (r, t)) := t$, $src(e, (r, t)) := (src(e), t+r)$, and $tgt(e, (r, t)) = (tgt(e), t)$. The associated hybrid type will have the desired behavior.

8.2 Delays

Intuitively, a machine acts as a delay of type A if it has type $A \times A$ and has the property that its second projection produces the same section as its first projection, except translated in time by d seconds, for some real $0 < d : \mathbb{R}$. This notion of "same" belies an implicit assumption that sections $a_1, a_2 : A$ can be compared for equality at different intervals of time. However, this is not the semantics of equality: only sections over the same interval can be compared for equality.

Instead of comparing for equality, one could start with a morphism $\phi \colon \mathtt{C} \to \mathtt{Prop}^A$, where $\mathtt{C}$ is a constant sheaf. Then for every $c : \mathtt{C}$, the predicate $\phi(c)$ is a translation-invariant predicate on A, so we may apply it at different times and compare the results. We can thus speak of ϕ-theoretic delays. Namely, a_2 is the ϕ-theoretic d delay of a_1 if, for all $c : \mathtt{C}$ and $r : \mathbb{R}$, the section a_1 satisfies $\phi(c)$ on the open set $\Uparrow[0, r]$ iff a_2 satisfies $\phi(c)$ on the translated open set $\Uparrow[d, d + r]$.

Definition 8.12 (Delay) Let A be a type, and suppose given a constant sheaf $\mathtt{C}$, a predicate $\phi : (C \times A) \to \mathtt{Prop}$, and a number $D : \mathbb{R}$ with $D > 0$. Fix $t : \mathtt{Time}$. We define $(a, a') : A \times A$ to be *ϕ-theoretically D-delayed* if the following holds:

$$\forall (c : \mathtt{C})(d, u : \mathbb{R}).\ (d < u) \Rightarrow \left(\downarrow^t_{[d,u+D]} \Uparrow^t_{[d,u]} \phi(c, a) \Leftrightarrow \downarrow^t_{[d,u+D]} \Uparrow^t_{[d+D,u+D]} \phi(c, a') \right). \tag{8.5}$$

Say that $\mathtt{C}$ *separates A (via ϕ)*, if the following formula holds:

$$\forall(a_1, a_2 : A). (\forall(c : \mathtt{C}). \phi(c, a_1) \Leftrightarrow \phi(c, a_2)) \Rightarrow (a_1 = a_2).$$

If $\mathtt{C}$ separates A, we may drop mention of ϕ and say that (a, a') is *D-delayed* iff it satisfies Eq. (8.5).

By the torsor Axiom 4, being D-delayed is not dependent on the choice of t : `Time`. One particularly intuitive case of delays is for numeric types.

Lemma 8.13 *Let $\mathbb{R}_j$ be the type of j-local real numbers, for a modality j. Then $\mathbb{Q} \times \mathbb{Q}$ separates R, via the predicate $\phi(q_1, q_2, r) := (q_1 < r < q_2)$. A similar statement holds when $\mathbb{R}_j$ is replaced by any of the other j-numeric types.*

Proof For any pair of lower reals, $\delta_1, \delta_2 : \underline{\mathbb{R}}_j$, it is obvious by propositional extensionality (see Axiom 0) that $\forall(q : \mathbb{Q}). (\delta_1 q \Leftrightarrow \delta_2 q)) \Rightarrow (\delta_1 = \delta_2)$. The same holds for the other numeric types. □

Example 8.14 Let's work out an example of the semantics of delays for the case $\mathbb{R}_\pi$, because its semantics are most familiar; the other numeric types are similar. Choose $D \in \mathbb{R} = [\![\mathbb{R}]\!](\ell)$ with $D > 0$ and $t = (d_t, u_t) \in [\![\mathtt{Time}]\!](\ell)$. Recall from Corollary 7.22 (or more generally Theorem 7.19) that elements $f, f' \in [\![\mathbb{R}_\pi]\!](\ell)$ can be identified with continuous real-valued functions $f, f' \colon (d_t, u_t) \to \mathbb{R}$.

We want to understand the semantics of the predicate ϕ defined in Lemma 8.13 saying that (f, f') are D-delayed:

$$\ell \Vdash \forall(q_1, q_2 : \mathbb{Q})(d, u : \mathbb{R}). (d < u) \Rightarrow \\ \left(\downarrow_{[d,u+D]}\uparrow_{[d,u]} (q_1 < f < q_2) \Leftrightarrow \downarrow_{[d,u+D]}\uparrow_{[d+D,u+D]} (q_1 < f' < q_2)\right). \tag{8.6}$$

Our *goal* is to show that Eq. (8.6) holds iff $f(x) = f'(x + D)$ for all $d_t < x < u_t - D$.

Equation (8.6) holds iff for every $q_1, q_2 \in \mathbb{Q}$ and $d < u$ in $\mathbb{R}$, we have

$$\ell \Vdash \left(\downarrow_{[d,u+D]}\uparrow_{[d,u]} (q_1 < f < q_2)\right) \Leftrightarrow \left(\downarrow_{[d,u+D]}\uparrow_{[d+D,u+D]} (q_1 < f' < q_2)\right).$$

One can show that this is the case iff we have

$$\ell \Vdash \downarrow^{(d_t,u_t)}_{[d,u+D]}\uparrow^{(d_t,u_t)}_{[d,u]} \left(q_1 < f\big|_{(d_t,u_t)} < q_2\right) \quad \text{iff}$$
$$\ell \Vdash \downarrow^{(d_t,u_t)}_{[d,u+D]}\uparrow^{(d_t,u_t)}_{[d+D,u+D]} \left(q_1 < f'\big|_{(d_t,u_t)} < q_2\right).$$

By the semantics of $\downarrow$ (Corollary 6.13), we see that if $d < d_t$ or $u_t < u + D$ then the above equivalence is vacuously satisfied. In other words, if $u_t - D \leq d_t$, then we already have achieved our goal. Thus we may suppose that $d_t \leq d < u \leq u_t - D$, and Eq. (8.6) has become

$$\ell \Vdash \uparrow^{(d_t,u_t)}_{[d,u]} \left(q_1 < f\big|_{(d_t,u_t)} < q_2\right) \quad \text{iff} \quad \ell \Vdash \uparrow^{(d_t,u_t)}_{[d+D,u+D]} \left(q_1 < f'\big|_{(d_t,u_t)} < q_2\right).$$

The left-hand side is equivalent to the following: for all $\langle r, s\rangle : \ell' \to \ell$, if $d < d_t + r < u_t - s < u$, then $q_1 < f\big|_{(d_t+r,u_t-s)} < q_2$. Similarly for the right, except d and u are replaced by $d + D$ and $u + D$. Thus Eq. (8.6) has become

$$\forall (d, u \in \mathbb{R}).\, (d_t \le d < u \le u_t - D) \Rightarrow \Big(q_1 < f\big|_{(d,u)} < q_2 \quad \text{iff} \quad q_1 < f'\big|_{(d+D,u+D)} < q_2\Big).$$

We are trying to show that this holds iff $f(x) = f'(x + D)$ for all $d_t < x < u_t - D$, and that follows from the continuity of f and f'.

8.3 Ordinary Differential Equations, Relations, and Inclusions

Consider a system of ordinary differential equations

$$\begin{aligned} \dot{x}_1 &= f_1(x_1, \ldots, x_n, a_1, \ldots, a_m) \\ \dot{x}_2 &= f_2(x_1, \ldots, x_n, a_1, \ldots, a_m) \\ &\vdots \\ \dot{x}_n &= f_n(x_1, \ldots, x_n, a_1, \ldots, a_m) \end{aligned} \tag{8.7}$$

In this setting, one generally considers each variable as a continuous function of time and the x_i as differentiable. Thus we may consider each variable as a π-local real, $x_i : \mathbb{R}_\pi$, $a_j : \mathbb{R}_\pi$, the semantics of which are continuous curves in $\mathbb{R}$; see Corollary 7.22. Our internal notion of differentiability is defined in Definition 7.31 and its semantics is indeed differentiability in the usual sense; see Corollary 7.45.

Thus we may interpret Eq. (8.7) entirely within the internal language of the topos $\mathcal{B}$. In fact, such systems can be generalized a great deal. For one thing, we can replace the system with a differential relation or a differential inclusion. Indeed, a differential relation is a predicate $P : (\mathbb{R}_\pi)^{2n+m} \to \texttt{Prop}$, whose solution set is

$$\{(x, a) : (\mathbb{R}_\pi)^n \times (\mathbb{R}_\pi)^m \mid P(x, \dot{x}, a)\}.$$

Requiring derivatives to remain within certain bounds is a common example of this notion, and—in case it was not clear—so is Eq. (8.7), when $P(x, x', a)$ is taken to be the proposition $x' = f(x, a) : \texttt{Prop}$.

In fact, the functions x_i, a_i, and f can even represent interval-valued functions. Because our internal definition of derivative is quite broadly defined, Eq. (8.7) will continue to make sense. It could also be generalized to something like $f(x, a) \sqsubseteq \dot{x}$, meaning the derivative is bounded by the interval $f(x, a)$; see Proposition 7.44. This may be the most natural form of differential inclusions, though differential relations as above cover the rest of them.

8.4 Systems, Components, and Behavior Contracts

8.4.1 Machines and Interfaces

We imagine machines—or systems—as inhabiting interfaces, which mark the boundary between system and environment. The interface is a collection of ports, each labeled with a behavior type. For example, here are pictures of some interfaces with varying numbers of ports:

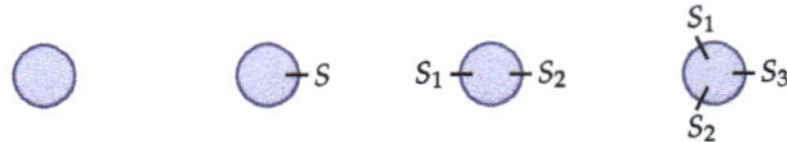

A machine inhabits an interface, and it has its own behavior type, which we call the *total behavior type*, as well as a map to the each port's behavior type. For example, a machine inhabiting the interface on the right might have total behavior type X and maps $p_i : X \to S_i$ for $i = 1, 2, 3$, which we call *port maps*. The tuple $\mathcal{X} = (X, p_1, p_2, p_3)$ defines the machine.

Example 8.15 Below in Sect. 8.4.4, we will show how one can interpret a labeled transition system as a machine. Here we show how to interpret a system of ODEs, e.g., Eq. (8.7), as a machine. For that system, there would be m ports, corresponding to the time-varying parameters $a_1, \ldots, a_m$. Each would have type $\mathbb{R}_\pi$, assuming we are expecting continuous signals. The machine would have total type

$$X = \{(x, a) : \mathbb{R}_\pi^{m+n} \mid \dot{x}_1 = f_1(x, a) \wedge \cdots \wedge \dot{x}_n = f_n(x, a)\}.$$

The port maps $p_i : X \to \mathbb{R}_\pi$ for $1 \leq i \leq m$ are the projections.

8.4.2 Systems and Behavior Contracts

A *behavior contract* on an interface $S_1, \ldots, S_n$ is a predicate on $S_1 \times \cdots \times S_n$. We prefer to think of it as a proposition in context $\Gamma = (s_1 : S_1, \ldots, s_n : S_n)$,

$$s_1 : S_1, \ldots, s_n : S_n \vdash \phi(s_1, \ldots, s_n) : \texttt{Prop}.^2$$

If the proposition $\forall(x : X).\,\phi\big(p_1(x), \ldots, p_n(x)\big)$ holds, we say that machine $\mathcal{X} = (X, p_1, \ldots, p_n)$ *validates* ϕ. However, for our work of proving properties of systems, the machine itself is often irrelevant; all that matters is its behavior contract it satisfies, namely $\Gamma \vdash \phi$.

A *system* consists of

- a finite number $n \in \mathbb{N}$ of *components*,
- for each $1 \le i \le n$ a context Γ_i, called the *interface* of the component,
- for each $1 \le i \le n$ a valid truth judgment $\Gamma_i \vdash \phi_i$, called the *behavior contract*,
- an *outer context* Γ',
- a *total context* Γ, and
- mappings (possibly with variable re-namings) $\rho_i : \Gamma_i \to \Gamma$, for each $1 \le i \le n$, and $\rho' : \Gamma' \to \Gamma$.

In other words, every variable $s : S$ in context Γ_i or Γ' is sent to some variable $t : S$ of the same type (S) in Γ. Let $S_1, \ldots, S_k$ be the set of types in Γ but not in the image of ρ'; we call these the *latent variables* of the system. Conversely, types in the image of ρ' are called *exposed variables*.

The variables in ϕ_i can be renamed via ρ_i, so we have $\Gamma \vdash \phi_i$ for each i. We say that the system satisfies outer contract $\Gamma' \vdash \phi'$ iff we have

$$\Gamma' \vdash \forall(s_1 : S_1) \cdots (s_k : S_k).\,[(\phi_1 \wedge \cdots \wedge \phi_n) \Rightarrow \phi']. \tag{8.8}$$

We will use this notion for our case study in Sect. 8.5. The strongest contract Φ' satisfied by the system is $\Phi' := \exists(s_1 : S_1) \cdots (s_k : S_k).\,\phi_1 \wedge \cdots \wedge \phi_n$. Indeed, we have $\Gamma' \vdash \forall(s_1 : S_1) \cdots (s_k : S_k).\,[(\phi_1 \wedge \cdots \wedge \phi_n) \Rightarrow \Phi']$ and if ϕ' also satisfies Eq. (8.8), then $\Gamma' \vdash \Phi' \Rightarrow \phi'$.

Example 8.16 Below is a picture of a system with four components:

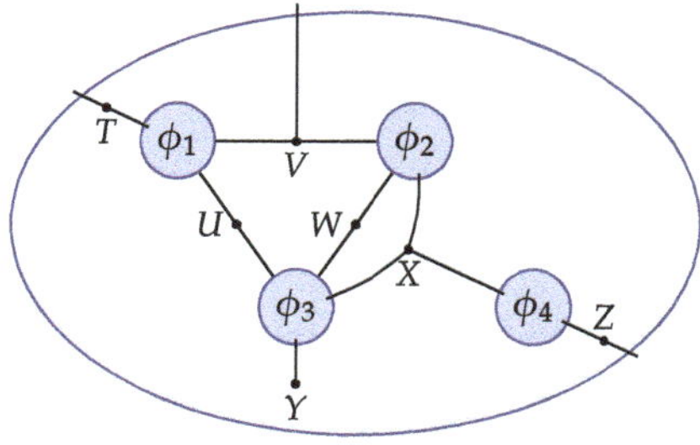

[2] A term in context, such as $s_1 : S_1, \ldots, s_n : S_n \vdash \phi(s_1, \ldots, s_n) : \texttt{Prop}$ is roughly the same as a formula $\phi : (S_1 \times \cdots \times S_n) \to \texttt{Prop}$. The main difference is that, in the former case, the variables have been named.

The four components have interfaces $\Gamma_1 = (t_1 : T, u_1 : U, v_1 : V)$; $\Gamma_2 = (v_2 : V, w_2 : W, x_2 : X)$; $\Gamma_3 = (u_3 : U, w_3 : W, x_3 : X, y_3 : Y)$, and $\Gamma_4 = (x_4 : X, z_4 : Z)$. The behavior contracts are $\Gamma_i \vdash \phi_i$ for each $1 \le i \le 4$. The outer context is $\Gamma' = (t' : T, v' : V, z' : Z)$. The total context is $\Gamma = (t : T, u : U, v : V, w : W, x : X, y : Y, z : Z)$. The mappings ρ_i and ρ' are the obvious "letter-preserving" functions; for example, $\rho_4(x_4) = x$ and $\rho_4(z_4) = z$. The strongest contract satisfied by the system is

$$t' : T, v' : V, z' : Z \vdash \exists (u : U)(v : V)(w : W)(x : X)(y : Y).$$
$$\phi_1(t', u, v) \wedge \phi_2(v', w, x) \wedge \phi_3(u, w, x, y) \wedge \phi_4(x, z').$$

8.4.3 *Control-Theoretic Perspective*

In a control-theoretic setting, some of the ports of an interface are designated as inputs and others are designated as outputs. In this setting, we might draw interfaces as follows:

where ports on the left of a box are inputs, and those on the right are outputs.

The idea is that input trajectories deterministically drive the total system. Slightly more precisely, let $i : X \to S$ be the input map. The condition is that for every tuple consisting of an initial state s of the system and a trajectory on each input port, each of which is compatible with s at the outset, there exists a unique trajectory of the system that extends the initial state. We might say that a machine satisfying the existence property is *total* and one satisfying the uniqueness property is *deterministic*.

Each of these can be stated in the internal language of $\mathcal{B}$. It is enough to choose $\epsilon > 0$ and ask that the total/deterministic conditions apply for extensions of length ϵ. Here is an internal notion of totalness:

$$\forall (t : \mathtt{Time})(P : X \to \mathtt{Prop})(s : S)(r : \mathbb{R}).\, r > 0 \Rightarrow$$
$$\Big(\uparrow_{[0,r]} \exists (x : X).\, (ix = s) \wedge Px\Big) \Rightarrow \uparrow_{[0,r+\epsilon]} \exists (x : X).\, (ix = s) \wedge \uparrow_{[0,r]} Px.$$

And here is an internal notion of determinacy:

$$\forall (t : \mathtt{Time})(x, x' : X)(r : \mathbb{R}).\, r > 0 \Rightarrow \big(\uparrow_{[0,r]}(x = x')\big)$$
$$\Rightarrow (ix = ix') \Rightarrow \uparrow_{[0,r+\epsilon]}(x = x')$$

In [SVS16] it is shown that when machines that are total, deterministic, and what the authors call "inertial"[3] are composed into systems by feeding output ports into input ports in arbitrary ways, the result is again total, deterministic, and inertial. This proof can be carried out within the internal language of $\mathcal{B}$, although we do not do so here.

8.4.4 Labeled Transition Systems

In this section, we apply the framework with the hybrid sheaf formalism of Sect. 8.1 to understand how labeled transition systems can be interpreted in the temporal type theory. Here is the definition of labeled transition system we use.

Definition 8.17 A *labeled transition system* consists of several components:

- a set Λ, elements of which are called *input tokens*;
- a set V, elements of which are called *states*;
- a set E, elements of which are called *transitions*;
- two function $s, t\colon E \to V$, called the *source* and *target* functions; and
- a function $\lambda\colon E \to \Lambda$, called the *edge label function*.

We want to interpret a labeled transition system as a machine, in the sense of Sect. 8.4.1. Its interface will look like this I-○-O or this I-▭-O. In order to construct a behavior type for the ports I and O, as well as the total behavior type of the machine, we must decide on what that behavior will be. In other words, we need to choose a way to incorporate the components of Definition 8.17 into a temporal story of the type of behavior it can exhibit. Consider the following story:

> Signals—in the form of input tokens—are appearing on the ports I of the machine. For example, something is barking "climb!" or "stay level!", and the machine is receiving these signals at arbitrary moments in time. The machine is almost always in some state $v \in V$; the only exception is that each time an input token $l \in \Lambda$ appears, the machine goes into an instantaneous transition e, namely one with label $\lambda(e) = l$ and source $s(e) = v$. Immediately afterwards, it is in the target state $t(e)$. Whatever state the machine is in, it outputs its state continuously.

[3]The inertiality condition for an output map $p : X \to S'$ says that every internal trajectory produces a guaranteed output trajectory of a slightly duration. This property can be stated internally as follows:

$$\forall(t:\texttt{Time})(P:X\to\texttt{Prop})(r:\mathbb{R}).\, r>0 \Rightarrow \left(\uparrow_{[0,r]}\exists(x:X).\,Px\right)$$
$$\Rightarrow \uparrow_{[0,r+\epsilon]}\exists(s':S').\,\left(\left(\uparrow_{[0,r]}\exists(x:X).\,Px\wedge(px=s')\right)\wedge\forall(x:X)\right.$$
$$\left.\,.\left(\uparrow_{[0,r]}Px\wedge(px=s')\right)\Rightarrow(px=s')\right).$$

From the above story, we should take the input behavior type I to be that of walks through the graph $E \rightrightarrows \{*\}$; see Sect. 8.1.2. Semantically, a walk of length ℓ is a finite path in this graph, which is just a list of elements of E, together with a positive duration of time between each. For example, if $E = \{\texttt{climb!}, \texttt{level!}\}$, then perhaps we could write

$$*(0.5) \quad \lightning_{\texttt{climb!}} \quad *(3.3) \quad \lightning_{\texttt{level!}} \quad *(0.1) \quad \lightning_{\texttt{level!}} \quad *(5.0) \quad \lightning_{\texttt{climb!}} \quad *(1.1)$$

to denote a certain walk of length 10. It rests for 0.5 s, instantly transitions along the edge $\texttt{climb!}$, rests for 3.3 s, instantly transitions along $\texttt{level!}$, etc. Each of the waiting periods occurs on the unique vertex, *, representing silence.

We should take the output behavior O to be that of walks through the graph $V \times V \rightrightarrows V$, i.e., the complete graph on V. A walk of length ℓ is a finite path in this graph, which is just a list of elements of V, together with a positive duration of time assigned to each. For example, if $V = \{a, b, c\}$, then perhaps we could write

$$a(2.4) \quad \lightning_{(a,b)} \quad b(1.1) \quad \lightning_{(b,b)} \quad b(2.0) \quad \lightning_{(b,a)} \quad a(3.3) \quad \lightning_{(a,c)} \quad c(0.8) \quad \lightning_{(c,b)} \quad b(0.4)$$

to denote a certain walk of length 10. It rests at a for 2.4 s, transitions along the unique edge to b, where it rests for 1.1 s, etc.

We take the total behavior X of the machine to be that of walks through the graph $s, t \colon E \rightrightarrows V$ given by the source and target functions. The last step needed to show that this machine inhabits the interface (see Sect. 8.4.1) is to give maps $X \to I$ and $X \to O$. By Proposition 8.11, we know that the random-walks construction is functorial. Hence, it suffices to give graph homomorphisms between them, and we do so below:

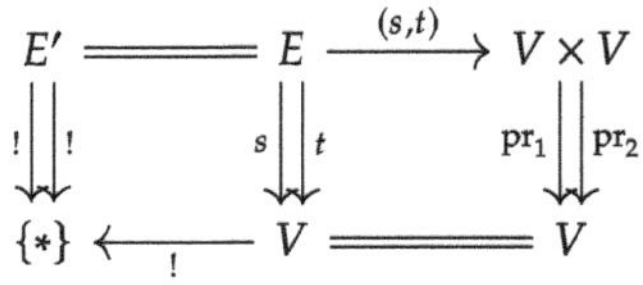

$$\begin{array}{ccccc} E' & = & E & \xrightarrow{(s,t)} & V \times V \\ {\scriptstyle !}\downdownarrows{\scriptstyle !} & & {\scriptstyle s}\downdownarrows{\scriptstyle t} & & {\scriptstyle \mathrm{pr}_1}\downdownarrows{\scriptstyle \mathrm{pr}_2} \\ \{*\} & \xleftarrow[!]{} & V & = & V \end{array}$$

Thus we have constructed a machine whose semantics matches the story above.

8.5 Case Study: The National Airspace System

In this section we will give a toy example, in order to see our temporal type theory—and its higher-order temporal logic—in action. Our goal is to describe very different sorts of systems—an ordinary differential equation, a time delay, and a labeled transition system—displaying behaviors that range from continuous to discrete, instantaneous to delayed. Each will be summarized by a behavior contract (a proposition in context), and we will combine them to prove a property of the system formed by their interaction.

The particular example we choose to model is that of "safe separation in the National Airspace System (NAS)" which we learned of by teaming up on a NASA grant with researchers at Honeywell Labs (see "Acknowledgements" section of Chap. 1). Airplanes, their pilots and onboard equipment, as well as radars and many other factors interact to form the NAS. One property of this system that needs to hold constantly is that of *safe separation*: that airplanes maintain a safe distance from each other. This is the system property we will prove from properties of components. Our proofs—as well as our descriptions of component properties—will take place in the higher-order logic of our sheaf topos $\mathcal{B} \cong \mathsf{Shv}(S_{\mathbb{IR}/\rhd})$, described in earlier chapters.

8.5.1 Background and Setup

We do a case study involving several disparate behavior contracts. In the National Airspace System, various systems like the Traffic Collision Avoidance System (TCAS) interact to keep airplanes safely separated. Here are wiring diagrams that represent some aspects of this system, both more globally and in a single airplane:

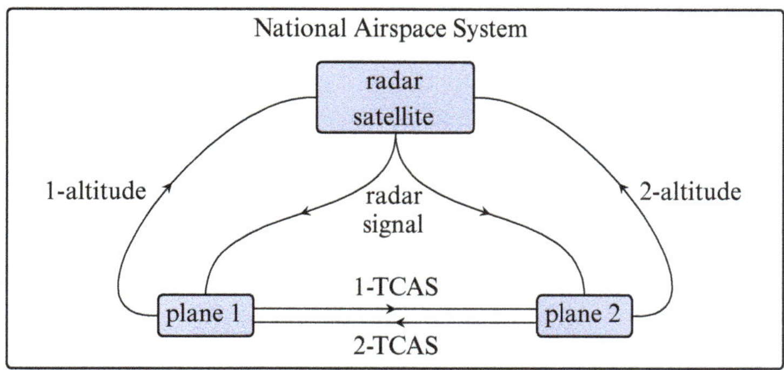

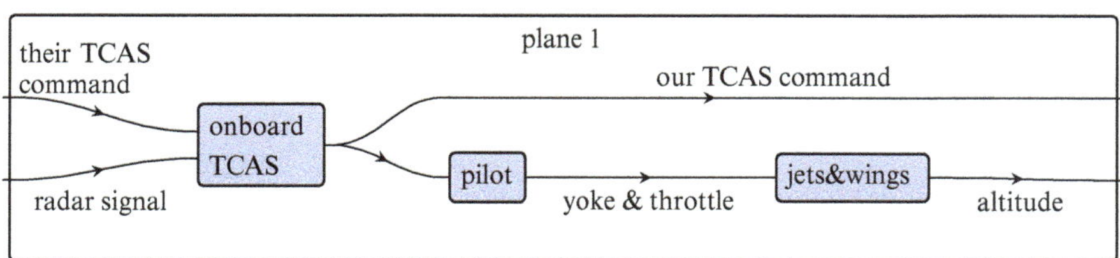

Both the radar and the other airplane's TCAS send a signal to our airplane, which is picked up by our TCAS. If there is any danger that safe separation will be violated, the TCAS alerts the pilot, who then uses the yoke (steering wheel) and throttle (gas pedal) to send a command to the plane. The jets and wings of the plane adjust accordingly to change the altitude.

Although this is already quite simplified, we will simplify even more. We imagine there is only one airplane, and it is trying to attain a safe altitude. To go from the scenario described above to this one, one could take the difference of the two airplane's altitudes. We pretend that the TCAS single-handedly measures the plane's altitude, tells the pilot if it is too low, and the pilot sends the message on to the plane's motors. In fact, we assume the pilot acts as a mere delay.

Here is the wiring diagram we will use:

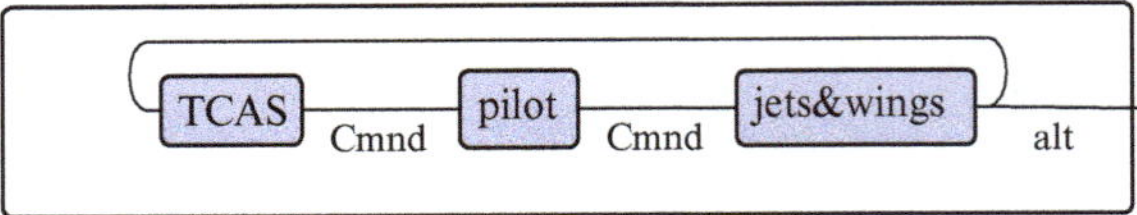

To each wire we will assign a sheaf, and two each box we will assign a behavior contract, or proposition, on its ports. We will then combine these contracts and prove that the whole system satisfies safe separation.

In our toy model, the altitude as a function of time is the only factor we will consider, and we denote it by a. The TCAS is tasked with controlling the plane so that it will be at a safe altitude, an altitude we denote by $\texttt{safe}$. We imagine that the plane can only do one of the two things: fly level or climb. The pilot commands the plane to either fly level or climb, and the plane responds instantly by assigning its velocity $\dot{a}$ to either 0 or a number called $\texttt{rate}$. This command by the pilot is in fact initiated by the TCAS, and the pilot delivers the command to the airplane after a delay of some number of seconds, a number we denote $\texttt{delay}$. The TCAS chooses its command according to the rule that if the altitude of the plane is less than some threshold, it sends the command to $\texttt{climb}$; if it is greater than this value it sends the command to stay $\texttt{level}$. The threshold is $\texttt{safe} + \texttt{margin}$, where $\texttt{margin} > 0$.

8.5.2 Toy Model and Proof of Safety

All of the proofs in this section take place in the temporal type theory laid out in Chaps. 4 and 5.

Take Cmnd $:= W_G$ to be the sheaf of walks (as in Sect. 8.1.2) through the graph G, drawn below, which has vertex set $\text{Cmnd}_0 = \{\texttt{level}, \texttt{climb}\}$ and edge set $E = \{\texttt{level!}, \texttt{climb!}\}$:

$G :=$ 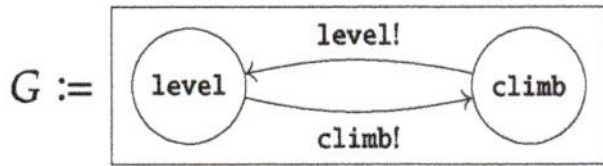

So Cmnd is the π-sheafification of a quotient of $\text{Cmnd}_0 + (E \times \texttt{Time})$, where Cmnd_0 is considered as the constant sheaf on two elements. In particular there is a map of sheaves $\text{Cmnd}_0 \to \text{Cmnd}$. The pilot will act as a delay that takes TCAS signals and returns commands to the yoke of the plane. The sheaf of altitudes is just the variable reals, $\mathbb{R}_\pi$.

Suppose we are given four positive rational numbers: $\texttt{safe}, \texttt{margin}, \texttt{del}, \texttt{rate} : \mathbb{Q}_{>0}$, corresponding to the safe altitude, the extra margin of safety provided by the TCAS, the pilot delay, and the ascent rate of the plane. Take Γ to be the type context

$$\Gamma := t : \texttt{Time}, T : \text{Cmnd}, P : \text{Cmnd}, a : \mathbb{R}_\pi, \texttt{safe} : \mathbb{Q}, \texttt{margin} : \mathbb{Q}, \texttt{del} : \mathbb{Q}, \texttt{rate} : \mathbb{Q}.$$

Behavior Contracts for the Toy Model

We will add four axioms $\Theta = \theta_1, \theta_2, \theta_3, \theta_4$ to the context Γ; these are the contracts satisfied by the various components. The first contract says that $\texttt{margin}$ is a positive rational number and altitude $a : \mathbb{R}_\pi$ is nonnegative:

$$\theta_1 := (\texttt{margin} > 0) \wedge (a \geq 0).$$

Second, we have the TCAS contract:

$$\theta_2 := (a > \texttt{safe} + \texttt{margin} \Rightarrow T = \texttt{level}) \wedge (a < \texttt{safe} + \texttt{margin} \Rightarrow T = \texttt{climb}).$$

Recall that in Definition 7.25 we defined the derivative for $a : \mathbb{R}_\pi$, denoted $\dot{a} : \bar{\underline{\mathbb{R}}}_\pi^\infty$. The airplane contract says that the instantaneous change in altitude is determined by the yoke:

$$\theta_3 := (P = \texttt{level} \Rightarrow \dot{a} = 0) \wedge (P = \texttt{climb} \Rightarrow \dot{a} = \texttt{rate}).$$

The delay contract is as in Definition 8.12 for $\phi : (\text{Cmnd}_0 \times \text{Cmnd}) \to \texttt{Prop}$ given by $\phi(v, c) := (v = c)$:

$$\theta_4 := \forall(d, u : \mathbb{R})(v : \text{Cmnd}_0).\, (d < u) \Rightarrow \Big(\downarrow_{[d, \texttt{delay}+u]} \uparrow_{[d,u]} (T = v) \Leftrightarrow \downarrow_{[d, \texttt{delay}+u]} \uparrow_{[d+\texttt{delay}, u+\texttt{delay}]} (P = v)\Big).$$

Proof of Safety

We take as a premise the four axioms $\Theta := \theta_1, \theta_2, \theta_3, \theta_4$ discussed above. We will prove that these four axioms are enough to guarantee safe separation. The rest of this section takes place in the logical context $\Gamma \mid \Theta$. In other words, we can freely refer to T,P,a,safe, margin, del, or rate as well as to axioms θ_1, θ_2, θ_3, and θ_4.

For any $a : \mathbb{R}_\pi$ and $r : \mathbb{R}$, recall the notation $a^@(r)$ from Eq. (7.2).

Lemma 8.18 *Given the contracts above, the plane's altitude $a : \mathbb{R}_\pi$ never decreases:*

$$\Gamma \mid \Theta \vdash \forall (t : \mathtt{Time})(r_1, r_2 : \mathbb{R}).\, r_1 < r_2 \Rightarrow a^@(r_1) \leq a^@(r_2).$$

It follows that for any t : Time and $r : \mathbb{R}$ we have the following:

$$q < a^@(r) \Rightarrow r < t \Rightarrow q < a \qquad \textit{and} \qquad a^@(r) < q \Rightarrow t < r \Rightarrow a < q.$$

Proof By Proposition 8.5, the command P : Cmnd is almost always a vertex,

$$\neg\neg(P = \mathtt{climb} \vee P = \mathtt{level}).$$

Since $(P = \mathtt{climb} \vee P = \mathtt{level}) \Rightarrow \dot{a} \geq 0$, by θ_3, and since $(\dot{a} \geq 0) = \neg(\dot{a} < 0)$ is a closed proposition for the $\neg\neg$ modality, we have shown $\dot{a} \geq 0$. By Proposition 7.37 we obtain $0 \leq a^@(r_2) - a^@(r_1)$. The result follows from arithmetic in $\mathbb{R}_{[r_1,r_2]}$; see Sect. 4.3.7.

The second statement follows from the first by Axiom 4 and Proposition 5.31. □

Recall from Proposition 6.12 that $\downarrow_0(P) = \downarrow_{[0,0]}(P)$ semantically means that for every interval of time $[d_t, u_t]$, if $d_t \leq 0 \leq u_t$, then P is true; we speak of this by saying "if time 0 is witnessed, P holds." We can now state and prove the safety of our system.

Proposition 8.19 *Let $M = \mathtt{delay} + \frac{\mathtt{safe}}{\mathtt{rate}}$. If time 0 is witnessed, then at all times $t > M$ the altitude of the airplane will be safe:*

$$\Gamma \mid \Theta \vdash \forall (t : \mathtt{Time}).\, \downarrow_0(t > M \Rightarrow a \geq \mathtt{safe}).$$

Proof Let $N = \frac{\mathtt{safe}}{\mathtt{rate}}$, so $M = N + \mathtt{delay}$. As a variable real, $a : \mathbb{R}_\pi$ is π-located (see Definition 4.21). Thus it is $@_N$-located, so since $\mathtt{margin} > 0$ (by θ_1), we have

$$@_N(\mathtt{safe} < a \vee a < \mathtt{safe} + \mathtt{margin})$$

The modality $@_N$ commutes with disjunction by Lemma 5.32, so we break into cases. If $@_N(\mathtt{safe} < a)$, then by Lemma 8.18 we get $N < t \Rightarrow \mathtt{safe} < a$ and hence $\downarrow_0(M < t \Rightarrow \mathtt{safe} \leq a)$.

For the second case, we assume $@_N(a < \mathtt{safe} + \mathtt{margin})$. Again by Lemma 8.18 we obtain $t < N \Rightarrow (a < \mathtt{safe}+\mathtt{margin})$. By θ_2 we have $t < N \Rightarrow T = \mathtt{climb}$, which implies $\uparrow_{[0,N]}(T = \mathtt{climb})$. By θ_4, since $0 < N$, this implies

$\downarrow_{[0,M]}\uparrow_{[\mathtt{delay},M]}(P = \mathtt{climb})$, which implies $\downarrow_{[0,M]}\uparrow_{[\mathtt{delay},M]}(\dot{a} = \mathtt{rate})$ by θ_3. By Proposition 7.37, this implies

$$\downarrow_{[0,M]}\big(\mathtt{safe} \leq a^{@}(M) - a^{@}(\mathtt{delay})\big)$$

because $\mathtt{safe} = \mathtt{rate}{*}(M - \mathtt{delay})$. Since $a \geq 0$ by θ_1, we have $a^{@}(\mathtt{delay}) \geq 0$ by Proposition 4.54, and hence $\downarrow_{[0,M]}@_M(a \geq \mathtt{safe})$. By Proposition 5.39, this is equivalent to $\downarrow_{[0,0]}\downarrow_{[M,M]}@_M(a \geq \mathtt{safe})$, which implies $\downarrow_0@_M(a \geq \mathtt{safe})$. One then uses Lemma 8.18 to conclude. □

8.6 Relation to Other Temporal Logics

Consider the temporal logic with the "until" and "since" operators, TL(Until, Since). This is the standard Boolean propositional logic augmented with two new connectives. Given a set Σ of atomic propositions, the set of propositional formulas is defined by the grammar

$$F := \top \mid \bot \mid P \mid \neg F \mid F_1 \wedge F_2 \mid F_1 \vee F_2 \mid F_1 \,\mathcal{U}\, F_2 \mid F_1 \,\mathcal{S}\, F_2,$$

where $P \in \Sigma$.

Intuitively, $P \,\mathcal{U}\, Q$ means that Q will be true at some time in the future and that P will be true until then. Likewise $P \,\mathcal{S}\, Q$ means that Q was true at some time in the past and that P has been true since then. To make descriptions like these precise, the meaning of temporal logic operators is often defined in terms of another—first-order—logic. For example, one could express the meaning of $P \,\mathcal{U}\, Q$ as follows:

$$(P \,\mathcal{U}\, Q)(t_0) := \exists t.\, (t_0 \leq t) \wedge Q(t) \wedge \forall t'.\, (t_0 \leq t' \leq t) \Rightarrow P(t'). \tag{8.9}$$

The first-order logic being used here is called the "First-Order Monadic Logic of Order,"[4] also denoted FO(<). This is the standard first-order logic with a single binary relation <, with the restriction that all predicates are unary ("monadic"). Then Eq. (8.9), together with the evident analogue for $\mathcal{S}$, gives an embedding of TL(Until, Since) into FO(<). In fact, Kamp's theorem—one of the earliest major results in the study of temporal logic (then called "tense logic")—shows that TL(Until, Since) is "expressively complete" for FO(<) in any Dedekind-complete linear-time semantics.

There is a—more or less obvious—embedding of FO(<) into the temporal type theory presented in this book. Each (unary) predicate P of FO(<) is represented by

[4]"Monadic" here refers to the restriction that all predicates must be unary, and has no connection to monads in the sense of category theory. The only time we use "monadic" in this sense is when discussing other temporal logics.

a term $P\colon \texttt{Time} \to \texttt{Prop}$. As an example of the embedding, Eq. (8.9) would be represented in our type theory as

$$(P\,\mathcal{U}\,Q)(t_0) := \exists(t : \texttt{Time}).\,(t_0 \leq t) \wedge Q(t) \wedge \forall(t' : \texttt{Time}).\,(t_0 \leq t' \leq t) \Rightarrow P(t'). \tag{8.10}$$

However, this embedding would not be sound for the *classical* TL(Until, Since), since the predicate given in (8.10) is not decidable. There are two options to fix this. First, one could be content with an embedding of a constructive version of TL(Until, Since). But as constructive logic is strictly more expressive than classical logic, there is no impediment to giving a sound embedding of classical TL(Until, Since)—our second option. To do this, we first require that all atomic propositions are represented as decidable predicates, e.g., we assume $\forall(t : \texttt{Time}).\,(P(t) \vee \neg P(t)) \wedge (Q(t) \vee \neg Q(t))$. Then we represent $\mathcal{U}$ as

$$\begin{aligned}(P\,\mathcal{U}\,Q)(t_0) &:= \neg\neg\exists(t : \texttt{Time}).\,(t_0 \leq t) \wedge Q(t) \wedge \forall(t' : \texttt{Time}).\,(t_0 \leq t' \leq t)\\ &\Rightarrow P(t') \Leftrightarrow \neg\forall(t : \texttt{Time}).\,((t_0 \leq t) \wedge \forall(u : \texttt{Time}).\,(t_0 \leq u \leq t)\\ &\Rightarrow P(u)) \Rightarrow \neg Q(t),\end{aligned}$$

where the second line is constructively equivalent to the first (hence the two are provably equivalent in our type theory).

The other standard temporal operators are definable in terms of $\mathcal{U}$, but we give explicit representations for some of them for concreteness:

$$\begin{aligned}(\bigcirc P)(t_0) &:= P(t_0 + 1)\\ (\Box P)(t_0) &:= \forall(t : \texttt{Time}).\,(t_0 \leq t) \Rightarrow P(t)\\ (\Diamond P)(t_0) &:= \neg\neg\exists(t : \texttt{Time}).\,(t_0 \leq t) \wedge P(t)\\ &\Leftrightarrow \neg\forall(t : \texttt{Time}).\,(t_0 \leq t) \Rightarrow \neg P(t)\\ (P\,\mathcal{R}\,Q)(t_0) &:= \forall(t : \texttt{Time}).\,((t_0 \leq t) \wedge \forall(u : \texttt{Time}).\,(t_0 \leq u \leq t)\\ &\Rightarrow \neg P(u)) \Rightarrow Q(t).\end{aligned}$$

The standard equivalences between these operators

$$\Diamond P \Leftrightarrow \neg(\Box\neg P) \qquad P\,\mathcal{U}\,Q \Leftrightarrow \neg(\neg P\,\mathcal{R}\,\neg Q).$$

are provable in the type theory.

In practice, one often wants to reason quantitatively about time, for example, to say that some proposition will be satisfied between 3 and 5 min from now. One well-studied quantitative temporal logic is metric temporal logic (MTL). In [HOW13], Kamp's theorem was extended to this quantitative setting by introducing the first-order monadic logic of order and metric (FO($<, +\mathbb{Q}$)), and showing that MTL is complete for FO($<, +\mathbb{Q}$). This augmented first-order logic simply adds functions

$+q$ to FO($<$) for all $q \in \mathbb{Q}$. For example, the metric "until" connective U_I of MTL, where $I = (d, u) \in \mathbb{Q} \times \mathbb{Q}$ is an interval, is encoded in FO($<, +\mathbb{Q}$) by

$$(P\ U_I\ Q)(t_0) := \exists t.\,((t_0 + d < t < t_0 + u) \wedge Q(t) \wedge \forall u.\,(t_0 < u < t \Rightarrow P(u)))\,.$$

The type `Time` in our system is a torsor over $\mathbb{R}$, and in particular there is an addition map $+\colon \mathtt{Time} \times \mathbb{Q} \to \mathtt{Time}$. Hence the embedding of FO($<$) is easily extended to FO($<, +\mathbb{Q}$). For example, the metric until connective U_I above is represented (in the classical embedding) by

$$\begin{aligned}(P\ U_I\ Q)(t_0) := \neg\neg\exists(t : \mathtt{Time}).\,((t_0 + d < t < t_0 + u) \wedge \\ Q(t) \wedge \forall(u : \mathtt{Time}).\,(t_0 < u < t \Rightarrow P(u)))\,.\end{aligned}$$

Appendix A
Predomains and Approximable Mappings

In Sect. 2.2 we introduced domains, which show up in many different disciplines, from order theory and topology to theoretical computer science. In this chapter, we discuss a notion of "basis" for domains, which we call *predomains*. They are more general than the well-known notion of abstract basis for domains [Gie+03, Definition III-4.15], and follow the work of Steve Vickers on what he calls "information systems" [Vic93]. Everything in this chapter is fully constructive, so it can be interpreted in the internal language of any topos $\mathcal{E}$.

In Appendix A.1 we define predomains and explain how they generate domains, as well as give examples of numeric predomains that show up throughout the book. In Appendix A.2 we give a notion of morphisms between predomains, called *approximable mappings*. We also prove there is an equivalence between the category of predomains and approximable mappings and that of domains and Scott-continuous functions. The benefit of considering domains is that they have nice semantics, constituting a full subcategory of topological spaces, whereas the benefit of considering predomains is that they consist of much less data in general. Finally in Appendix A.3 we discuss how the above work relates with subtoposes and modalities. By doing so, we were able to treat the various Dedekind numeric objects in various modalities (see Sect. 4.3) in a unified way.

While it may sometimes be difficult for a domain-theory novice to find intuition for the notions in this chapter, the proofs are fairly elementary.

A.1 Predomains and Their Associated Domains

In this section we define predomains and then discuss how to obtain a domain from a predomain. In Appendix A.2 we will show that this construction is part of an equivalence of categories.

P. Schultz, D. I. Spivak, *Temporal Type Theory*, Progress in Computer Science and Applied Logic 29, https://doi.org/10.1007/978-3-030-00704-1

A.1.1 *Introduction to Predomains*

We begin with the definition of predomain, which one will recognize is self-dual.

Definition A.1 (Predomain) A *predomain* is an inhabited set B together with a binary relation $\prec$ satisfying

$$b_1 \prec b_2 \;\Leftrightarrow\; \exists(b \in B).\, b_1 \prec b \prec b_2.$$

We sometimes denote the predomain simply by B if the order is clear from context.

To avoid uses of the axiom of choice when working internally to a topos, we assume a predomain comes with a specified function sending any b_1, b_2 such that $b_1 \prec b_2$ to a $\{b_1|b_2\}$ such that $b_1 \prec \{b_1|b_2\} \prec b_2$. The notation is inspired by Conway's surreal numbers, and can be thought of as the "simplest" element between b_1 and b_2. All our work below is completely independent of this choice.

Say that $(B, \prec)$ is *rounded* if for any b there exists b_1, b_2 with $b_1 \prec b \prec b_2$. Again we may assume B comes with specified choices, which we may denote $\{\varnothing|b\} \prec b \prec \{b|\varnothing\}$.

Any predomain B has an opposite, B^{op}, with the same underlying set and opposite order. Given any two predomains B_1 and B_2, there is a "product" predomain $B_1 \times B_2$ with underlying set the cartesian product, and with the component-wise order.

Remark A.2 When working in the internal language of a topos, replace the word "set" with "type." From that point of view, a predomain is a type B together with a predicate $\prec\colon B \times B \to \texttt{Prop}$, satisfying $\forall(b_1, b_2, b_3 : B).\,(b_1 \prec b_2) \wedge (b_2 \prec b_3) \Rightarrow (b_1 \prec b_3)$, and a function $\{-|-\} : \{(b_1, b_2) : B \times B \mid b_1 \prec b_2\} \to B$, satisfying $\forall(b_1, b_3 : B).\, b_1 \prec b_3 \Rightarrow (b_1 \prec \{b_1|b_3\}) \wedge (\{b_1|b_3\} \prec b_3)$.

In this chapter, we generally use the set-theoretic rather than topos/type-theoretic language, with hope that readers will be able to use this chapter independently of the rest of the book.

Example A.3 The pair $(\mathbb{Q}, <)$, where $<$ is the usual order on rational numbers, is a predomain. While perhaps cryptic now, we denote this predomain by $\underline{\mathbb{R}}_{\text{pre}} := (\mathbb{Q}, <)$; the reason will become clear in Example A.9. Its opposite is $\overline{\mathbb{R}}_{\text{pre}} := (\mathbb{Q}, >)$.

We refer to the predomain $\underline{\mathbb{R}}_{\text{pre}} \times \overline{\mathbb{R}}_{\text{pre}}$ as that of *improper intervals* and denote it $\underline{\overline{\mathbb{R}}}_{\text{pre}}$. Its order is obviously given by $(q_1, q_2) \prec (q_1', q_2')$ iff $q_1 < q_1'$ and $q_2' < q_2$. The predomain of *proper intervals*, denoted $\mathbb{IR}_{\text{pre}}$, is given by $\{(q_1, q_2) \in \underline{\overline{\mathbb{R}}}_{\text{pre}} \mid q_1 < q_2\}$ with the induced order.

It will also be useful to define $\underline{\mathbb{R}}^{\infty}_{\text{pre}}$ to be the predomain $\mathbb{Q} \sqcup \{-\infty\}$ equipped with the relation $<$ given by

$$a < b \Leftrightarrow (a, b \in \mathbb{Q} \wedge a < b) \vee (a = -\infty),$$

in particular $-\infty < -\infty$. Define $\bar{\mathbb{R}}^{\infty}_{\text{pre}} := (\mathbb{Q} \sqcup \{\infty\}, >)$ to be its opposite. Again, $\bar{\underline{\mathbb{R}}}^{\infty}_{\text{pre}} := \underline{\mathbb{R}}^{\infty}_{\text{pre}} \times \bar{\mathbb{R}}^{\infty}_{\text{pre}}$ and $\mathbb{IR}^{\infty}_{\text{pre}} := \{(q_1, q_2) \in \bar{\underline{\mathbb{R}}}^{\infty}_{\text{pre}} \mid q_1 < q_2\}$, where $-\infty < \infty$.

We sometimes use the following shorthand notation. For any finite set $F \subseteq B$ and element $b \in B$, write $F \prec b$ to denote $\forall (b' \in F).\, b' \prec b$.

Definition A.4 (Up/Down Closure, Open Subsets, Specialization Order, Joins, Meets) Let $(B, \prec)$ be a predomain. For a subset $X \subseteq B$, define the subsets

$$\twoheaduparrow X := \{b \in B \mid \exists (x \in X).\, x \prec b\} \qquad \text{and} \qquad \twoheaddownarrow X := \{b \in B \mid \exists (x \in X).\, b \prec x\}$$

called the *up-closure* and the *down-closure* of X, respectively. Despite the name "closure," one may have $X \not\subseteq \twoheaduparrow X$ or $X \not\subseteq \twoheaddownarrow X$; however, both operations $\twoheaduparrow$ and $\twoheaddownarrow$ are idempotent. A subset X is called *rounded upper* if $X = \twoheaduparrow X$ and *rounded lower* if $X = \twoheaddownarrow X$. We may refer to rounded upper subsets as *open subsets* and denote the set of such by $\Omega(B)$.

A predomain can be equipped with two canonical preorders, which we call the *lower specialization order* and *upper specialization order*. The lower and upper specialization orders are defined

$$b \leqslant b' := \twoheaddownarrow b \subseteq \twoheaddownarrow b' \qquad \text{and} \qquad b \eqslantless b' := \twoheaduparrow b' \subseteq \twoheaduparrow b.$$

Clearly, $b \prec b'$ implies both $b \leqslant b'$ and $b \eqslantless b'$. Note also that any open set $X = \twoheaduparrow X$ is an upper set in $(B, \leqslant)$, i.e. $x \in X$ and $x \leqslant x'$ implies $x' \in X$.

Given two elements $b_1, b_2 \in B$, we say that they have a *join* iff there exists an element $b_1 \curlyvee b_2 \in B$ that is a join in both specialization orders $(B, \leqslant)$ and $(B, \eqslantless)$; similarly with meets $b_1 \curlywedge b_2$. We say that B has *conditional joins* if, whenever b_1 and b_2 are such that there exists a b' with $\{b_1, b_2\} \leqslant b'$, then b_1 and b_2 have a join $b_1 \curlyvee b_2$. Dually, say that B has *conditional meets* if $b' \eqslantless \{b_1, b_2\}$ implies they have a meet $b_1 \curlywedge b_2$. If any two elements have a meet (resp. join), we say B has *binary meets* (resp. binary joins).

Example A.5 Let B be any of the following predomains from Example A.3: $\underline{\mathbb{R}}_{\text{pre}}$, $\bar{\mathbb{R}}_{\text{pre}}$, $\bar{\underline{\mathbb{R}}}_{\text{pre}}$, $\underline{\mathbb{R}}^{\infty}_{\text{pre}}$, $\bar{\mathbb{R}}^{\infty}_{\text{pre}}$, or $\bar{\underline{\mathbb{R}}}^{\infty}_{\text{pre}}$. Then the two specializations orders for B coincide; e.g., in $\bar{\mathbb{R}}_{\text{pre}}$ we have $b_1 \leqslant b_2$ iff $b_1 \le b_2$ iff $b_1 \eqslantless b_2$. Moreover, B is rounded and has binary meets and binary joins.

The predomains $\mathbb{IR}_{\text{pre}}$ and $\mathbb{IR}^{\infty}_{\text{pre}}$ are rounded and have binary meets and conditional joins, but not binary joins. For example, the join of (q_1, q_2) and (q_1', q_2') in $\bar{\underline{\mathbb{R}}}_{\text{pre}}$ is $(\max(q_1, q_1'), \min(q_2, q_2'))$, and this is not always a join in $\mathbb{IR}_{\text{pre}}$ because $q_1 < q_2$ and $q_1' < q_2'$ does not imply $\max(q_1, q_1') \overset{?}{<} \min(q_2, q_2')$.

The way we obtain domains from predomains is through rounded ideals, so the following—and its equivalent formulation in Lemma A.7—will be fundamental.

Definition A.6 (Rounded Ideals, Rounded Filters) Let $(B, \prec)$ be a predomain. A subset $I \subseteq B$ is called a *rounded ideal* if, for any finite set $F \subseteq B$, we have

$$F \subseteq I \;\Leftrightarrow\; \exists(b \in I).\, F \prec b.$$

Write $\mathrm{RId}(B)$ for the set of rounded ideals in B.

A *rounded filter* in B is a rounded ideal in B^{op}. Denote the set of rounded filters in B by $\Omega_{\mathrm{Filt}}(B)$. Note that any rounded filter is open, $\Omega_{\mathrm{Filt}}(B) \subseteq \Omega(B)$.

We spell out Definition A.6 in Lemma A.7.

Lemma A.7 *A subset $I \subseteq B$ is a rounded ideal iff the following three conditions hold:*

(nonempty)	$\exists b \in I,$
(down-closed)	*$b \in I$ and $b' \prec b$ implies $b' \in I$,*
(up-directed)	*$\{b_1, b_2\} \subseteq I$ implies $\exists b' \in I, \{b_1, b_2\} \prec b'$.*

If B has conditional joins, the up-directed condition may be replaced by the following pair of conditions:

(rounded)	*$b \in I$ implies $\exists b' \in I, b \prec b'$,*
(up-directed')	*$\{b_1, b_2\} \subseteq I$ implies $b_1 \curlyvee b_2$ exists and is in I.*

Dually, a subset $U \subseteq B$ is a rounded filter if it is nonempty, up-closed, and down-directed.

Proof This is all straightforward, except the statement about replacing up-directed with rounded and up-directed' in the case that B has conditional joins. Clearly, up-directed always implies rounded. Supposing B has conditional joins, up-directed and down-closed also implies up-directed'.

For the converse, take $b_1, b_2 \in I$ such that their join $b := b_1 \curlyvee b_2$ exists with $b \in I$. We *do not* necessarily have $\{b_1, b_2\} \prec^? b$; however, we do have $b_1 \preccurlyeq b$ and $b_2 \preccurlyeq b$ by definition, and by roundedness there exists $b' \in I$ with $b \prec b'$, so I is up-directed as desired. □

As mentioned in the proof of Lemma A.7, roundedness follows from up-directedness, and from the structure of the lemma, it may appear to take a subsidiary role. However, as the name "rounded ideal" suggests, roundedness is an essential aspect and—as we will see in the following proof—a very useful technical condition.

Proposition A.8 *Let B and B' be predomains and $B \times B'$ their product. We have isomorphisms*

$$\mathrm{RId}(B) \times \mathrm{RId}(B') \xrightarrow{\cong} \mathrm{RId}(B \times B') \qquad \textit{and} \qquad \Omega_{\mathrm{Filt}}(B) \times \Omega_{\mathrm{Filt}}(B') \xrightarrow{\cong} \Omega_{\mathrm{Filt}}(B \times B')$$

Proof The two statements are dual, so it suffices to prove the first. We begin by giving the maps in either direction. We send a pair of ideals $I \subseteq B$ and $I' \subseteq B'$ to the ideal $I \times I' \subseteq B \times B'$, and this clearly a rounded ideal if I and I' are. Given some $J \subseteq \mathrm{RId}(B \times B')$, define

$$I := \{b \in B \mid \exists(b' \in B').\,(b, b') \in J\} \quad \text{and} \quad I' := \{b' \in B' \mid \exists(b \in B).\,(b, b') \in J\}$$

It is easy to see that I is nonempty, rounded, and up-directed. To see it is down-closed, choose $b \in I$ and $b_0 \prec b$. We have some b' with $(b, b') \in J$ so by roundedness we can choose some (b_1, b_1') with $(b, b') \prec (b_1, b_1')$. Then we have $(b_0, b') \prec (b_1, b_1')$, so $b_0 \in I$. We have shown that $I \subseteq B$, and similarly $I' \subseteq B'$, are rounded ideals. Thus we have defined mappings in either direction, and they are evidently mutually inverse. □

We are ready to motivate the "cryptic" notation for the predomains in Example A.3.

Example A.9 Let $\underline{\mathbb{R}}_{\text{pre}} = (\mathbb{Q}, <)$ be as in Example A.3. In the category of sets, $\mathrm{RId}(\underline{\mathbb{R}}_{\text{pre}})$ can be identified with the set of *lower reals*, $\mathbb{R} \sqcup \{+\infty\}$. Indeed if r is an lower real, then one sees that $\{q \in \mathbb{Q} \mid q < r\}$ is a rounded ideal by simply checking the conditions of Lemma A.7. Conversely, if $I \subseteq \mathbb{Q}$ is a rounded ideal, then one can take its supremum $\sup(I) \in \mathbb{R} \sqcup \{\infty\}$; it is easy to see these functions are mutually inverse. The order on rounded ideals corresponds to the usual $\leq$ order on $\mathbb{R} \sqcup \{\infty\}$.

Recall that $\bar{\underline{\mathbb{R}}}_{\text{pre}} \cong \bar{\mathbb{R}}_{\text{pre}} \times \underline{\mathbb{R}}_{\text{pre}}$, where $\underline{\mathbb{R}}_{\text{pre}} = (\bar{\mathbb{R}}_{\text{pre}})^{\text{op}}$. By Proposition A.8, a rounded ideal $I \in \mathrm{RId}(\bar{\underline{\mathbb{R}}}_{\text{pre}})$ can be identified with a pair (D_I, U_I) of "cuts," i.e. rounded ideals $D_I \subseteq \bar{\mathbb{R}}_{\text{pre}}$ and $U_I \subseteq \underline{\mathbb{R}}_{\text{pre}}$.

We will see in Theorem A.18 that $\underline{\mathbb{R}} := \mathrm{RId}(\underline{\mathbb{R}}_{\text{pre}})$, $\bar{\mathbb{R}} := \mathrm{RId}(\bar{\mathbb{R}}_{\text{pre}})$, and $\bar{\underline{\mathbb{R}}} := \mathrm{RId}(\bar{\underline{\mathbb{R}}}_{\text{pre}})$ naturally have the structure of a domain. The last of these is the domain of improper intervals, e.g. from [Kau80]. The domain $\mathbb{IR} := \mathrm{RId}(\mathbb{IR}_{\text{pre}})$ consisting of rounded ideals in the sub-predomain $\mathbb{IR}_{\text{pre}}$ is the standard interval domain, and is important throughout the main body of this book. An element $I \in \mathbb{IR}$ corresponds to a pair of disjoint cuts, $D_I \cap U_I = \emptyset$.

We similarly define domains corresponding to the other predomains from Example A.3:

$$\underline{\mathbb{R}}^{\infty} :- \mathrm{RId}(\underline{\mathbb{R}}^{\infty}_{\text{pre}}), \quad \bar{\mathbb{R}}^{\infty} :- \mathrm{RId}(\bar{\mathbb{R}}^{\infty}_{\text{pre}}), \quad \bar{\underline{\mathbb{R}}}^{\infty} := \mathrm{RId}(\bar{\underline{\mathbb{R}}}^{\infty}_{\text{pre}}). \quad \mathbb{IR}^{\infty} := \mathrm{RId}(\mathbb{IR}^{\infty}_{\text{pre}}),$$

Lemma A.10 *If B is a rounded predomain and has conditional joins, then for any $b \in B$, its down-closure $\downarrow b$ is a rounded ideal.*

Proof $\downarrow b$ is obviously down-closed, and it is nonempty since B is rounded (Definition A.1). The (rounded) and (up-directed') conditions of Lemma A.7 are obvious too. □

If $(B, \prec)$ is a predomain, so is $(\mathrm{RId}(B), \subseteq)$.

Proposition A.11 *If B has conditional joins, then so does* $\mathrm{RId}(B)$. *If B has binary meets, then so does* $\mathrm{RId}(B)$.

Proof For the first, suppose $I_1, I_2, I \in \mathrm{RId}(B)$ with $\{I_1, I_2\} \subseteq I$. Then for any $b_1 \in I_1$ and $b_2 \in I_2$, the join $b_1 \curlyvee b_2$ exists and is in I. One checks that the set

$$I_1 \curlyvee I_2 = \downarrow\!\{ b_1 \curlyvee b_2 \mid b_1 \in I_1, b_2 \in I_2 \}$$

is a rounded ideal and that it is the join of I_1 and I_2 in $\mathrm{RId}(B)$. For example, to see that $I_1 \curlyvee I_2$ is inhabited, take $b_1 \in I_1$ and $b_2' \prec b_2 \in I_2$, note $b_2 \leqslant b_1 \curlyvee b_2$, and find $b_2' \in I_1 \curlyvee I_2$.

Now suppose B has binary meets, and let $I_1, I_2 \in \mathrm{RId}(B)$ be rounded ideals. The intersection $I_1 \cap I_2$ is clearly down-closed. It is nonempty because there exist $b_1 \in I_1$ and $b_2 \in I_2$, hence $b_1 \curlywedge b_2 \in I_1 \cap I_2$. For roundedness, if $b \in I_1 \cap I_2$, then there exist $b_1 \in I_1$ and $b_2 \in I_2$ with $b \prec \{b_1, b_2\}$, hence $b \prec b_1 \curlywedge b_2 \in I_1 \cap I_2$. Directedness is similar. □

A.1.2 Domains from Predomains

Recall the notion of domains, directed sets, and the way-below relation $\ll$ from Sect. 2.2. In this section, we use the traditional notation $\sqsubseteq$ for the order relation on a domain.

Proposition A.12 *If $(D, \sqsubseteq)$ is a domain, with way-below relation $\ll$, then $(D, \ll)$ is a predomain. Moreover,*

1. *The domain order $\sqsubseteq$ and the predomain's lower specialization order $\leqslant$ coincide.*
2. *For every $x \in D$, the set $\downarrow x$ is a rounded ideal in the predomain, and the down-closure function $\downarrow : D \to \mathrm{RId}(D, \ll)$ is an order isomorphism $(D, \sqsubseteq) \cong (\mathrm{RId}(D), \subseteq)$.*
3. *A subset $U \subseteq D$ is open in the predomain $(D, \ll)$ iff it is Scott open in the domain $(D, \sqsubseteq)$.*

Proof We know that $\ll$ is transitive and interpolative from Remark 2.8 and Proposition 2.14, so $(D, \ll)$ is a predomain. Item 1 is straightforward (e.g., if $x \ll a \Rightarrow x \ll b$, then $a = \bigvee \downarrow a \leq \bigvee \downarrow b = b$), and so is Item 2 (the inverse is $\bigvee : \mathrm{RId}(D) \to D$).

For Item 3, suppose $U \subseteq D$ is open in the predomain, i.e. $U = \uparrow U$, and that $\bigvee X = u \in U$. There exists $u' \ll u$ with $u' \in U$, and there exists $x \in X$ such that $u' \leq x$. Then use any $u'' \ll u' \leq x$ to see $x \in U$. Conversely, if U is open in the domain, it is easy to check that $U \subseteq \uparrow U$, and the fact that $\uparrow U \subseteq U$ follows from Item 1. □

Example A.13 The two specialization orders for the predomain underlying a domain $(D, \sqsubseteq)$ need not coincide. The lower specialization order $\leqslant$ always agrees with the domain order $\sqsubseteq$, but the upper specialization order can be quite different.

Indeed, one can see this difference in the domain $\mathbb{IR} = \text{RId}(\mathbb{IR}_{\text{pre}})$ from Example A.9 (see also Proposition A.17 and Theorem A.18), where the $\leqslant$ relation is better behaved than the $\lessdot$ relation. We denote the elements of $\mathbb{IR}$ by $[d, u]$, where $d, u \in \mathbb{R}$ and $d \leq u$. Then $[d_1, u_1] \leqslant [d_2, u_2]$ iff $d_1 \leq d_2 \leq u_2 \leq u_1$, whereas $[d_1, u_1] \lessdot [d_2, u_2]$ iff either $d_1 \leq d_2 \leq u_2 \leq u_1$ or $d_2 = u_2$. For example, if we identify the set $\mathbb{R}$ of real numbers with intervals of the form $[r, r]$, then $[r_1, r_1] \leqslant [r_2, r_2]$ iff $r_1 = r_2$ whereas $[r_1, r_1] \lessdot [r_2, r_2]$ for all $r_1, r_2 \in \mathbb{R}$.

One can think of a predomain as a sort of basis for a domain, and Proposition A.12 is analogous to the statement "the domain is a basis for itself." We next explain how predomains present domains. This culminates in Theorem A.18, where we show that for any predomain B, the set $\text{RId}(B)$ of rounded ideals ordered by inclusion is a domain. Hence $\text{RId}(B)$ is a topological space, and we will show in Theorem A.25 that its frame of opens is isomorphic to $\Omega(B)$. It is clear that the opens of the form $\twoheaduparrow b$ form a basis for the topology. The following lemma shows that in fact $\Omega_{\text{Filt}}(B)$ is also a basis. Of course, if B is rounded and has conditional meets, then $\twoheaduparrow b$ is always a rounded filter, in which case this is trivial.

Lemma A.14 *Let $(B, \prec)$ be a predomain. For any $b, b' \in B$,*

$$b \prec b' \Rightarrow \exists (U \in \Omega_{\text{Filt}}(B)).\, b' \in U \subseteq \twoheaduparrow b,$$

and dually,

$$b' \prec b \Rightarrow \exists (I \in \text{RId}(B)).\, b' \in I \subseteq \twoheaddownarrow b.$$

Proof If B has conditional meets, we can simply take $U := \twoheaduparrow\{b|b'\}$.

Otherwise, inductively construct a descending sequence $b \prec \cdots \prec b_n \prec b_{n-1} \prec \cdots \prec b_1 = b'$, with $b_n := \{b|b_{n-1}\}$. Let $U = \bigcup_n \twoheaduparrow b_n$. Clearly, $U \subseteq \twoheaduparrow b$, and $b' \in \twoheaduparrow b_2 \subseteq U$. It remains to check that U is a rounded filter. If $x \in U$ and $x \prec y$, then $y \in U$, since $b_n \prec x$ for some n hence $b_n \prec y$. And if $F \subset U$ is a finite subset, then there is some n for which $b_n \prec F$, and $b_{n+1} \prec b_n$ implies $b_n \in U$. □

The next two lemmas lead up to Proposition A.17, which characterizes the way below relation in the poset $(\text{RId}(B), \subseteq)$.

Lemma A.15 *For any $b \in I \in \text{RId}(B)$, there exist $\bar{b} \in I$ and $J \in \text{RId}(B)$ such that $b \in J \subseteq \twoheaddownarrow \bar{b} \subseteq I$.*

Proof Since I is rounded, there is some $\bar{b} \in I$ with $b \prec \bar{b}$. Then by Lemma A.14 there is a $J \in \text{RId}(B)$ with $b \in J \subseteq \twoheaddownarrow \bar{b}$. Since I is down-closed, $\bar{b} \in I$ implies $\twoheaddownarrow \bar{b} \subseteq I$. □

Lemma A.16 *For any $I \in \text{RId}(B)$, the set $X = \{ J \in \text{RId}(B) \mid \exists (j \in I).\, J \subseteq \twoheaddownarrow j \}$ is directed, and $I = \bigcup X$.*

Proof That $I = \bigcup X$ is clear from Lemma A.15. To show directedness, suppose $J_1, J_2 \in X$, so $J_1 \subseteq \downarrow j_1$ and $J_2 \subseteq \downarrow j_2$ for some $j_1, j_2 \in I$. Then because I is an ideal, there exists a $j \in I$ with $\{j_1, j_2\} \prec j$. By Lemma A.15, there exists a $\bar{j} \in I$ and $J \in \mathrm{RId}(B)$ with $j \in J \subseteq \downarrow \bar{j}$. Hence $J \in X$, and $J_1 \subseteq \downarrow j_1 \subseteq \downarrow j \subseteq J$ and similarly $J_2 \subseteq J$. □

Proposition A.17 *For any $I, I' \in \mathrm{RId}(B)$, $I' \ll I$ if and only if $I' \subseteq \downarrow i$ for some $i \in I$. Dually, for any $U, U' \in \Omega_{\mathrm{Filt}}(B)$, $U' \ll U$ iff $U' \subseteq \uparrow u$ for some $u \in U$.*

Proof Suppose $I' \ll I$. By Lemma A.16, I is the directed union of the set $\{ J \in \mathrm{RId}(B) \mid \exists (j \in I).\, J \subseteq \downarrow j \}$, hence there exist $J \in \mathrm{RId}(B)$ and $j \in I$ such that $I' \subseteq J \subseteq \downarrow j$.

Conversely, suppose $i \in I$ and $I' \subseteq \downarrow i$, and suppose $I \subseteq \bigcup X$ for some directed subset $X \subseteq \mathrm{RId}(B)$. Then $i \in J$ for some $J \in X$, and $I' \subseteq \downarrow i \subseteq J$, hence $I' \ll I$. □

Theorem A.18 *If B is a predomain, then the poset* $(\mathrm{RId}(B), \subseteq)$ *is a domain.*

Proof It is easy to see that rounded ideals are closed under directed unions, so $\mathrm{RId}(B)$ is a dcpo. Combining Lemma A.16 and Proposition A.17, every $I \in \mathrm{RId}(B)$ is the union of the directed set $\{ J \in \mathrm{RId}(B) \mid J \ll I \}$. □

Proposition A.19 *Let $(D, \sqsubseteq)$ be a domain and $A \subseteq D$ a subset. Then the following are equivalent:*

1. *For all $d \in D$, the set $\downarrow d \cap A$ is directed and $d = \sup(\downarrow d \cap A)$.*
2. *For all $d_1, d_2 \in D$, if $d_1 \ll d_2$, then there exists $a \in A$ such that $d_1 \ll a \ll d_2$.*
3. *For all $d_1, d_2 \in D$, if $d_1 \ll d_2$, then there exists $a \in A$ such that $d_1 \sqsubseteq a \ll d_2$.*

Proof The equivalence of the first three, as well as other conditions, is shown in [Gie+03, III-4.2]. □

Definition A.20 Let D be a domain. A *basis* for D is a subset $A \subseteq D$ satisfying any of the equivalent conditions of Proposition A.19

Any domain D is clearly a basis for itself. It is easy to see that if A is a basis and $A \subseteq A'$, then A' is a basis. It is also easy to check that if A is a basis then $d_1 \sqsubseteq d_2$ iff $(a \ll d_1) \Rightarrow (a \ll d_2)$ for all $a \in A$.

Proposition A.21 *Suppose B is rounded and has conditional joins. Then there is an embedding $\downarrow\colon B \to \mathrm{RId}(B)$, where $b \prec b'$ iff $\downarrow b \ll \downarrow b'$. In particular, B serves as a basis for* $\mathrm{RId}(B)$.

Proof By Lemma A.10, $\downarrow b$ is a rounded ideal. The fact that $b \prec b'$ iff $\downarrow b \ll \downarrow b'$ follows from Proposition A.17. To see that the set $\{\downarrow b\}_{b \in B}$ serves as a basis, one checks using Proposition A.17 that $\downarrow I \cap \{\downarrow b\}_{b \in B} = \downarrow I$ for any $I \in \mathrm{RId}(B)$. □

Remark A.22 In Proposition A.21, sufficient conditions for B to embed as a basis of $\mathrm{RId}(B)$ were given, but in fact a necessary and sufficient condition is that B have the *finite interpolation property*, $F \prec b$ iff $\exists (b' \in B).\, F \prec b' \prec b$. Predomains with this property are well known in the domain literature: they are called *abstract bases*; see [Gie+03, Definition III-4.15].

Next we aim to prove that $\Omega(B)$ is a frame, isomorphic to the frame of Scott-open subsets of $\mathrm{RId}(B)$. We introduce the following notation for $I \in \mathrm{RId}(B)$ and $U \in \Omega(B)$:

$$I \models^B U := \exists(b : B).\,(b \in I) \wedge (b \in U). \tag{A.1}$$

We will see in Lemma A.23 that $I \models^B U$ iff I is contained in the Scott-open subset corresponding to U. By a *rounded filter* in a domain D (e.g., $D = \mathrm{RId}(B)$), we mean in the predomain $(D, \ll)$; see Proposition A.12.

Lemma A.23 *Let $(B, \prec)$ be a predomain. For any $U \in \Omega(B)$, the set $\mathcal{U}_U = \{\, I \in \mathrm{RId}(B) \mid I \models^B U \,\}$ is a Scott-open subset $\mathcal{U}_U \in \Omega(\mathrm{RId}(B))$. If $U \in \Omega_{\mathrm{Filt}}(B)$, then $\mathcal{U}_U \in \Omega_{\mathrm{Filt}}(\mathrm{RId}(B))$.*

Proof It is easy to see that $\mathcal{U}_U$ is Scott-open: it is clearly an upper set, and if $X \subseteq \mathrm{RId}(B)$ is a directed set with $\bigcup X \in \mathcal{U}_U$, then there is a $b \in U \cap \bigcup X$ so $b \in J$ for some $J \in X$, hence $J \models^B U$ and $J \in \mathcal{U}_U$.

Suppose U is filtered; $\mathcal{U}_U$ is obviously up-closed and nonempty. Let $I_1, I_2 \in \mathcal{U}_U$ and let $i_1 \in U \cap I_1$ and $i_2 \in U \cap I_2$. Then there are $i, i' \in U$ with $i \prec i' \prec \{i_1, i_2\}$. By Lemma A.14, there is a $J \in \mathrm{RId}(B)$ with $i \in J \subseteq {\downarrow\!\!\downarrow} i'$. Hence $i \in U \cap J$, so $J \in \mathcal{U}_U$, and $J \subseteq {\downarrow\!\!\downarrow} i' \subseteq I_1 \cap I_2$ implies $J \ll \{I_1, I_2\}$ by Proposition A.17, showing that $\mathcal{U}_U$ is directed, hence filtered. □

Lemma A.24 *For any Scott open $\mathcal{U} \in \Omega(\mathrm{RId}(B))$, the set $U_{\mathcal{U}} = \{\, b \in B \mid \exists(J \in \mathcal{U}).\, J \subseteq {\downarrow\!\!\downarrow} b \,\}$ is an open of B. For any $I \in \mathcal{U}$, we have $I \models^B U_{\mathcal{U}}$. Moreover, if $\mathcal{U} \in \Omega_{\mathrm{Filt}}(\mathrm{RId}(B))$, then $U_{\mathcal{U}} \in \Omega_{\mathrm{Filt}}(B)$.*

Proof Choose $I \in \mathcal{U}$. Since $\mathcal{U}$ is rounded (Proposition A.12 and Lemma A.7) there is a $J \in \mathcal{U}$ with $J \ll I$. By Proposition A.17 there is a $j \in I$ with $J \subseteq {\downarrow\!\!\downarrow} j$. But then $j \in U_{\mathcal{U}} \cap I$, so $I \models^B U_{\mathcal{U}}$.

We now show that $U_{\mathcal{U}}$ is open. Suppose $b' \in U_{\mathcal{U}}$, so $J \subseteq {\downarrow\!\!\downarrow} b'$ for some $J \in \mathcal{U}$. If $b' \prec b$, then $J \subseteq {\downarrow\!\!\downarrow} b' \subseteq {\downarrow\!\!\downarrow} b$, hence $b \in U_{\mathcal{U}}$. Conversely, since $J \models^B U_{\mathcal{U}}$ there exists some $b'' \in U_{\mathcal{U}} \cap J$, hence $b'' \in U_{\mathcal{U}}$, and $b'' \in J \subseteq {\downarrow\!\!\downarrow} b'$ shows $b'' \prec b'$.

Finally, suppose $\mathcal{U}$ is a rounded filter. If $b_1, b_2 \in U_{\mathcal{U}}$, then $I_1 \subseteq {\downarrow\!\!\downarrow} b_1$ and $I_2 \subseteq {\downarrow\!\!\downarrow} b_2$ for some $I_1, I_2 \in \mathcal{U}$. Because $\mathcal{U}$ is a rounded filter, there exist $I, J \in \mathcal{U}$ such that $J \ll I \ll \{I_1, I_2\}$, hence by Proposition A.17 there is some $j \in I$ with $J \subseteq {\downarrow\!\!\downarrow} j$. So $j \in U_{\mathcal{U}}$, and $j \in I \subseteq I_1 \cap I_2 \subseteq {\downarrow\!\!\downarrow} b_1 \cap {\downarrow\!\!\downarrow} b_2$ implies $j \prec \{b_1, b_2\}$, completing the proof that $U_{\mathcal{U}}$ is a rounded filter. □

Theorem A.25 *The constructions of Lemmas A.23 and A.24 determine order isomorphisms*

$$\Omega(\mathrm{RId}(B)) \cong \Omega(B) \qquad \textit{and} \qquad \Omega_{\mathrm{Filt}}(\mathrm{RId}(B)) \cong \Omega_{\mathrm{Filt}}(B).$$

Proof Both constructions are clearly monotonic, so we show they are mutually inverse.

If $\mathcal{U} \in \Omega(\mathrm{RId}(B))$, then $\mathcal{U}_{U_{\mathcal{U}}} = \mathcal{U}$ if $I \models^B U_{\mathcal{U}} \Leftrightarrow I \in \mathcal{U}$. The $\Leftarrow$ implication holds by Lemma A.24. In the other direction, suppose $b \in I \cap U_{\mathcal{U}}$. Then by definition of $U_{\mathcal{U}}$, there is a $J \in \mathcal{U}$ such that $J \subseteq \downarrow b$. But then by Proposition A.17, $J \ll I$, so $I \in \mathcal{U}$.

If $U \in \Omega(B)$, then to show $U_{\mathcal{U}_U} = U$ we need to prove $b \in U$ if and only if there exists an $J \in \mathrm{RId}(B)$ such that $J \models^B U$ and $J \subseteq \downarrow b$. For the forward direction, if $b \in U$, then there is a $b' \in U$ with $b' \prec b$, and so by Lemma A.14 there is an $J \in \mathrm{RId}(B)$ with $b' \in J \subseteq \downarrow b$. In the other direction, if $J \subseteq \downarrow b$ and $b' \in J \cap U$, then $b' \prec b$ so $b \in U$. □

Corollary A.26 *By duality, Theorem A.25 implies* $\Omega_{\mathrm{Filt}}(\Omega_{\mathrm{Filt}}(B)) \cong \mathrm{RId}(B)$.

Corollary A.27 *For any domain D, there is an isomorphism of domains* $D \cong \mathrm{RId}(D, \ll)$.

Proof This follows from Theorem A.25—$\Omega(\mathrm{RId}(D, \ll)) \cong \Omega(D)$—and the fact that domains are sober and thus determined by their frame of opens. □

A.2 Approximable Mappings

Domains are, in particular, topological spaces (see Remark 2.17) and predomains give a sort of basis for them. In this section we define a sort of mapping $B \to B'$ between predomains, called *approximable mappings*, whose purpose is to induce a continuous function between the corresponding domains $\mathrm{RId}(B) \to \mathrm{RId}(B')$. In fact, all morphisms between domains arise in this way—RId is an equivalence between the category of predomains and the category of domains—as we show in Corollary A.35. Again, the work in this section was adapted from [Vic93].

A.2.1 Morphisms of Predomains

Soon we will define a category **Predom** whose objects are predomains (see Definition A.1) and whose morphisms are approximable mappings, defined as follows.

Definition A.28 (Approximable Mapping) Given two predomains B and B', an *approximable mapping* $H: B \to B'$ is a relation, written $H(b, b')$ for $b \in B$ and $b' \in B'$, satisfying

1. $b_1 \prec b_2$ and $H(b_1, b')$ implies $H(b_2, b')$
2. $H(b_2, b')$ implies there exists a $b_1 \prec b_2$ with $H(b_1, b')$
3. $b'_1 \prec b'_2$ and $H(b, b'_2)$ implies $H(b, b'_1)$
4. $H(b_1, b'_i)$ for all $1 \leq i \leq n$ and $b_1 \prec b_2$ implies there exists a $b' \in B'$ such that $H(b_2, b')$ and $b'_i \prec b'$ for all $1 \leq i \leq n$, for any $n \geq 0$.

The last (4) may be replaced by the following two:

4'a if $b_1 \prec b_2$, then there exists a $b' \in B'$ with $H(b_2, b')$
4'b if $b_1 \prec b_2$, $H(b_1, b'_1)$, and $H(b_1, b'_2)$, then there exists a $b' \in B'$ with $H(b_2, b')$ and $\{b'_1, b'_2\} \prec b'$.

If B is rounded, and B and B' have conditional joins, then (4) be instead be replaced by

4"a $\forall(b \in B). \exists(b' \in B'). H(b, b')$
4"b $H(b, b'_1)$ implies there exists a $b'_2 \in B'$ with $b'_1 \prec b'_2$ and $H(b, b'_2)$
4"c if $H(b, b'_1)$ and $H(b, b'_2)$, then $b'_1 \curlyvee b'_2$ exists and $H(b, b'_1 \curlyvee b'_2)$.

The composite of two approximable mappings $H: B \to B'$ and $H': B' \to B''$ is simply the relational composite:

$$(H; H')(b, b'') \Leftrightarrow \exists(b' \in B'). H(b, b') \wedge H'(b', b''),$$

and the identity on B is $\mathrm{id}_B(b, b') \Leftrightarrow (b' \prec b)$.

There is certainly redundancy in the definition of approximable mapping. For example, given conditions 2 and 4, condition 1 holds iff condition 3 holds.

Lemma A.29 *If $H: B \to B'$ is an approximable mapping, then—with no additional constraints on B and B'—condition 4"b holds. In particular, $H(b, b')$ iff $b \in \Omega(H)(\Uparrow b')$.*

Proof Suppose $H(b, b'_1)$. Then by 2. there is some $b_1 \prec b$ with $H(b_1, b'_1)$, and by 4. with $n = 1$, there exists a b'_2 with $H(b, b'_2)$ with $b'_1 \prec b'_2$. □

Proposition A.30 *Predomains and approximable mappings form a category* **Predom**.

Proof Unitality of identities follows from condition 2 and its dual, condition 4" (see Lemma A.29). Suppose given approximable mappings $H: A \to B$ and $H': B \to C$. The composite clearly satisfies conditions 1, 2, and 3, so we check 4.

Let $a_1 \prec a_2$ and for all $1 \leq i \leq n$, $H(a_1, b_i)$ and $H'(b_i, c_i)$. Then there is an $a' \in A$ with $a_1 \prec a' \prec a_2$, a $b \in B$ with $H(a', b)$ and $b_i \prec b$ for all i, and a $b' \in B$ with $H(a_2, b')$ and $b \prec b'$. Then $H'(b, c_i)$ for all i, so there is a c with $H'(b', c)$ and $c_i \prec c$ for all i. Hence $(H; H')(a_2, c)$. □

A.2.2 Approximable Mappings to Scott-Continuous Functions

In Proposition A.34 we will prove that an approximable mapping between predomains induces a continuous morphism between the corresponding domains. We are already able to say what it does on open sets.

Lemma A.31 *Any approximable mapping* $H\colon B \to B'$ *defines a function* $\Omega(H)\colon \Omega(B') \to \Omega(B)$*, defined on* $U' \in \Omega(B')$ *by*

$$\Omega(H)(U') := \{u \in B \mid \exists u' \in U' \text{ such that } H(u, u')\}.$$

Proof Suppose $U' \subseteq B'$ is any subset and let $U := \Omega(H)(U')$ be defined as above. If $b \in \uparrow U$ then there exists $u \in U$ with $u \prec b$, and $b \in U$ follows from condition 1.

For the converse, we need the assumption $U' \in \Omega(B')$, i.e. $U' = \uparrow U'$. For any $b \in U$, there exists $u \in U$ such that $u \prec b$ by condition 2. so $b \in \uparrow U$, as desired. □

Lemma A.32 *Any approximable mapping* $H\colon B \to B'$ *defines a function* $\mathrm{RId}(H)\colon \mathrm{RId}(B) \to \mathrm{RId}(B')$*, defined on an ideal* $I \in \mathrm{RId}(B)$ *by*

$$\mathrm{RId}(H)(I) := \{\, b' \in B' \mid \exists (b \in I).\, H(b, b')\,\}. \tag{A.2}$$

Proof If $I \in \mathrm{RId}(B)$, then it is easy to check that $\mathrm{RId}(H)(I)$ is a rounded ideal (this uses 1, 3, 4, and the roundedness of I). □

We want to show that $\mathrm{RId}(H)$ is continuous. To do so, we will begin by proving the following general fact, reminiscent of the theory of Chu spaces.

Lemma A.33 *Suppose that* $\mathcal{X} = (X, \Omega(X), \in_X)$ *and* $\mathcal{Y} = (Y, \Omega(Y), \in_Y)$ *are topological spaces. Then there is a natural bijection between the set* $\mathbf{Top}(\mathcal{X}, \mathcal{Y})$ *of continuous maps* $\mathcal{X} \to \mathcal{Y}$ *and the set*

$$\{(f : X \to Y,\, f^* : \Omega(Y) \to \Omega(X)) \mid \forall (x : X)(U : \Omega(Y)).\, f(x) \in_Y U \Leftrightarrow x \in_X f^*(U)\}.$$

Recall from Eq. (A.1) the relation $\models^B$ on $\mathrm{RId}(B) \times \Omega(B)$. Using the isomorphism $\Omega(B) \cong \Omega(\mathrm{RId}(B))$ from Theorem A.25, we consider $(\mathrm{RId}(B), \Omega(B), \models^B)$ as a topological space. Then Lemma A.33 says that $\mathrm{RId}(H)$ is continuous iff $(\mathrm{RId}(H)(I) \models^{B'} U') \Leftrightarrow (I \models^B \Omega(H)(U'))$ for any $I \in \mathrm{RId}(B)$ and $U' \in \Omega(B')$.

Proposition A.34 *For any approximable mapping* $H\colon B \to B'$*, the map* $\mathrm{RId}(H)\colon \mathrm{RId}(B) \to \mathrm{RId}(B')$ *from Eq.* (A.2) *is continuous. Its inverse-image map for open sets is given by* $\Omega(H)\colon \Omega(B') \to \Omega(B)$*. Explicitly, for any* $I \in \mathrm{RId}(B)$ *and* $U' \in \Omega(B')$*,*

$$\mathrm{RId}(H)(I) \models^{B'} U' \quad\Leftrightarrow\quad I \models^B \Omega(H)(U'). \tag{A.3}$$

Conversely, for any continuous map $f\colon \mathrm{RId}(B) \to \mathrm{RId}(B')$*, represented by the induced frame homomorphism* $f^{-1}\colon \Omega(B') \to \Omega(B)$*, there is a unique approximable mapping* $H\colon B \to B'$ *such that* $f = \mathrm{RId}(H)$*, given by defining* $H(b, b') := (b \in f^{-1}(\uparrow b'))$.

Proof The fact that $\mathrm{RId}(H)$ is continuous then follows from Lemma A.33 and Eq. (A.3), the latter of which is easily checked:

$$\mathrm{RId}(H)(I) \models^{B'} U' \;\Leftrightarrow\; \exists(b \in I)(b' \in U').\, H(b, b') \;\Leftrightarrow\; I \models^{B} \Omega(H)(U').$$

If $f\colon \mathrm{RId}(B) \to \mathrm{RId}(B')$ is any continuous map, then $f^{-1}\colon \Omega(\mathrm{RId}(B')) \to \Omega(\mathrm{RId}(B))$ is a frame homomorphism, i.e. an order preserving map which preserves finite meets and arbitrary joins. By Theorem A.25 we may regard this as a frame homomorphism $f^{-1}\colon \Omega(B') \to \Omega(B)$. Then we must check that $H(b, b') := b \in f^{-1}(\uparrow b')$ defines an approximable mapping. Openness of $f^{-1}(\uparrow b')$ implies 1 and 2, and 3 follows because f^{-1} is order preserving; we next show 4.

Suppose $b_1 \prec b_2$. In any frame, let $\top$ denote the top element and $\curlywedge$ denote the meet. To see 4'a, we have

$$b_2 \in \uparrow b_1 \subseteq \top = f^{-1}(\top) = f^{-1}\Big(\bigcup_{b' \in B'} \uparrow b'\Big) = \bigcup_{b' \in B'} f^{-1}(\uparrow b').$$

For 4'b, if $b_1 \in f^{-1}(\uparrow b'_1)$ and $b_1 \in f^{-1}(\uparrow b'_2)$, then since open sets are up-closed, we have $\uparrow b_1 \subseteq f^{-1}(\uparrow b'_1)$ and $\uparrow b_1 \subseteq f^{-1}(\uparrow b'_2)$, and hence,

$$b_2 \in \uparrow b_1 \subseteq f^{-1}(\uparrow b'_1) \curlywedge f^{-1}(\uparrow b'_2) = f^{-1}(\uparrow b'_1 \curlywedge \uparrow b'_2).$$

This again implies there exists a basic open $\uparrow b' \subseteq \uparrow b'_1 \curlywedge \uparrow b'_2$ such that $b_2 \in f^{-1}(\uparrow b')$.

To see that $f = \mathrm{RId}(H)$, it suffices to check that $f^{-1} = \Omega(H)$ because domains are sober spaces (see Remark 2.17). We have

$$f^{-1}(U') = f^{-1}\Big(\bigcup_{b' \in U'} \uparrow b'\Big) = \bigcup_{b' \in U'} f^{-1}(\uparrow b'),$$

so $b \in f^{-1}(U')$ if and only if $\exists(b' \in U').\, b \in f^{-1}(\uparrow b')$, i.e. $b \in \Omega(H)(U')$.

Finally, if H' is any other approximable mapping with $\mathrm{RId}(H') = f$, then $H = H'$ follows from Lemma A.29. □

Corollary A.35 *Let* **Predom** *be as in Proposition A.30, and let* **Top** *be the category of topological spaces. Then* RId *defines a fully faithful functor* **Predom** $\to$ **Top**. *Its essential image is the full subcategory of continuous dcpos, i.e. domains, with their Scott-open topology.*

Proof The fact that RId is bijective on hom-sets was shown in Proposition A.34, and the fact that its essential image is the domains follows from Corollary A.27. All that remains to be shown is that RId preserves composition and identity, but this is a simple check. □

We record now a few special kinds of approximable mappings which will be needed later: dense, filtered, and Lawson.

Definition A.36 (Dense Approximable Mapping) Let B and B' be rounded predomains. Say an approximable mapping $H: B \to B'$ is *dense* if it satisfies the formula $\forall(b' : B').\, \exists(b : B).\, H(b, b')$.

Say that a map $h: R \to R'$ between topological spaces is *dense* if, for every inhabited open $U \subseteq R'$, the preimage $h^{-1}(U)$ is inhabited. This is the case iff the image subset $h(R) \subseteq R'$ is dense in the usual sense.

Proposition A.37 *Let B, B' be predomains such that B' is rounded, and $H: B \to B'$ an approximable mapping. Then H is dense iff the morphism of topological spaces* $\mathrm{RId}(H): \mathrm{RId}(B) \to \mathrm{RId}(B')$ *is dense.*

Proof For the forward direction, suppose that H is dense and that $U' \in \Omega(B')$ is inhabited. By Proposition A.34 it suffices to show that $U := \Omega(H)(U')$ is inhabited. We have $b' \in U'$, so there exists $b \in B$ such that $H(b, b')$, which precisely says $b \in U$.

For the backward direction, suppose $\Omega(H)$ sends inhabited sets to inhabited sets, and take any $b_1' \in B'$. Since B' is rounded, there exists $b_2' \in B'$ with $b_1' \prec b_2'$. Let $U' := \uparrow b_1'$ and note $b_2' \in U'$. Then $U := \Omega(H)(U')$ is inhabited, so there exists $b \in U$, i.e. there exists $b' \in U'$ such that $H(b, b')$. Hence $b_1' \prec b'$, and by condition 3. $H(b, b_1')$ as desired. □

Definition A.38 (Lawson, Filtered Approximable Mapping) Say an approximable mapping $H: B \to B'$ is a *Lawson approximable mapping* (for short, "H is Lawson") if H satisfies the strengthening of condition 2 of Definition A.28, which is dual to 4: for any $n \geq 0$, if $H(b_i, b_1')$ for all $1 \leq i \leq n$ and $b_2' \prec b_1'$, then there exists a $b \in B$ such that $H(b, b_2')$ and $b \prec b_i$ for all $1 \leq i \leq n$.

If B is rounded, and both B and B' have conditional meets, this is equivalent to the duals of 4"a, 4"b, 4"c. Thus in this case, H is Lawson iff H is dense (Definition A.36) and satisfies

$$\big(H(b_1, b') \wedge H(b_2, b')\big) \Rightarrow H(b_1 \curlywedge b_2, b'). \tag{A.4}$$

If H is any (not necessarily dense) approximable mapping and B, B' are any predomains, we say H is *filtered* if it satisfies (A.4).

More generally, say that an approximable mapping $H: B_1 \times B_2 \to B_3$ is *filtered in the first variable* if

$$\Big(H\big((b_1, b_2), b_3\big) \wedge H\big((b_1', b_2), b_3\big)\Big) \Rightarrow H\big((b_1 \curlywedge b_1', b_2), b_3\big).$$

A.2.3 Approximable Mappings from Functions

Most of the continuous maps we will need to construct will in fact be induced by *functions* of predomains—meaning functions between underlying sets—in the following way.

Proposition A.39 *Let $f\colon B \to B'$ be a function between predomains, and assume B' is rounded and has conditional joins. Then the relation*

$$f^*(b, b') :\equiv b' \prec fb$$

is an approximable mapping if and only if

1. *$b_1 \prec b_2$ implies $fb_1 \leqslant fb_2$,*
2. *$b' \prec fb_2$ implies there exists some $b_1 \prec b_2$ with $b' \prec fb_1$.*

If these conditions hold, it follows that $b_1 \leqslant b_2$ implies $fb_1 \leqslant fb_2$.

Moreover, if B is also rounded, then f^* is dense if and only if $\forall(b' : B').\exists(b : B).\, b' \prec fb$. If B and B' also have conditional meets, then f^* is filtered if and only if f preserves meets: $fb_1 \wedge fb_2 = f(b_1 \wedge b_2)$.

Proof Suppose $f^*(b, b')$ defines an approximable mapping. If $b_1 \prec b_2$, then

$$b' \prec fb_1 \Leftrightarrow f^*(b_1, b') \Rightarrow f^*(b_2, b') \Leftrightarrow b' \prec fb_2,$$

hence $fb_1 \leqslant fb_2$. And the second condition is just part 2. of Definition A.28.

Conversely, suppose f satisfies the conditions of the proposition. By Lemma A.10, $\downarrow b' \in \mathrm{RId}(B')$ for all $b' \in B'$. Conditions 3. and 4. of Definition A.28 follow, and conditions 1. and 2. follow easily from 1. and 2. from the proposition.

To see that f preserves the specialization order, let $b_1 \leqslant b_2$ and $b' \prec fb_1$. Then by 2. there is a b_0 with $b' \prec fb_0$ and $b_0 \prec b_1$. But then $b_0 \prec b_1$ and $b_1 \leqslant b_2$ implies $b_0 \prec b_2$, hence $fb_0 \leqslant fb_2$ by 1., which together with $b' \prec fb_0$ implies $b' \prec fb_2$.

The characterizations of when f^* is dense or filtered follow directly from Definitions A.36 and A.38. □

Corollary A.40 *Any inclusion $B \subseteq B'$ of predomains defines an approximable mapping.*

Proposition A.41 *Let $B_1, \ldots, B_n$ and C be predomains, and let $f\colon B_1 \times \cdots \times B_n \to C$ be a function on underlying sets. Assume C is rounded and has conditional joins. Then the relation*

$$f^*((b_1, \ldots, b_n), c) :\equiv c \prec f(b_1, \ldots, b_n)$$

is an approximable mapping if and only if it satisfies the conditions of Proposition A.39 in each variable separately, i.e. for each $1 \leq i \leq n$,

1. $b_i \prec b'_i$ *implies* $f(b_1, \ldots, b_i, \ldots, b_n) \leqslant f(b_1, \ldots, b'_i, \ldots, b_n)$, *and*
2. $c \prec f(b_1, \ldots, b_n)$ *implies there exists a* $b'_i \prec b_i$ *with* $c \prec f(b_1, \ldots, b'_i, \ldots, b_n)$.

Moreover, if B is also rounded, then f^* *is dense if and only if*

$$\forall(b' : B'). \exists(b_1, \ldots, b_n : B). b' \prec f(b_1, \ldots, b_n).$$

For any $1 \leq i \leq n$, *if* B_i *and* C *also have conditional meets, then* f^* *is filtered in variable i if and only if*

$$f(b_1, \ldots, b_i, \ldots, b_n) \curlywedge f(b_1, \ldots, b'_i, \ldots, b_n) = f(b_1, \ldots, b_i \curlywedge b'_i, \ldots, b_n).$$

Proof Assume $f^*((b_1, \ldots, b_n), c)$ defines an approximable mapping. Then condition 1. follows exactly as in the proof of Proposition A.39. For the condition 2., let $b = (b_1, \ldots, b_n)$ and suppose $c \prec f(b)$. By Definition A.28 condition 2., there exists $b' = (b'_1, \ldots, b'_n)$ such that $b'_i \prec b_i$ for each i and $c \prec f(b')$. We obtain $c \prec f(b_1, \ldots, b'_i \ldots, b_n)$ by $n-1$ applications of condition 1, one for each $j \neq i$.

The converse direction is just as in the proof of Proposition A.39: Definition A.28 conditions 3 and 4 follow from Lemma A.10, and conditions 1 and 2 are n-applications of conditions 1 and 2 from the proposition.

The characterizations of when f^* is dense or filtered follow directly from Definitions A.36 and A.38. □

When f^* is the approximable mapping associated to a function $f : B_1 \times \cdots \times B_n \to C$ by Proposition A.41, the induced continuous map $\mathrm{RId}(f^*) : \mathrm{RId}(B_1) \times \cdots \times \mathrm{RId}(B_n) \to \mathrm{RId}(C)$ has a simple form:

$$\mathrm{RId}(f^*)(I_1, \ldots, I_n) := \{c \in C \mid \exists(b_1 \in I_1) \cdots (b_n \in I_n). c \prec f(b_1, \ldots, b_n)\}. \tag{A.5}$$

Here we are using the isomorphism $\mathrm{RId}(B_1 \times \cdots \times B_n) \cong \mathrm{RId}(B_1) \times \cdots \times \mathrm{RId}(B_n)$ from Proposition A.8.

The following is easy to prove and has an obvious analogue for multi-variable functions.

Proposition A.42 *Suppose* $f_1 : B_1 \to B_2$ *and* $f_2 : B_2 \to B_3$ *are functions satisfying the conditions of Proposition A.39. Then so does* $f_2 \circ f_1$, *and we have* $(f_2 \circ f_1)^* = f_2^* \circ f_1^*$.

Our main use of predomains is to define continuous functions between various sorts of real numbers objects in our topos, and various subtoposes. We deal with subtoposes in Sect. A.3, but before doing so we give an example that will be useful for defining multiplication.

Example A.43 Let $\underline{\mathbb{R}}^+_{\mathrm{pre}}$ denote the predomain whose carrier set is $\mathbb{Q}_{\geq 0} = \{q \in \mathbb{Q} \mid q \geq 0\}$ and where $q_1 \prec q_2$ iff $q_1 < q_2$ or $q_1 = q_2 = 0$. Let $\bar{\mathbb{R}}^+_{\mathrm{pre}}$ denote the predomain has the same carrier set, but with $q_1 \prec q_2$ iff $q_1 < q_2$. Defining $\underline{\mathbb{R}}^+ = \{r \in \underline{\mathbb{R}} \mid 0 \leq r\}$ and $\bar{\mathbb{R}}^+ := \{r \in \bar{\mathbb{R}} \mid 0 \leq r\}$, then classically there are isomorphisms

$$\underline{\mathbb{R}}^+ \cong \mathrm{RId}(\underline{\mathbb{R}}^+_{\mathrm{pre}}) \qquad \text{and} \qquad \bar{\mathbb{R}}^+ \cong \mathrm{RId}(\bar{\mathbb{R}}^+_{\mathrm{pre}})$$

We want to show that various operations on real numbers are continuous. For example, the subtraction operation $(q_1, q_2) \mapsto q_1 - q_2$ reverses the order in the second variable, so it defines a function $-\colon \underline{\mathbb{R}}_{\mathrm{pre}} \times \bar{\mathbb{R}}_{\mathrm{pre}} \to \underline{\mathbb{R}}_{\mathrm{pre}}$ and a function $-\colon \bar{\mathbb{R}}_{\mathrm{pre}} \times \underline{\mathbb{R}}_{\mathrm{pre}} \to \bar{\mathbb{R}}_{\mathrm{pre}}$. One checks easily that it satisfies the conditions of Proposition A.41, and hence these are approximable mappings.

Similarly, the function $(q_1, q_2) \mapsto \max(q_1, q_2)$ satisfies the Proposition A.41 and hence defines approximable mappings $\max\colon \underline{\mathbb{R}}_{\mathrm{pre}} \times \underline{\mathbb{R}}_{\mathrm{pre}} \to \underline{\mathbb{R}}_{\mathrm{pre}}$ and $\max\colon \bar{\mathbb{R}}_{\mathrm{pre}} \times \bar{\mathbb{R}}_{\mathrm{pre}} \to \bar{\mathbb{R}}_{\mathrm{pre}}$. We also need two auxiliary notations: $q^+ := \max(q, 0)$ and $q^- := \max(0 - q, 0)$.

Finally, the product of nonnegative lower (resp. upper) reals is unproblematic. That is, the function $(q_1, q_2) \mapsto q_1 \times q_2$ satisfies the Proposition A.41 and hence defines approximable mappings $\times\colon \underline{\mathbb{R}}^+_{\mathrm{pre}} \times \underline{\mathbb{R}}^+_{\mathrm{pre}} \to \underline{\mathbb{R}}^+_{\mathrm{pre}}$ and $\times\colon \bar{\mathbb{R}}^+_{\mathrm{pre}} \times \bar{\mathbb{R}}^+_{\mathrm{pre}} \to \bar{\mathbb{R}}^+_{\mathrm{pre}}$. Summarizing, we have defined approximable mappings

$$\begin{aligned}
-\colon \underline{\mathbb{R}}_{\mathrm{pre}} \times \bar{\mathbb{R}}_{\mathrm{pre}} &\to \underline{\mathbb{R}}_{\mathrm{pre}} & -\colon \bar{\mathbb{R}}_{\mathrm{pre}} \times \underline{\mathbb{R}}_{\mathrm{pre}} &\to \bar{\mathbb{R}}_{\mathrm{pre}} \\
\max\colon \underline{\mathbb{R}}_{\mathrm{pre}} \times \underline{\mathbb{R}}_{\mathrm{pre}} &\to \underline{\mathbb{R}}_{\mathrm{pre}} & \max\colon \bar{\mathbb{R}}_{\mathrm{pre}} \times \bar{\mathbb{R}}_{\mathrm{pre}} &\to \bar{\mathbb{R}}_{\mathrm{pre}} \\
.^+\colon \underline{\mathbb{R}}_{\mathrm{pre}} &\to \underline{\mathbb{R}}^+_{\mathrm{pre}} & .^+\colon \bar{\mathbb{R}}_{\mathrm{pre}} &\to \bar{\mathbb{R}}^+_{\mathrm{pre}} \\
.^-\colon \underline{\mathbb{R}}_{\mathrm{pre}} &\to \bar{\mathbb{R}}^+_{\mathrm{pre}} & .^-\colon \bar{\mathbb{R}}_{\mathrm{pre}} &\to \underline{\mathbb{R}}^+_{\mathrm{pre}}
\end{aligned}$$

Taking products, we also obtain approximable mappings $-\colon \bar{\underline{\mathbb{R}}}_{\mathrm{pre}} \times \bar{\underline{\mathbb{R}}}_{\mathrm{pre}} \to \bar{\underline{\mathbb{R}}}_{\mathrm{pre}}$ and $\max\colon \bar{\underline{\mathbb{R}}}_{\mathrm{pre}} \times \bar{\underline{\mathbb{R}}}_{\mathrm{pre}} \to \bar{\underline{\mathbb{R}}}_{\mathrm{pre}}$. Since approximable mappings induce continuous functions between domains, one may of course remove the $_{\mathrm{pre}}$ subscripts in all of the above cases.

Multiplication $*\colon \bar{\underline{\mathbb{R}}}_{\mathrm{pre}} \times \bar{\underline{\mathbb{R}}}_{\mathrm{pre}} \to \bar{\underline{\mathbb{R}}}_{\mathrm{pre}}$ is then defined by $(d_1, u_1) * (d_2, u_2) = (d', u')$ where

$$\begin{aligned}
d' &:= \max(d_1^+ d_2^+, u_1^- u_2^-) - \max(u_1^+ d_2^-, u_2^+ d_1^-) \\
u' &:= \max(u_1^+ u_2^+, d_1^- d_2^-) - \max(d_1^+ u_2^-, d_2^+ u_1^-)
\end{aligned} \tag{A.6}$$

This is a composite of approximable mappings, and hence is approximable, so it defines a continuous map of domains

$$*\colon \bar{\underline{\mathbb{R}}} \times \bar{\underline{\mathbb{R}}} \to \bar{\underline{\mathbb{R}}}$$

which satisfies the usual properties of Kaucher multiplication; see [Kau80].

A.3 Predomains in Subtoposes

In this section we interpret predomains and rounded ideals inside a subtopos that is given by some modality j : $\texttt{Prop} \to \texttt{Prop}$ on a topos $\mathcal{E}$. We also consider what happens when we change the modality.

We continue to think of a predomain as an object B in $\mathcal{E}$ together with a map $\prec : B \times B \to \texttt{Prop}$, eliding the extra data of the map $(b_1, b_2) \mapsto \{b_1|b_2\}$; see Definition A.1.

Definition A.44 (*j*-**Rounded Ideal,** *j*-**Open Subset**) Let $(B, \prec)$ be an internal predomain in $\mathcal{E}$ and j a modality. Define the type of *j-rounded ideals*, denoted $\mathrm{RId}_j(B)$, to be the subtype of $B \to \texttt{Prop}$ given by those I satisfying

(j-closed)	$\forall (b : B).\, j(Ib) \Rightarrow Ib$
(down-closed)	$\forall (b_1, b_2 : B).\, (b_1 \prec b_2) \Rightarrow Ib_2 \Rightarrow Ib_1$
(up-directed)	$\forall (b_1, b_2 : B).\, Ib_1 \Rightarrow Ib_2 \Rightarrow j$ $\big(\exists b_3.\, (b_1 \prec b_3) \wedge (b_2 \prec b_3) \wedge Ib_3\big).$
(nonempty)	$j(\exists (b : B).\, Ib)$

Similarly, define the type of *j-open subsets*, denoted $\Omega_j(B)$, by

$$\Omega_j(B) := \{U : B \to \texttt{Prop} \mid \forall (b : B).\, Ub \Leftrightarrow j\exists (b' : B).\, Ub' \wedge (b' \prec b)\}$$

and the relation $\models^B_j$ on $\mathrm{RId}_j(B) \times \Omega_j(B)$ given by

$$I \models^B_j U \Leftrightarrow j\exists (b : B).\, Ib \wedge Ub.$$

Example A.45 For each of the predomains in Example A.3 we can define the corresponding domain of j-rounded ideals, as we did in Example A.9. Except for two exceptions, each of the resulting domains models a kind of Dedekind numeric objects, e.g. the lower reals, the upper reals, the improper intervals, etc. The exceptions are the j-rounded ideals in $\mathbb{IR}_{\mathrm{pre}} = \{(q_1, q_2) \in \bar{\underline{\mathbb{R}}}_{\mathrm{pre}} \mid q_1 < q_2\}$ and similarly $\mathbb{IR}^{\infty}_{\mathrm{pre}}$, which do not appropriately handle disjointness. To correct this, we define

$$\mathbb{IR}_{j,\mathrm{pre}} := \{(q_1, q_2) \in \bar{\underline{\mathbb{R}}}_{\mathrm{pre}} \mid j(q_1 < q_2)\} \quad \text{and}$$
$$\mathbb{IR}^{\infty}_{j,\mathrm{pre}} := \{(q_1, q_2) \in \bar{\underline{\mathbb{R}}}^{\infty}_{\mathrm{pre}} \mid j(q_1 < q_2)\}.$$

With these definitions in hand, we can define the following domains:

$$\underline{\mathbb{R}}_j := \mathrm{RId}_j(\underline{\mathbb{R}}_{\mathrm{pre}}) \quad \bar{\mathbb{R}}_j := \mathrm{RId}_j(\bar{\mathbb{R}}_{\mathrm{pre}}) \quad \bar{\underline{\mathbb{R}}}_j := \mathrm{RId}_j(\bar{\underline{\mathbb{R}}}_{\mathrm{pre}}) \quad \mathbb{IR}_j := \mathrm{RId}_j(\mathbb{IR}_{j,\mathrm{pre}})$$

$$\underline{\mathbb{R}}^{\infty}_j := \mathrm{RId}_j(\underline{\mathbb{R}}^{\infty}_{\mathrm{pre}}) \quad \bar{\mathbb{R}}^{\infty}_j := \mathrm{RId}_j(\bar{\mathbb{R}}^{\infty}_{\mathrm{pre}}) \quad \bar{\underline{\mathbb{R}}}^{\infty}_j := \mathrm{RId}_j(\bar{\underline{\mathbb{R}}}^{\infty}_{\mathrm{pre}}) \quad \mathbb{IR}^{\infty}_j := \mathrm{RId}_j(\mathbb{IR}^{\infty}_{j,\mathrm{pre}})$$

In Proposition 4.30 it is shown that these definitions agree with our definition (4.21) of Dedekind j-numeric objects in any topos.

Corollary A.46 *For any topos $\mathcal{E}$, if $(B, \prec)$ is an internal predomain, then* $\mathrm{RId}_j(B)$ *is an internal domain in $\mathcal{E}_j$. Given $I, I' : B \to$ `Prop`, the domain order is given by $I' \sqsubseteq I \Leftrightarrow \forall(b : B).\, I'b \Rightarrow Ib$, and the way-below relation is given by $I' \ll I \Leftrightarrow j\exists(b : B).\, Ib \wedge \forall(b' : B).\, I'b' \Rightarrow j(b' \prec b)$.*

Proof The first part is Theorem A.18 and the second part is Proposition A.17, applied in the topos $\mathcal{E}_j$. □

In the following propositions, we explain how Definition A.44 is natural with respect to changing either the modality j or the predomain B.

Proposition A.47 *Fix a predomain $(B, \prec)$. Suppose j' and j are modalities, such that $j'P \Rightarrow jP$ for any P :* `Prop`*. Then applying j pointwise (i.e. by composing $X : B \to$* `Prop` *with j :* `Prop` $\to$ `Prop`*) defines a monotonic map on rounded ideals and on open sets, which we denote*

$$j\colon \mathrm{RId}_{j'}(B) \to \mathrm{RId}_j(B) \qquad \textit{and} \qquad j\colon \Omega_{j'}(B) \to \Omega_j(B)$$

Proof Suppose $I : B \to$ `Prop` is in $\mathrm{RId}_{j'}(B)$; we will show jI satisfies the four conditions necessary to be in $\mathrm{RId}_j(B)$. The first follows because j is a modality, and likewise the third follows because $Ib_2 \Rightarrow Ib_1$ implies $jIb_2 \Rightarrow jIb_1$. For the second, assuming $j'(\exists b.\, Ib)$ we get $j(\exists b.\, Ib)$ by assumption, and thus $j(\exists b.\, jIb)$. For the fourth, assuming jIb_1 and jIb_2 we get $j'j(\exists b_3.\, (b_1 \prec b_3) \wedge (b_2 \prec b_3) \wedge Ib_3)$, hence $j(\exists b_3.\, (b_1 \prec b_3) \wedge (b_2 \prec b_3) \wedge jIb_3)$ using $j' \Rightarrow j$ to remove the j', and $Ib_3 \Rightarrow jIb_3$ to add the j. This is clearly monotonic, because for any b, $Ib \Rightarrow I'b$ implies $jIb \Rightarrow jI'b$.

The proof that $U \in \Omega'_j(B)$ implies $jU \in \Omega_j(B)$ is similar but easier. □

Remark A.48 In general the map $j\colon \mathrm{RId}_{j'}(B) \to \mathrm{RId}_j(B)$ from Proposition A.47 is open (as shown), but *not* continuous. Indeed, for $I \in \mathrm{RId}_{j'}(B)$ and $U \in \Omega_j(B) \subseteq \Omega_{j'}(B)$, the condition Lemma A.33 that $(jI \models_j U) \Leftrightarrow^? (I \models_{j'} U)$ translates to

$$(j\exists b.\, Ub \wedge jIb) \Leftrightarrow^? (j'\exists b.\, Ub \wedge Ib)$$

and there is no reason to expect that to hold.

Proposition A.49 *Suppose given an approximable mapping $H\colon B \to B'$, and let j be a modality. For any $I \in \mathrm{RId}_j(B)$ and $U' \in \Omega_j(B')$, define $\mathrm{RId}_j(H)(I) := I'$ and $\Omega_j(H)(U') := U$ where*

$$I'b' \Leftrightarrow j\exists(b : B).\, Ib \wedge H(b, b') \qquad \textit{and} \qquad Ub \Leftrightarrow j\exists(b' : B').\, U'b' \wedge H(b, b').$$

This defines a morphism of domains $\mathrm{RId}_j(H)\colon \mathrm{RId}_j(B) \to \mathrm{RId}_j(B')$.

Proof Take $I : \mathrm{RId}_j(B)$ and let I' be as above; we must first show I' satisfies the conditions of Definition A.44. The first is obvious, so consider the second. Since I is rounded, condition 4 of Definition A.28 implies $\exists(b : B)(b' : B').\, Ib \wedge H(b, b')$, which directly implies $j(\exists b'.\, I'b')$. For the third, we are given $b'_1 \prec b'_2$ with $I'b'_2$ and need to show $I'b'_1$, but this is just condition 3. For the fourth, we assume $I'b'_1$ and $I'b'_2$, and hence obtain

$$j\big(\exists(b_1, b_2 : B).\, Ib_1 \wedge Ib_2 \wedge H(b_1, b'_1) \wedge H(b_2, b'_2)\big)$$

Because I is directed we have $j\big(\exists b.\, Ib \wedge H(b, b'_1) \wedge H(b, b'_2)\big)$, and because it is rounded condition 4 implies

$$j\big(\exists(b : B).\, Ib \wedge \exists(b' : B').\, (b'_1 \prec b') \wedge (b'_2 \prec b') \wedge H(b, b')\big).$$

Rearranging, we obtain $j\big(\exists(b' : B').\, (b_1 \prec b') \wedge (b_2 \prec b') \wedge \big(\exists(b : B).\, Ib \wedge H(b, b')\big)\big)$, which implies $j(\exists b'.\, b_1 \prec b' \wedge b_2 \prec b' \wedge I'b')$, as desired.

It remains to show that $\mathrm{RId}_j(H)$ is continuous, and by Lemma A.33, it suffices to take $I \in \mathrm{RId}_j(B)$ and $U' \in \Omega_j(B')$, let $I' := \mathrm{RId}_j(H)(I)$ and $U := \Omega_j(H)(U)$, and show $(I \models^B_j U) \Leftrightarrow (I' \models^{B'}_j U')$. Using the fact that B and B' are inhabited, it is easy to check the following equivalences, completing the proof:

$$\begin{aligned}(I \models^B_j U) &\Leftrightarrow j\exists b.\, Ib \wedge j\exists b'.\, U'b' \wedge H(b, b')\\ &\Leftrightarrow j\exists b \exists b'.\, Ib \wedge U'b' \wedge H(b, b')\\ &\Leftrightarrow j\exists b'.\, U'b' \wedge j\exists b.\, Ib \wedge H(b, b') \Leftrightarrow (I' \models^{B'}_j U').\end{aligned}$$

□

As in Eq. (A.5), when f^* is the approximable mapping associated to a function $f\colon B_1 \times \cdots \times B_n \to C$ by Proposition A.41, the induced continuous map $\mathrm{RId}_j(f^*)\colon \mathrm{RId}_j(B_1) \times \cdots \times \mathrm{RId}_j(B_n) \to \mathrm{RId}_j(C)$ has a simple form:

$$\mathrm{RId}_j(f^*)(I_1, \ldots, I_n) := \{\, c \in C \mid j\exists(b_1 \in I_1) \cdots (b_n \in I_n).\, c \prec f(b_1, \ldots, b_n) \,\}. \tag{A.7}$$

Proposition A.50 *Suppose that $H\colon B \to B'$ is an approximable mapping and that j_1, j_2 are modalities such that $j_1 P \Rightarrow j_2 P$ for all P. Then the following squares, where the sides are as defined in Propositions A.47 and A.49, commute:*

$$\begin{array}{ccc} \mathrm{RId}_{j_1}(B) & \xrightarrow{\mathrm{RId}_{j_1}(H)} & \mathrm{RId}_{j_1}(B') \\ {\scriptstyle j_2}\downarrow & & \downarrow{\scriptstyle j_2} \\ \mathrm{RId}_{j_2}(B) & \xrightarrow[\mathrm{RId}_{j_2}(H)]{} & \mathrm{RId}_{j_2}(B') \end{array} \qquad \begin{array}{ccc} \Omega_{j_1}(B) & \xleftarrow{\Omega_{j_1}(H)} & \Omega_{j_1}(B') \\ {\scriptstyle j_2}\downarrow & & \downarrow{\scriptstyle j_2} \\ \Omega_{j_2}(B) & \xleftarrow[\Omega_{j_2}(H)]{} & \Omega_{j_2}(B') \end{array}$$

Proof For the first square, choose I in $\mathrm{RId}_{j_1} B$ and $b' \in B'$. It is easy to check that indeed

$$j_2 j_1\big(\exists(b : B).\, Ib \wedge H(b, b')\big) \Leftrightarrow j_2\big(\exists(b : B).\, j_2 Ib \wedge H(b, b')\big).$$

The proof that the second square commutes is similar. □

A.3.1 Closed Modalities

The map $j \colon \mathrm{RId}_{j'} \to \mathrm{RId}_j$ from Proposition A.47 has a section if j is a closed modality. Recall that j is closed if there is some proposition $\phi : \mathtt{Prop}$ such that $jP = \phi \vee P$ for all $P : \mathtt{Prop}$.

Proposition A.51 *Let B be a predomain, let j be a closed modality, and let j' be a modality such that $j' \Rightarrow j$. Then $\Omega_j(B) \subseteq \Omega_{j'}(B)$, and if B has binary joins then* $\mathrm{RId}_j(B) \subseteq \mathrm{RId}_{j'}(B)$.

Proof Suppose $jP = \phi \vee P$ and consider the second claim by taking a j-rounded ideal $I \in \mathrm{RId}_j(B)$. We may assume $j' = \mathrm{id}$, so we need to show that I is nonempty and up-directed. We can easily prove $\phi \Rightarrow \exists(b : B).\, Ib$, using that B is inhabited and that I is j-closed (so that $\phi \Rightarrow Ib$ for any b). Hence I being j-nonempty implies that I is nonempty.

To show that I is j'-up-directed, let $b_1, b_2 : B$ be elements such that Ib_1 and Ib_2. Using again that I is j-closed, and taking $b_3 = b_1 \vee b_2$, we can prove $\phi \Rightarrow \exists(b_3 : B).\, (b_1 \prec b_3) \wedge (b_2 \prec b_3) \wedge Ib_3$. Hence I being j-up-directed implies that I is up-directed.

The proof of the first claim is similar, but easier. □

Proposition A.52 *Let B and B' be predomains, let j be a closed modality, and let j' be a modality such that $j' \Rightarrow j$. Let $H \colon B \to B'$ be an approximable mapping. If B and B' have binary joins and H is dense, then the left-hand diagram commutes; if B is rounded, then the right-hand diagram commutes:*

$$\begin{array}{ccc} \mathrm{RId}_j(B) & \xrightarrow{\mathrm{RId}_j(H)} & \mathrm{RId}_j(B') \\ \big\downarrow & & \big\downarrow \\ \mathrm{RId}_{j'}(B) & \xrightarrow[\mathrm{RId}_{j'}(H)]{} & \mathrm{RId}_{j'}(B') \end{array} \qquad \begin{array}{ccc} \Omega_j(B) & \xleftarrow{\Omega_j(H)} & \Omega_j(B') \\ \big\downarrow & & \big\downarrow \\ \Omega_{j'}(B) & \xleftarrow[\Omega_{j'}(H)]{} & \Omega_{j'}(B') \end{array}$$

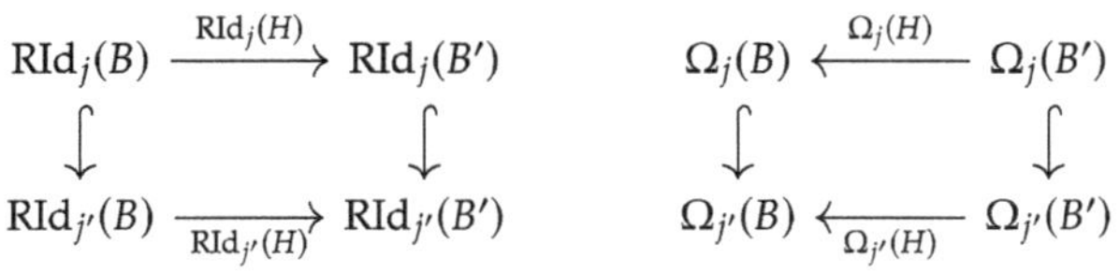

Proof Suppose $jP = \phi \vee P$; it suffices to prove the result for $j' = \mathrm{id}$. For the first diagram, given $I \in \mathrm{RId}_j(B)$, we want to show that given any $b' \in B'$ the following holds:

$$j\exists(b : B).\, Ib \wedge H(b, b') \Leftrightarrow \exists(b : B).\, Ib \wedge H(b, b')$$

It suffices to show $\phi \Rightarrow \exists(b : B).\, Ib \wedge H(b, b')$. Since H is dense, we have $\exists b.\, H(b, b')$, and since I is j-closed, we have $\forall b.\, \phi \Rightarrow Ib$, and the result follows.

For the second diagram take $U' \in \Omega_j(B')$ and $b \in B$. We want to show

$$j\exists(b' : B').\, U'b' \wedge H(b, b') \Leftrightarrow \exists(b' : B').\, U'b' \wedge H(b, b')$$

Again it suffices to show $\phi \Rightarrow \exists(b' : B').\, U'b' \wedge H(b, b')$, and again since U' is j-closed, we have $\forall b'.\, \phi \Rightarrow U'b'$. Since B is rounded, $\exists b'.\, H(b, b')$, and the result follows. □

One proves the following proposition similarly.

Proposition A.53 *Let B_1, B_2, and B_3 be predomains with binary joins. Let ϕ_1, ϕ_2, ϕ_3 : `Prop` be such that $(\phi_1 \vee \phi_2) \Rightarrow \phi_3$, and let j_1, j_2, and j_3 be the corresponding closed modalities. Let j' be any modality such that $j'P \Rightarrow (j_1 P \wedge j_2 P \wedge j_3 P)$ for all P. Let $H\colon B_1 \times B_2 \to B_3$ be a dense approximable mapping. Then the dotted lift exists in the diagram:*

$$\begin{array}{ccc}
\mathrm{RId}_{j_1}(B_1) \times \mathrm{RId}_{j_2}(B_2) & \dashrightarrow & \mathrm{RId}_{j_3}(B_3) \\
\downarrow & & \downarrow \\
\mathrm{RId}_{j'}(B_1) \times \mathrm{RId}_{j'}(B_2) & \xrightarrow[\mathrm{RId}_{j'}(H)]{} & \mathrm{RId}_{j'}(B_3)
\end{array}$$

Lemma A.54 *Suppose B is any predomain, j is a closed modality with $jP := \phi \vee P$, and take any $I \in \mathrm{RId}(B)$ and $U \in \Omega_j(B)$. Then $(j\exists b.\, Ib \wedge Ub) \Leftrightarrow (\exists b.\, Ib \wedge Ub)$.*

Proof It suffices to show $\phi \Rightarrow \exists b.\, Ib \wedge Ub$. Since ideals are inhabited, we have $\exists b.\, Ib$, and since U is j-closed, $\forall b.\, \phi \Rightarrow Ub$. The result follows directly from there. □

Proposition A.55 *Let B be a predomain, let j be a closed modality, and let j' be a modality such that $j' \Rightarrow j$. Then the inclusion $\mathrm{RId}_j(B) \subseteq \mathrm{RId}_{j'}$ from Proposition A.51 is continuous.*

Proof To see that $\mathrm{RId}_j(B) \subseteq \mathrm{RId}_{j'}$ is continuous, it suffices by Lemma A.33 to show that $(I \models_j jU) \Leftrightarrow (I \models_{j'} U)$ for any $I \in \mathrm{RId}_j(B)$ and $U \in \Omega_{j'}(B)$. This translates to showing $(j\exists b.\, Ib \wedge jUb) \Leftrightarrow^? (j'\exists b.\, Ib \wedge Ub)$, and it suffices to show $(j\exists b.\, Ib \wedge Ub) \Rightarrow^? \exists b.\, Ib \wedge Ub$, which is Lemma A.54. □

A.3.2 Constant Predomains and Ideals

In the main body of this book, we consider a sheaf topos $\mathcal{B}$ for which constant sheaves C have a number of special properties. One of the most useful is Eq. (A.8) shown below, which Johnstone calls the "dual Frobenius rule"; see [Joh02, Example C.1.1.16e].[1]

$$\forall (P : \texttt{Prop})(P' : \texttt{C} \to \texttt{Prop}).\, \big(\forall (c : \texttt{C}).\, P \vee P'(c)\big) \Rightarrow \big(P \vee \forall (c : \texttt{C}).\, P'(c)\big). \tag{A.8}$$

This property does not hold for constant sheaves in an arbitrary topos $\mathcal{E}$. However it does hold if either

- $\texttt{C}$ is finite (i.e., a finite coproduct $\texttt{C} = 1 + 1 + \cdots + 1$ of copies of the terminal object) or
- there is a defining site $(\mathcal{S}, \chi)$ for $\mathcal{E}$ such that every covering family in χ is filtered.

For example, the dual Frobenius property holds for constant objects in any presheaf topos. It also holds in $\mathcal{B} = \mathsf{Shv}(S_{\mathbb{IR}/\rhd})$ because every covering family in $S_{\mathbb{IR}/\rhd}$ is filtered; see Definition 3.5. Since our main application of the material in this section is to the specific topos $\mathcal{B}$, we use the name "constant predomains" to refer to predomains having this property.

Definition A.56 (Constant Predomains, Ideals, Decidable Mappings) In a topos $\mathcal{E}$, we say that a predomain $(B, \prec)$ is *constant* if B satisfies Eq. (A.8) and $\prec\colon B \times B \to \texttt{Prop}$ is a decidable predicate, i.e. $\forall (b, b' : B).\, (b' \prec b) \vee \neg(b' \prec b)$.

If B and B' are constant predomains, say that an approximable mapping $H\colon B \to B'$ is *decidable* if $H(b, b') \vee \neg H(b, b')$ holds for all $b : B$ and $b' : B'$.

Finally, suppose B is a constant predomain and j is a modality. Say that a j-rounded ideal $I \in \mathrm{RId}_j(B)$ is *j-constant* if it satisfies $\forall (b : B).\, Ib \vee (Ib \Rightarrow j\bot)$. We denote the type of j-constant rounded ideals in B by $\mathrm{cRId}_j(B)$.

Example A.57 The relation $<$ on $\mathbb{Q}$ is decidable in any topos $\mathcal{E}$. Thus if $\mathbb{Q}$ satisfies Eq. (A.8), then the predomain $(\mathbb{Q}, <)$, as well as all its variants from Example A.3), are constant in $\mathcal{E}$. For example, this is the case in $\mathcal{B}$. Each of the variants is also rounded and has conditional joins in the sense of Definition A.4. In Proposition A.58, we show that elements of the predomain are constant as elements of the associated domain.

Proposition A.58 *Suppose that $(B, \prec)$ is a constant predomain that is rounded and has conditional joins, and let j be a modality. For any $b : B$, the predicate $j(b' \prec b)$ defines a map $B \to \mathrm{RId}_j(B)$, which we denote $\downarrow_j$, and every rounded ideal of the form $\downarrow_j b$ is j-constant:*

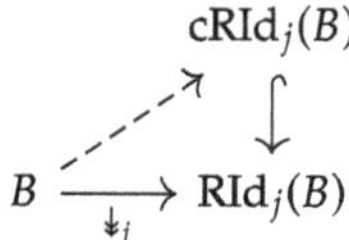

[1] The dual Frobenius rule shows up in the main body of this book as Axiom 2.

In particular, the constant elements form a basis of the domain.

Proof The first claim is obvious; see Lemma A.10. To check the second claim, take $b, b' : B$. From the fact that $\prec$ is decidable, we get $(b' \prec b) \vee ((b' \prec b) \Rightarrow \bot)$, hence $j(b' \prec b) \vee ((b' \prec b) \Rightarrow j\bot)$ as desired. For the third claim, we use the fact that any superset of a basis is a basis; see Proposition A.19. □

Proposition A.59 *Let B be a constant predomain and let j' and j be modalities such that $j' \Rightarrow j$. Then the dotted lift exists in the diagram:*

$$\begin{array}{ccc} \mathrm{cRId}_{j'}(B) & \dashrightarrow & \mathrm{cRId}_{j}(B) \\ \downarrow & & \downarrow \\ \mathrm{RId}_{j'}(B) & \xrightarrow[j]{} & \mathrm{RId}_{j}(B) \end{array}$$

Proof It is straightforward to check that $Ib \vee (Ib \Rightarrow j'\bot)$ implies $jIb \vee (jIB \Rightarrow j\bot)$. □

Proposition A.60 *Let B and B' be constant predomains, let j be a modality, and let $H : B \to B'$ be a decidable approximable mapping. Then the dotted lift exists in the diagram:*

$$\begin{array}{ccc} \mathrm{cRId}_{j}(B) & \dashrightarrow & \mathrm{cRId}_{j}(B') \\ \downarrow & & \downarrow \\ \mathrm{RId}_{j}(B) & \xrightarrow[\mathrm{RId}_j(H)]{} & \mathrm{RId}_{j}(B') \end{array}$$

Proof Choose $I \in \mathrm{cRId}_j(B)$ and let $I' := \mathrm{RId}_j(H)(I)$, so $I'b' \Leftrightarrow j\exists(b : B).\, Ib \wedge H(b, b')$. We want to show that $I'b' \vee (I'b' \Rightarrow j\bot)$ for any $b' : B'$, so fix b'. For any $b : B$, since $Ib \vee (Ib \Rightarrow j\bot)$ and $H(b, b') \vee \neg H(b, b')$, it follows that $(Ib \wedge H(b, b')) \vee ((Ib \wedge H(b, b')) \Rightarrow j\bot)$, and thus

$$\forall(b : B).\, I'b' \vee ((Ib \wedge H(b, b')) \Rightarrow j\bot).$$

By Eq. (A.8), we get $I'b' \vee \forall(b : B).\, (Ib \wedge H(b, b')) \Rightarrow j\bot$, and this implies the result. □

Proposition A.61 *Let B_1, B_2, and B_3 be constant predomains with binary joins. Let ϕ_1, ϕ_2, ϕ_3 : `Prop` be such that $(\phi_1 \vee \phi_2) \Rightarrow \phi_3$, and let j'_1, j'_2, and j'_3 be the corresponding closed modalities. Let j be any modality such that $jP \Rightarrow (j'_1 P \wedge j'_2 P \wedge j'_3 P)$ for all P. Let $H\colon B_1 \times B_2 \to B_3$ be a decidable dense approximable mapping. Then the dotted lifts exist in the diagram:*

$$\begin{array}{ccc}
\mathrm{cRId}_{j_1'}(B_1) \times \mathrm{cRId}_{j_2'}(B_2) & \dashrightarrow & \mathrm{cRId}_{j_3'}(B_3) \\
\downarrow & & \downarrow \\
\mathrm{RId}_{j_1'}(B_1) \times \mathrm{RId}_{j_2'}(B_2) & \dashrightarrow & \mathrm{RId}_{j_3'}(B_3) \\
\downarrow & & \downarrow \\
\mathrm{RId}_j(B_1) \times \mathrm{RId}_j(B_2) & \xrightarrow[\mathrm{RId}_j(H)]{} & \mathrm{RId}_j(B_3)
\end{array}$$

Proof The bottom lift exists by Proposition A.53, so we check the top one. Choose $I_1 \in \mathrm{cRId}_{j_1'}(B_1)$ and $I_2 \in \mathrm{cRId}_{j_2'}(B_2)$, and let $I_3 := \mathrm{RId}_j(H)(I_1, I_2)$. Then by using the bottom lift we have

$$I_3 b_3 \Leftrightarrow \phi_3 \vee \exists (b_1 : B_1)(b_2 : B_2).\, I_1 b_1 \wedge I_2 b_2 \wedge H(b_1, b_2, b_3)$$

for any $b_3 : B_3$. We want to show $I_3 b_3 \vee (I_3 b_3 \Rightarrow \phi_3)$.

For any $b_1 : B_1$ and $b_2 : B_2$, we have $I_1 b_1 \vee (I_1 b_1 \Rightarrow \phi_1)$ and $I_2 b_2 \vee (I_2 b_2 \Rightarrow \phi_2)$, as well as $H(b_1, b_2, b_3) \vee \neg H(b_1, b_2, b_3)$, and it follows that

$$\forall (b_1, b_2).\, (I_1 b_1 \wedge I_2 b_2 \wedge H(b_1, b_2, b_3)) \vee ((I_1 b_1 \wedge I_2 b_2 \wedge H(b_1, b_2, b_3)) \Rightarrow \phi_3)$$

Clearly, $(I_1 b_1 \wedge I_2 b_2 \wedge H(b_1, b_2, b_3)) \Rightarrow I_3 b_3$. By Eq. (A.8) we have $I_3 b_3 \vee \forall (b_1, b_2).\, (I_1 b_1 \wedge I_2 b_2 \wedge H(b_1, b_2, b_3) \Rightarrow \phi_3)$, which immediately implies the result. □

A.3.3 Dense Modalities

Definition A.62 (Dense Modality) A modality j is called *dense* if $j\bot \Rightarrow \bot$.

Proposition A.63 *If j is a dense modality, then any decidable proposition P is j-closed.*

Proof Suppose $P \vee \neg P$. If jP and $\neg P$, then $j\bot$, hence $\bot$. Thus we have $(jP \wedge P) \Rightarrow P$ and $(jP \wedge \neg P) \Rightarrow P$, giving $jP \Rightarrow P$. □

A.3.4 Proper Modalities

Definition A.64 (Directed Preorder, Directed Join, Continuous Inclusion) A *directed preorder* is a pair $(D, \leq)$ satisfying

- $\exists (d : D).\, \top$,
- $\forall (d : D).\, d \leq d$,
- $\forall (d_1, d_2, d_3 : D).\, (d_1 \leq d_2) \Rightarrow (d_2 \leq d_3) \Rightarrow (d_1 \leq d_3)$, and
- $\forall (d_1, d_2 : D).\, \exists (d' : D).\, (d_1 \leq d') \wedge (d_2 \leq d')$.

Recall that a predicate $F : D \to \texttt{Prop}$ is monotonic if $(d_1 \le d_2) \Rightarrow Fd_1 \Rightarrow Fd_2$. Given a monotonic predicate F, we refer to the proposition $\exists(d : D).\, Fd$ as the *(directed) join of F, indexed by D.*

Let j be a modality. We will say the inclusion $\texttt{Prop}_j \subseteq \texttt{Prop}$ is *continuous* if it preserves directed joins indexed by constant types. More precisely, the inclusion is continuous if, for every directed preorder $(D, \le)$ such that D is constant in the sense of Appendix A.3.2 and every monotonic j-closed predicate $F : D \to \texttt{Prop}_j$, the following implication holds: $(j\exists(d : D).\, Fd) \Rightarrow (\exists(d : D).\, Fd)$.

Proposition A.65 *Let j be a modality. The following are equivalent:*

1. *The inclusion $\texttt{Prop}_j \subseteq \texttt{Prop}$ is continuous in the above sense.*
2. *If C is any inhabited constant object and $P : C \to \texttt{Prop}$ is a subset such that*

$$\forall(c_1, c_2 : C).\, \exists(c' : C).\, (P(c_1) \Rightarrow P(c')) \wedge (P(c_2) \Rightarrow P(c')). \tag{A.9}$$

then $(j\exists(c : C).\, P(c)) \Rightarrow \exists(c : C).\, jP(c)$.

Proof 1 $\Rightarrow$ 2) Define an order on C by $(c_1 \le c_2) \Leftrightarrow (P(c_1) \Rightarrow P(c_2))$. Then clearly C is directed and $P\colon C \to \texttt{Prop}$ is monotonic, hence jP is monotonic and j-closed, therefore $j\exists(c : C).\, P(c)$ implies $j\exists(c : C).\, jP(c)$, which implies $(\exists(c : C).\, jP(c))$.

2 $\Rightarrow$ 1) Let D be a constant, directed preorder, and $F\colon D \to \texttt{Prop}_j$ a monotonic map. Then D clearly satisfies (A.9), hence $j\exists(d : D).\, Fd$ implies $\exists(d : D).\, jFd$, which implies $\exists(d : D).\, Fd$. □

Definition A.66 (Proper) A modality j satisfying the equivalent conditions of Proposition A.65 will be called *proper*.

Proposition A.67 *For any proper modality j, and any constant predomain $(B, \prec)$ with binary meets, there is an inclusion* $\mathrm{RId}_j(B) \subseteq \mathrm{RId}(B)$.

Proof Let $I \in \mathrm{RId}_j(B)$; we want to show that $I \in \mathrm{RId}(B)$. By Lemma A.7 we need to show that $\exists(b : B).\, Ib$, and that if Ib_1 and Ib_2, then $\exists(b' : B).\, (b_1 \prec b') \wedge (b_2 \prec b') \wedge Ib'$. Both easily follow from (A.9) by letting $C := B$ in the first case and $C := \{b : B \mid (b_1 \prec b) \wedge (b_2 \prec b)\}$ in the second, and in both cases using $P := I$ and letting $c' := c_1 \curlywedge c_2$ be the meet. □

Example A.68 We saw in Proposition 5.72 that π is a proper modality. By Corollary 5.53 it is also dense in the sense of Definition A.62, so Proposition A.69 applies.

Proposition A.69 *For any modality j which is proper and dense, and any constant predomain B with binary meets, the inclusion* $\mathrm{cRId}(B) \subseteq \mathrm{RId}(B)$ *factors through* $\mathrm{RId}_j(B) \subseteq \mathrm{RId}(B)$*, and in fact* $\mathrm{cRId}(B) = \mathrm{cRId}_j(B)$.

Proof The first claim follows immediately from Propositions A.63 and A.67, and the second claim is then obvious, since $j\bot \Leftrightarrow \bot$. □

We've seen in Proposition A.51 statements about when inclusions of the form $\mathrm{RId}_j(B) \subseteq \mathrm{RId}_{j'}(B)$ commute with approximable mappings, where our primary interest is in the case $j' = \mathrm{id}$. The analogous statements about proper modalities turn out to be somewhat more subtle.

Proposition A.70 *Let j be a proper modality, and let $H\colon B \twoheadrightarrow B'$ be any approximable mapping, where B and B' are constant predomains with binary meets. Then the square in the diagram below need not commute, but it does upon restriction to* $\mathrm{cRId}(B)$*:*

$$\begin{array}{ccccc} \mathrm{cRId}(B) & \hookrightarrow & \mathrm{RId}_j(B) & \xrightarrow{\mathrm{RId}_j(H)} & \mathrm{RId}_j(B') \\ & & \downarrow & & \downarrow \\ & & \mathrm{RId}(B) & \xrightarrow[\mathrm{RId}(H)]{} & \mathrm{RId}(B') \end{array}$$

Proof Let $I \in \mathrm{cRId}(B)$. We want to show that for any $b' \in B'$, the proposition $\exists(b : B).\, Ib \wedge H(b, b')$ is j-closed. The type $\{b : B \mid Ib\}$ is constant and, with the lower-specialization order $\leqslant$ on B, it is also directed. Moreover, it follows directly from Definition A.28 that if $b_1 \leqslant b_2$ then $H(b_1, b') \Rightarrow H(b_2, b')$. Therefore, because j is proper, $\exists(b : B \mid Ib).\, H(b, b')$ is j-closed. □

We say that $H\colon B \twoheadrightarrow B'$ is *decidable* if it is decidable as a relation, i.e. $\forall b, b'.\, H(b, b') \vee \neg H(b, b')$. Recall the notion of filtered approximable mappings from Definition A.38.

Proposition A.71 *Let j be a proper modality, let B and B' be constant predomains with binary meets, and let $H\colon B \twoheadrightarrow B'$ be a decidable filtered approximable mapping. Then the diagram commutes:*

$$\begin{array}{ccc} \mathrm{RId}_j(B) & \xrightarrow{\mathrm{RId}_j(H)} & \mathrm{RId}_j(B') \\ \downarrow & & \downarrow \\ \mathrm{RId}(B) & \xrightarrow[\mathrm{RId}(H)]{} & \mathrm{RId}(B') \end{array}$$

Proof Let $I \in \mathrm{RId}_j(B)$. We want to show that for any $b' \in B'$, the proposition $\exists(b : B).\, Ib \wedge H(b, b')$ is j-closed. Because H is decidable, the type $\{b : B \mid H(b, b')\}$ is constant, and because H is filtered, $H(b_1, b')$ and $H(b_2, b')$ implies $H(b_1 \curlywedge b_2, b')$. We also have $I(b_1) \Rightarrow I(b_1 \curlywedge b_2)$ and $I(b_2) \Rightarrow I(b_1 \curlywedge b_2)$, so because j is proper, $\exists(b : B \mid H(b, b')).\, I(b)$ is j-closed. □

Propositions A.70 and A.71 can be usefully combined, as in the following proposition, whose proof is similar to the previous two and is left to the reader.

Proposition A.72 *Let j be a proper modality, let B_1, B_2, and B_3 be constant predomains with binary meets, and let $H\colon B_1 \times B_2 \twoheadrightarrow B_3$ be a decidable* *approximable mapping which is filtered in the second variable, i.e.* $H((b_1, b_2), b_3)$ *and* $H((b_1, b_2'), b_3)$ *implies* $H((b_1, b_2 \curlywedge b_2'), b_3)$. *Then the square in the diagram below need not commute, but it does upon restriction to* $\mathrm{cRId}(B_1) \times \mathrm{RId}_j(B_2)$*:*

$$\begin{array}{ccccc} \mathrm{cRId}(B_1) \times \mathrm{RId}_j(B_2) & \hookrightarrow & \mathrm{RId}_j(B_1) \times \mathrm{RId}_j(B_2) & \xrightarrow{\mathrm{RId}_j(H)} & \mathrm{RId}_j(B_3) \\ & & \big\downarrow & & \big\downarrow \\ & & \mathrm{RId}(B_1) \times \mathrm{RId}(B_2) & \xrightarrow[\mathrm{RId}(H)]{} & \mathrm{RId}(B_3) \end{array}$$

We can also establish analogues of the previous propositions for opens, rather than approximable mappings, either by using essentially the same proofs or by making use of the observation that opens in a predomain B are "the same as" approximable mappings $B \to 2$ to the two element linear poset. We state the analogue of Proposition A.72 for the record.

Proposition A.73 *Let* j *be a proper modality, let* B_1 *and* B_2 *be constant predomains with binary meets, and let* $U \in \Omega(B_1 \times B_2)$ *be a decidable open which is filtered in the second variable, i.e.* $U(b_1, b_2)$ *and* $U(b_1, b_2')$ *implies* $U(b_1, b_2 \curlywedge b_2')$. *Then the square in the diagram below need not commute, but it does upon restriction to* $\mathrm{cRId}(B_1) \times \mathrm{RId}_j(B_2)$*:*

$$\begin{array}{ccccc} \mathrm{cRId}(B_1) \times \mathrm{RId}_j(B_2) & \hookrightarrow & \mathrm{RId}_j(B_1) \times \mathrm{RId}_j(B_2) & \xrightarrow{\vDash_j U} & \mathtt{Prop}_j \\ & & \big\downarrow & & \big\downarrow \\ & & \mathrm{RId}(B_1) \times \mathrm{RId}(B_2) & \xrightarrow[\vDash U]{} & \mathtt{Prop} \end{array}$$

Appendix B
$\mathbb{IR}_{/\rhd}$ as a Continuous Category

Recall the category $\mathbb{IR}_{/\rhd}$ from Definition 3.1, which is a the translation-invariant version of the interval domain $\mathbb{IR}$. Whereas $\mathbb{IR}$ is a continuous poset, which we and other authors call a domain, $\mathbb{IR}_{/\rhd}$ is a continuous category. We begin this chapter with a review of continuous categories.

B.1 Review of Continuous Categories

Throughout the book, we have made significant use of the fact that $\mathbb{IR}$ is a continuous poset—or domain—in understanding the semantics of the topos $\mathcal{B}$. There is a strong relationship between $\mathbb{IR}$ and its quotient $\mathbb{IR}_{/\rhd}$ on which $\mathcal{B}$ is the topos of sheaves. Unlike $\mathbb{IR}$, the category $\mathbb{IR}_{/\rhd}$ is not a poset, let alone a domain, but we will be able to generalize much of the story about $\mathbb{IR}$ to $\mathbb{IR}_{/\rhd}$, making use of the theory of *continuous categories*. Continuous categories are a generalization—first described in [JJ82]—of continuous posets. This Sect. (B.1) is largely a review of [JJ82], or of [Joh02, Ch. C4], so we only sketch the proofs in this section.

Definition B.1 (Filtered Categories) A category I is *filtered* if every finite diagram in I has a cocone in I.

Let C be a category. The category $\mathsf{Ind}\text{-}C$ has as objects pairs (I, F), where I is a filtered category and $F\colon I \to C$ is a functor. The morphisms are given by

$$\mathsf{Ind}\text{-}C\big((I, F), (J, G)\big) := \lim_{i\in I^{\mathrm{op}}} \operatorname*{colim}_{j\in J} C(Fi, Gj).$$

Although the above definition of morphisms in $\mathsf{Ind}\text{-}(C)$ may look convoluted, it is justified by seeing Ind-objects as presheaves, as we now explain.

For any object $c \in C$, the over-category C/c is filtered, so the forgetful functor $C/c \to C$ is an object of $\mathsf{Ind}\text{-}(C)$. Its opposite can be identified with the category of elements of the representable functor, $(C/c) \cong \mathrm{el}(yc)^{\mathrm{op}}$, where

P. Schultz, D. I. Spivak, *Temporal Type Theory*, Progress in Computer Science and Applied Logic 29, https://doi.org/10.1007/978-3-030-00704-1

$yc = C(-, c)$. Noting that any object $(I, F) \in \mathsf{Ind}\text{-}(C)$ can be regarded as a presheaf by $\operatorname{colim}(I \xrightarrow{F} C \xrightarrow{y} \mathsf{Psh}(C))$ leads to the following.

Proposition B.2 *For any category C, the following categories are equivalent:*

- *the category* $\mathsf{Ind}\text{-}C$,
- *the full subcategory of* $\mathsf{Psh}(C)$ *spanned by filtered colimits of representables,*
- *the full subcategory of* $\mathsf{Psh}(C)$ *spanned by functors* $F\colon C^{\mathrm{op}} \to \mathbf{Set}$ *for which* $\mathrm{el}(F)^{\mathrm{op}}$ *is filtered.*
- *the full subcategory of* $\mathsf{Psh}(C)$ *spanned by flat functors* $F\colon C^{\mathrm{op}} \to \mathbf{Set}$.

The third and fourth are exactly the same; the former is the definition of flat in this case. Recall from Definition 2.5 the notion of ideals in a poset, namely nonempty, down-closed, up-directed subsets. The following is [JJ82, Lemma 1.1].

Lemma B.3 *If P is a poset, then* $\mathsf{Ind}\text{-}P$ *is equivalent to the category of ideals in* P.

Recall from Propositions 2.6 and 2.9 that a poset P is directed-complete iff the canonical map $P \to \mathrm{Id}(P)$ has a left adjoint sup, and that in this case P is continuous iff sup has a further left adjoint. In light of Lemma B.3, Proposition B.4 and Definition B.5 generalize the definition of continuous posets to categories.

Proposition B.4 *A category C has filtered colimits iff the Yoneda embedding* $y\colon C \to \mathsf{Ind}\text{-}C$ *has a left adjoint, which we denote* $\operatorname{colim}\colon \mathsf{Ind}\text{-}C \to C$.

Definition B.5 (Continuous Categories) A category C is *continuous* if it has filtered colimits and the functor $\operatorname{colim}\colon \mathsf{Ind}\text{-}C \to C$ has another left adjoint

$$C \underset{W}{\overset{y}{\rightrightarrows}} \mathsf{Ind}\text{-}C, \qquad \mathsf{Ind}\text{-}C \xrightarrow{\operatorname{colim}} C$$

From the embedding $\mathsf{Ind}\text{-}C \to \mathsf{Psh}(C)$, we can think of $W\colon C \to \mathsf{Ind}\text{-}C$ as defining a profunctor, which we denote by

$$\langle\!\langle -, - \rangle\!\rangle\colon C^{\mathrm{op}} \times C \to \mathbf{Set}, \tag{B.1}$$

i.e., $\langle\!\langle c, d \rangle\!\rangle := W(d)(c)$.

It is easy to see that $c \cong \operatorname{colim}(yc)$ for any $c \in C$, so we obtain a map $i\colon W(c) \to yc$. Thus for any $c, c' \in C$ and $f \in \langle\!\langle c, c' \rangle\!\rangle$, following [JJ82], we write $f\colon c \rightsquigarrow c'$ and call it a *wavy arrow*. The map i is not a monomorphism in general, so a wavy arrow is additional structure on—not just a property of—an arrow. We will see later in Theorem B.21 that in $\mathbb{IR}_{/\rhd}$, this notion of wavy arrow coincides with the one we defined earlier in Definition 3.4.

Example B.6 Any accessible category C is continuous. The extra left adjoint $W\colon C \to \mathsf{Ind}\text{-}C$ sends any object to the diagram of compact objects over it.

Remark B.7 The 2-functor $\mathsf{Ind}\colon \mathbf{Cat} \to \mathbf{Cat}$ is the free cocompletion under directed colimits, and like all free cocompletions it has the structure of a *lax idempotent 2-monad* (also known as a *KZ-monad* or *KZ-doctrine*) [Koc95]. In particular this means that the structure map of an algebra $\mathrm{colim}_C\colon \mathsf{Ind}\text{-}C \to C$ is always left adjoint to the unit of the monad $y\colon C \to \mathsf{Ind}\text{-}C$, as in Proposition B.4. The category $\mathsf{Ind}\text{-}C$ has filtered colimits, and they give the components of the monad multiplication, $\mathrm{colim}_{\mathsf{Ind}\text{-}C}\colon \mathsf{Ind}\text{-}\mathsf{Ind}\text{-}C \to \mathsf{Ind}\text{-}C$.

When the structure map colim of an algebra has a further left adjoint W, it follows that $W\colon C \to \mathsf{Ind}\text{-}C$ is an algebra homomorphism (in this case, preserves directed colimits), and a section of colim [Joh02, B.1.1.15]. That is,

$$\begin{aligned} \mathrm{colim}_{\mathsf{Ind}\text{-}C} \circ \mathsf{Ind}\text{-}W &\cong W \circ \mathrm{colim}_C \\ \mathrm{colim}_C \circ W &\cong \mathrm{id}_C\,. \end{aligned} \tag{B.2}$$

In a continuous poset, the way-below relation is interpolative. In [JJ82] Joyal and Johnstone prove the analogous fact for any continuous category C, namely that the following natural map is an isomorphism

$$\int^{\ell'} \langle\!\langle \ell, \ell' \rangle\!\rangle \times \langle\!\langle \ell', \ell'' \rangle\!\rangle \xrightarrow{\cong} \langle\!\langle \ell, \ell'' \rangle\!\rangle. \tag{B.3}$$

Thus we have an idempotent comonad on C in the category of profunctors.

B.2 The (Connected, Discrete Bifibration) Factorization System

In order to prove that $\mathbb{IR}_{/\rhd}$ is a continuous category, it will be useful to use the (connected, discrete bifibration) orthogonal factorization system on **Cat**, which we learned about from André Joyal's (semi-defunct) "CatLab."[1] Since we can find neither a reference for this material nor proofs in the literature, we describe it below, assuming the reader is aware of the orthogonal factorization systems (final, discrete fibration) and (initial, discrete opfibration) on **Cat** due to Street and Walters [SW73].

Recall that the fully faithful functor $U\colon \mathbf{Grpd} \to \mathbf{Cat}$ has a left adjoint, given on a category C by adding a formal inverse to each morphism in C. An explicit description is given in Lemma B.8. We denote both this groupoidification functor and the corresponding idempotent monad by $\Gamma\colon \mathbf{Cat} \to \mathbf{Grpd} \subseteq \mathbf{Cat}$, and we denote the unit of the adjunction by $\eta\colon \mathrm{id}_{\mathbf{Cat}} \to \Gamma$.

Lemma B.8 *For any category C, the unit $\eta_C\colon C \to \Gamma C$ is identity on objects. A morphism $c \to d$ in ΓC can be identified with a zigzag $Z_n \to C$, up to the*

[1] See https://ncatlab.org/joyalscatlab/published/Factorisation+systems#in_cat.

congruence defined below. A zigzag is a diagram in C of the following form for some $n \in \mathbb{N}$:

$$c = z_0 \xrightarrow{f_1} z_1' \xleftarrow{f_1'} \cdots \xrightarrow{f_n} z_n' \xleftarrow{f_n'} z_n = d$$

which we denote by the list $(f_n', f_n, \ldots, f_1', f_1)$. The identity zigzag is $()$, i.e. the one of length $n = 0$, and composition is given by list concatenation. The congruence is generated by the relations

$$(f_2', f_2, \mathrm{id}, f_1) \sim (f_2', f_2 \circ f_1) \qquad (f_2', \mathrm{id}, f_1', f_1) \sim (f_2' \circ f_1', f_1)$$
$$(f, f) \sim () \qquad (f_2', f, f, f_1) \sim (f_2', f_1)$$

The inverse of a zigzag is given by reversing it.

The unit η_C sends $f : c \to d$ to (id_d, f). If G is a groupoid, then η_G is an isomorphism.

Definition B.9 (Connected) A functor between groupoids is called *connected* if it is essentially surjective and full. A functor F between categories is called *connected* if ΓF is connected as a functor of groupoids. A groupoid or category is *connected* if the unique functor to the terminal category is connected.

Remark B.10 Note that a groupoid G is connected if and only if it has at least one object, and for any two objects $a, b \in G$ there exists a morphism $f : a \to b$. Thus a category C is connected if and only if it has at least one object, and for any two objects $a, b \in C$ there exists a zig-zag from a to b.

Recall that a groupoid is called contractible if it is nonempty and each of its hom-sets has exactly one element. The following is straightforward.

Proposition B.11 *Let G be a groupoid. The following are equivalent*

- *G is contractible.*
- *There exists a connected functor $1 \to G$.*
- *G is nonempty and every functor $1 \to G$ is connected.*

Proposition B.12 *If I is a filtered category, then there exists a connected functor $1 \to I$.*

Proof By Proposition B.11 it suffices to show that ΓI is contractible. Since I is nonempty, so is ΓI, so take $i \in \Gamma I$. It suffices to show that i has only one endomorphism, i.e., that any zigzag $(f_n', f_n, \ldots, f_1', f_1) : i \to i$ is equivalent to $()$. Note that for any commutative diagram of the following form,

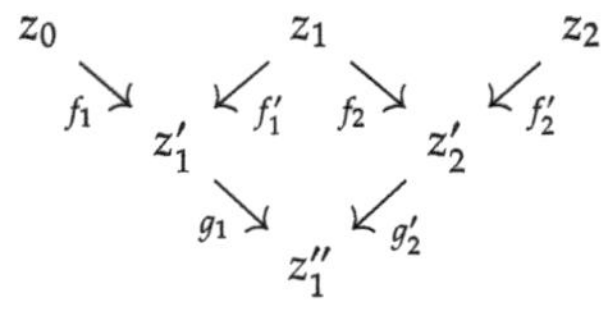

$$\begin{array}{ccccccccc}
z_0 & & & & z_1 & & & & z_2 \\
& \overset{f_1}{\searrow} & & \overset{f_1'}{\swarrow} & & \overset{f_2}{\searrow} & & \overset{f_2'}{\swarrow} & \\
& & z_1' & & & & z_2' & & \\
& & & \overset{g_1}{\searrow} & & \overset{g_2'}{\swarrow} & & & \\
& & & & z_1'' & & & &
\end{array}$$

we have $(f_2', f_2, f_1', f_1) = (g_2' \circ f_2', g_1 \circ f_1)$. The result then follows by induction on n, using the fact that I is filtered. □

Proposition B.13 *Let $F: G \to G'$ be a functor between groupoids.*

- *F is a discrete bifibration iff it is a discrete fibration iff it is a discrete opfibration.*
- *F is connected iff it is initial iff it is final.*

It follows that there is a (connected, discrete bifibration) orthogonal factorization system on **Grpd**.

Proof To prove the first claim, it suffices by duality to show that if F is a discrete fibration then it is a discrete opfibration, and this is straightforward. The third claim follows from the first and second by the more general result for categories [SW73]. For the second claim, F is initial iff final by duality, so it remains to show it is final iff connected.

By definition, F is final iff for all $g' \in G'$ the comma category $(g' {\downarrow} F)$ is nonempty and every two objects in it have a morphism between them. It is easy to check that F is essentially surjective iff, for all $g' \in G'$ the comma category $(g' {\downarrow} F)$ is nonempty. It is also easy to check that F is full iff every two objects in $(g' {\downarrow} F)$ have a morphism between them. □

Our first goal is to extend the (connected, discrete bifibration) orthogonal factorization system, given in Proposition B.13, from **Grpd** to **Cat**; see Theorem B.16.

Lemma B.14 *A bifibration over a groupoidification is a groupoidification.*

More precisely, let D be a category and suppose given a bifibration of groupoids $p: G \to \Gamma D$. Let $p': C \to D$ be its pullback along η_D as shown,

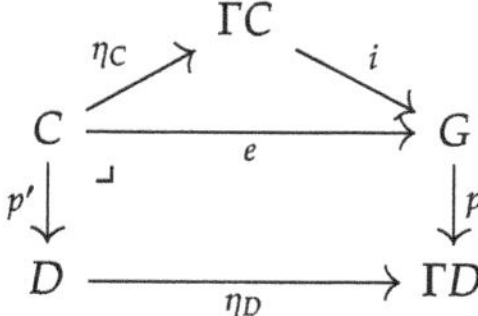

and let i be the universal arrow as shown. Then i is an isomorphism.

Proof We will find a functor $j: G \to \Gamma C$ inverse to i. The functors η_D and η_C are identity on objects, so e and i are too. Hence we need to only define j on morphisms and prove that i and j are mutually inverse.

Given a morphism $f: g \to g'$ in G, we may identify $p(f)$ with a zigzag $z: Z_n \to D$. Its image $\eta_D \circ z$ can be lifted uniquely along p to a zigzag $z': Z_n \to G$, because p is a discrete bifibration, and it is easy to see that $z' \mapsto f$ under the isomorphism $\Gamma G \to G$. Since $p(f) = \eta_D \circ z$ and C is a pullback, there is a unique functor $f': Z_n \to C$ making the diagram commute: $e(f') = f$ and $p'(f') = z$. We define $j(f) := \eta_C \circ f'$, and note that $(i \circ j)(f) = e(f') = f$. It is also easy to show that $j \circ i$ is the identity, completing the proof. □

Proposition B.15 *Let $p\colon C \to D$ be a functor and $\Gamma p\colon \Gamma C \to \Gamma D$ its groupoidification. Then p is a discrete bifibration iff Γp is a discrete bifibration and the η-naturality square below is a pullback:*

$$\begin{array}{ccc} C & \xrightarrow{\eta_C} & \Gamma C \\ {\scriptstyle p}\downarrow & \lrcorner & \downarrow{\scriptstyle \Gamma p} \\ D & \xrightarrow[\eta_D]{} & \Gamma D \end{array}$$

Proof The pullback of a discrete bifibration is a discrete bifibration, so one direction is obvious.

Suppose p is a discrete bifibration, and consider the induced functor $P\colon D \to$ **Set** whose category of elements is p. Note that for every morphism f in D, the induced function $P(f)$ is a bijection. Hence P factors as $D \xrightarrow{\eta_D} \Gamma D \xrightarrow{P'}$ **Set** for a unique functor P', whose category of elements a discrete bifibration $p'\colon G \to \Gamma D$. The pullback of p' along η_D is p, and the result follows by Lemma B.14. □

Theorem B.16 *There is an orthogonal factorization system $(\mathcal{L}, \mathcal{R})$ on* **Cat***, where $\mathcal{L}$ consists of the connected functors and $\mathcal{R}$ consists of the discrete bifibrations.*

Proof For any functor $f : C \to D$, form the following diagram from η's naturality square

$$\begin{array}{ccccc} C & & \xrightarrow{\eta_C} & & \Gamma(C) \\ & \searrow^{c} & & \downarrow{\scriptstyle \Gamma(f)} & & \searrow^{c'} \\ {\scriptstyle f}\downarrow & & \widetilde{f} & \xrightarrow{e} & & \widetilde{\Gamma(f)} \\ & \swarrow_{b} & & & & \swarrow_{b'} \\ D & & \xrightarrow[\eta_D]{} & & \Gamma(D) \end{array}$$

by first factoring $\Gamma(f) = b' \circ c'$, where c' is connected and b' is a bifibration (see Proposition B.13), then forming the pullback of b' along η_D to give b and e, and finally adding the universal morphism c making the diagram commute. Clearly, b is a bifibration because it is the pullback of one.

By Lemma B.14 the functor e factors through an isomorphism $\Gamma(\widetilde{f}) \cong \widetilde{\Gamma(f)}$. Hence we can identify c' with $\Gamma(c)$, and by Definition B.9 c is connected. This defines the (connected, discrete bifibration) factorization: $f = b \circ c$.

It remains to show the orthogonality, so consider a solid-arrow square as below:

$$\begin{array}{ccc} X & \xrightarrow{f} & C \\ {\scriptstyle c}\downarrow & \nearrow & \downarrow{\scriptstyle b} \\ Y & \xrightarrow[g]{} & D \end{array}$$

where c is connected and b is a discrete bifibration; we want to show there is a unique dotted-arrow $Y \to C$ making the diagram commute. By Proposition B.15, the functor Γb is a bifibration and the front face in the commutative diagram below is a pullback:

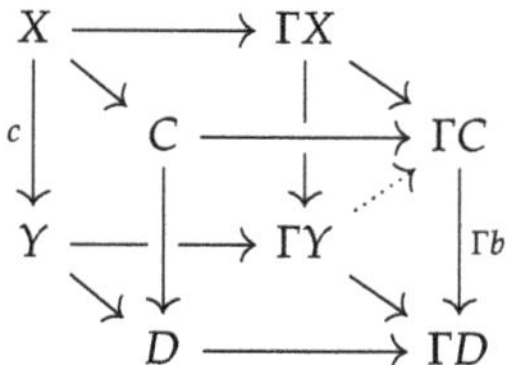

where the back and right-hand faces are given by universality of Γ. The functor Γc is connected by definition, so there is a unique dotted-arrow lift as shown. Since the front face is a pullback, we obtain a map $Y \to C$ making the diagram commute, as required. □

For any two categories C, D, let $C \star D$ denote their join. Recall the notation of linear order categories—in particular 0, 1, and 2—from Notation 1.1. It follows easily from Definition B.9 that if C is a connected category, then so are the inclusions $j: C \to (C \star 1)$ and $j': C \to (1 \star C)$.

Proposition B.17 *Suppose that $p: C \to D$ is a discrete bifibration. Then it preserves and reflects connected limits and colimits.*

Proof We consider the case for colimits, the case for limits being dual. Let $2 = 1 \star 1$ be the arrow category, with the two maps $1 \to 2$ denoted $d = \mathrm{id} \star !: 1 \star 0 \to 1 \star 1$ and $c = ! \star \mathrm{id}: 0 \star 1 \to 1 \star 1$. Also let $j = \mathrm{id} \star !: I \star 0 \to I \star 1$. We can represent the colimit cocone of a diagram $F: I \to C$ as a functor $\operatorname{colim} F: I \star 1 \to C$ such that $(\operatorname{colim} F) \circ j = F$, and such that for any $\phi: I \star 1 \to C$ satisfying $\phi \circ j = F$ there exists a unique $\phi': I \star 2 \to C$ such that the two right triangles of the following diagram commute:

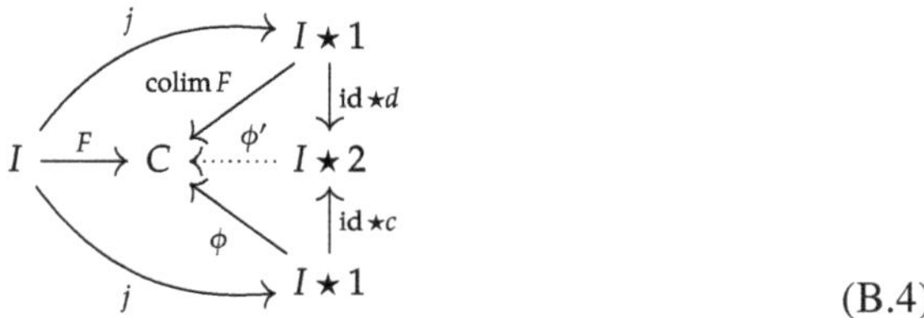

(B.4)

Let I be a small connected category and suppose that $F: I \to C$ has a colimit, $\operatorname{colim} F$. Then $p \circ (\operatorname{colim} F)$ is a candidate colimit of $p \circ F$, so to see that it has the universal property, suppose we have a solid-arrow diagram

$$\begin{array}{ccc} I & \xrightarrow{F} & C \\ {\scriptstyle j}\downarrow & \overset{\phi}{\nearrow} & \downarrow{\scriptstyle p} \\ I \star 1 & \xrightarrow[\psi]{} & D \end{array} \tag{B.5}$$

Since j is connected, we obtain a unique lift ϕ as shown. Then letting ϕ' be as in Eq. (B.4), we claim $p \circ \phi'$ is the unique functor such that $(p \circ \phi') \circ (\mathrm{id} \star d) = p \circ (\operatorname{colim} F)$ and $(p \circ \phi') \circ (\mathrm{id} \star c) = \psi$. To prove uniqueness, suppose $\psi' \colon I \star 2 \to D$ is another functor such that $\psi' \circ (\mathrm{id} \star d) = p \circ (\operatorname{colim} F)$ and $\psi' \circ (\mathrm{id} \star c) = \psi$. Again we find a unique lift $\hat{\psi}'$,

$$\begin{array}{ccc} I \star 1 & \xrightarrow{\operatorname{colim} F} & C \\ {\scriptstyle \mathrm{id}\star d}\downarrow & \overset{\hat{\psi}'}{\nearrow} & \downarrow{\scriptstyle p} \\ I \star 2 & \xrightarrow[\psi']{} & D. \end{array} \tag{B.6}$$

But then $\hat{\psi}' \circ (\mathrm{id} \star c)$ is also a lift for the square (B.5), so $\hat{\psi}' \circ (\mathrm{id} \star c) = \phi$. Since also $\hat{\psi}' \circ (\mathrm{id} \star d) = \operatorname{colim} F$, we have $\hat{\psi}' = \phi'$ by the uniqueness of ϕ' in Eq. (B.4). Hence $\psi' = p \circ \hat{\psi}' = p \circ \phi'$ as desired.

The proof that p reflects connected colimits is easier: given a diagram (B.5) such that ψ is a colimiting cocone in D, one checks that the unique lift ϕ is a colimit cocone in C. □

Corollary B.18 *If $p \colon C \to D$ is a discrete bifibration and p is surjective on objects, then $\mathsf{Ind}\text{-}(p)$ is surjective on objects. In particular, if C has all filtered colimits (resp. filtered limits), then so does D.*

Proof Let $F \colon I \to D$ be a filtered diagram. By Proposition B.12, there exists a connected functor $i \colon 1 \to I$, and because p is surjective on objects there is a map $c \colon 1 \to C$ such that $p \circ c = F \circ i$. Thus we obtain a lift $G \colon I \to C$ such that $p \circ G = F$. If it has a colimit in C, then by Proposition B.17 this colimit is preserved by p. □

Lemma B.19 *If $(\mathcal{L}, \mathcal{R})$ is an orthogonal factorization system on* **Cat** *and $p \colon C \to D$ is in the right class, then so is $p^K \colon C^K \to D^K$ for any small category K.*

Proof This follows from the well-known fact that $\mathcal{L}$ is closed under limits in the arrow category, because if c is in the left class then so is $c \times \mathrm{id}_K$. □

Lemma B.20 *If $p \colon C \to D$ is a discrete fibration (resp. discrete opfibration), then so is the functor $\mathsf{Ind}\text{-}(p) \colon \mathsf{Ind}\text{-}(C) \to \mathsf{Ind}\text{-}(D)$.*

Proof We consider only the discrete fibration (d-fib) case, the other being similar. Note that $p \colon C \to D$ is a d-fib iff $C^K \to D^K$ is a d-fib for all small categories K by Lemma B.19. We need to show that $\mathsf{Ind}\text{-}(p)$ has the right lifting property with respect to the inclusion of the codomain, $c \colon 1 \to 2$, as well as the codiagonal $c \sqcup_1 c \to 2$, which we denote $\delta \colon 2 \sqcup_1 2 \to 2$.

For the first, suppose given filtered diagrams $F \colon I \to C$ and $G \colon J \to D$ and a map $f \colon \operatorname{colim}(p \circ F) \to \operatorname{colim}(G)$ in $\mathsf{Ind}\text{-}(D)$. By [KS06, Corollary 6.1.14] there is a filtered category K, final functors $i \colon K \to I$ and $j \colon K \to J$, and a natural transformation $\varphi \colon p \circ F \circ i \to G \circ j$ such that $\operatorname{colim} \varphi$ is isomorphic to f; see the left-hand diagram below.

$$\begin{array}{ccccc} & \overset{i}{\nearrow} & I & \xrightarrow{F} & C \\ K & & \varphi\Downarrow & & \downarrow p \\ & \underset{j}{\searrow} & J & \xrightarrow[G]{} & D \end{array} \qquad \begin{array}{ccc} K & \xrightarrow{Fi} & C \\ \| & pFI & \downarrow p \\ K & \underset{Gj}{\overset{\varphi\Downarrow}{\rightrightarrows}} & D \end{array} \qquad \begin{array}{ccc} 1 & \xrightarrow{Fi} & C^K \\ c\downarrow & & \downarrow p^K \\ 2 & \xrightarrow[\varphi]{} & D^K \end{array}$$

Since precomposition with final functors preserves colimits, the original lifting problem can be replaced by the one shown in the middle, or equivalently the diagram on the right. It has a lift because p, and hence p^K, is a discrete fibration

To lift diagrams of the shape δ, one reasons similarly. Namely, using [KS06, Corollary 6.1.14], one finds a single K such that the lifting problem can be rewritten as follows:

$$\begin{array}{ccc} K & \overset{\varphi_1\Downarrow}{\underset{\varphi_2\Uparrow}{\longrightarrow}} & C \\ \| & & \downarrow p \\ K & \overset{\varphi\Downarrow}{\rightrightarrows} & D \end{array} \qquad \begin{array}{ccc} 2\sqcup_1 2 & \xrightarrow{(\varphi_1,\varphi_2)} & C^K \\ \delta\downarrow & & \downarrow p \\ 2 & \xrightarrow[\varphi]{} & D^K \end{array}$$

This again has a lift because p is a discrete fibration. We leave the details to the reader. □

B.3 Proof that $\mathbb{IR}_{/\triangleright}$ Is Continuous

Recall the definition of $\mathbb{IR}_{/\triangleright}$, the category of the translation-invariant intervals, from Definition 3.1. It has objects $\ell \in \mathbb{R}_{\geq 0}$ and morphisms $\langle r, s\rangle : \ell \to \ell'$ for $r, s \geq 0$, with composition given by component-wise sum. In Definition 3.4 we defined such a morphism to be a *wavy arrow* if $r > 0$ and $s > 0$, in which case we write $\langle r, s\rangle : \ell \rightsquigarrow \ell'$. We can thus define a profunctor $\langle\!\langle -, -\rangle\!\rangle : \mathbb{IR}_{/\triangleright}^{\text{op}} \times \mathbb{IR}_{/\triangleright} \to \mathbf{Set}$ sending $\ell, \ell' \in \mathbb{IR}_{/\triangleright}$ to the set

$$\langle\!\langle \ell, \ell'\rangle\!\rangle := \big\{\langle r, s\rangle \in \mathbb{IR}_{/\triangleright}(\ell, \ell') \mid r > 0, s > 0\big\}. \tag{B.7}$$

Clearly, this is a subprofunctor of the hom-profunctor $\mathbb{IR}_{/\triangleright}(-,-)$. In order for it to agree with our definition of wavy arrows from Definition B.5, we need to show that induces a left adjoint to colim. Indeed, for any $\ell' \in \mathbb{IR}_{/\triangleright}$, define a presheaf

$$W(\ell') : \mathbb{IR}_{/\triangleright}^{\text{op}} \to \mathbf{Set}, \qquad \ell \mapsto \langle\!\langle \ell, \ell'\rangle\!\rangle. \tag{B.8}$$

W is functorial in ℓ', lands in $\mathsf{Ind}\text{-}(\mathbb{IR}_{/\triangleright})$, and is left adjoint to colim as we now show.

Theorem B.21 *The category* $\mathbb{IR}_{/\triangleright}$ *is a continuous category. In particular, the functor* W *defined in Eq.* (B.8) *is left adjoint to* colim: $\mathsf{Ind}\text{-}(\mathbb{IR}_{/\triangleright}) \to \mathbb{IR}_{/\triangleright}$.

Proof Consider the following diagram of categories:

$$\begin{array}{ccc} \mathbb{IR} & \underset{\bigvee}{\overset{\downarrow}{\rightleftarrows}} & \mathsf{Ind}\text{-}(\mathbb{IR}) \\ {\scriptstyle p}\downarrow & & \downarrow{\scriptstyle \mathsf{Ind}\text{-}p} \\ \mathbb{IR}_{/\rhd} & \underset{\text{colim}}{\overset{W}{\rightleftarrows}} & \mathsf{Ind}\text{-}(\mathbb{IR}_{/\rhd}) \end{array} \tag{B.9}$$

On the top, we have that $\downarrow$ is left adjoint to $\bigvee$ and that $\bigvee \circ \downarrow = \mathrm{id}_{\mathbb{IR}}$ by Proposition 2.9. Our goal is to fill out the bottom half. That is, we want to show that $\mathbb{IR}_{/\rhd}$ has filtered colimits and to construct a left adjoint W to colim: $\mathsf{Ind}\text{-}(\mathbb{IR}_{/\rhd}) \to \mathbb{IR}_{/\rhd}$, such that colim $\circ W$ is the identity and such that both the leftward and the rightward squares in (B.9) commute.

We showed in Lemma 3.3 that $p\colon \mathbb{IR} \to \mathbb{IR}_{/\rhd}$ is a discrete bifibration and that it is surjective on objects. Thus $\mathbb{IR}_{/\rhd}$ has filtered colimits and p preserves them by Corollary B.18 and Proposition B.17; hence the leftward diagram commutes.

For $\ell, \ell' \in \mathbb{IR}_{/\rhd}$, let $\langle\!\langle \ell, \ell' \rangle\!\rangle$ be as in Eq. (B.7), and let $W(\ell')\colon \mathbb{IR}^{\mathrm{op}}_{/\rhd} \to \mathbf{Set}$ be as in Eq. (B.8). The latter is a flat functor because any pair of objects $\langle r_1, s_1\rangle \in \langle\!\langle \ell_1, \ell' \rangle\!\rangle$ and $\langle r_2, s_2\rangle \in \langle\!\langle \ell_2, \ell' \rangle\!\rangle$ have a coproduct $\langle \max(r_1, r_2), \max(s_1, s_2)\rangle$ in $(\mathbb{IR}^{\mathrm{op}}_{/\rhd}/\ell')^{\mathrm{op}}$. It is easy to check that $W \circ p = \mathsf{Ind}\text{-}(p) \circ \downarrow$, so the rightward diagram commutes.

Note that every isomorphism in $\mathbb{IR}_{/\rhd}$ is an identity. We next see that $\mathrm{colim}(W(\ell)) = \ell$ for any $\ell \in \mathbb{IR}_{/\rhd}$. This follows from the analogous fact for domains D, that $\bigvee \downarrow d = d$ for any $d \in D$. Indeed choose $\tilde{\ell}$ with $p(\tilde{\ell}) = \ell$ and we have

$$\mathrm{colim}\, W(\ell) = \mathrm{colim}\, W(p\tilde{\ell}) = \mathrm{colim}\, \mathsf{Ind}\text{-}(p)(\downarrow\tilde{\ell}) = p\bigvee\downarrow\tilde{\ell} = p\tilde{\ell} = \ell.$$

We show that W is left adjoint to colim by constructing a unit $\eta\colon \mathrm{id} \to \mathrm{colim} \circ W$ and counit $\epsilon\colon W \circ \mathrm{colim} \to \mathrm{id}$ and checking the triangle identities $(\mathrm{colim}\,\epsilon) \cdot (\eta\, \mathrm{colim}) = \mathrm{id}_{\mathrm{colim}}$ and $(\epsilon W) \cdot (W\eta) = \mathrm{id}_W$. The unit η is the identity, so the triangle identities simplify to $\mathrm{colim}\,\epsilon = \mathrm{id}_{\mathrm{colim}}$ and $\epsilon W = \mathrm{id}_W$.

We want to construct a counit $\epsilon_F\colon W(\mathrm{colim}\, F) \to F$ for any $F \in \mathsf{Ind}\text{-}\mathbb{IR}_{/\rhd}$. Since p is a bifibration and surjective on objects, $\mathsf{Ind}\text{-}(p)$ is surjective on objects by Corollary B.18. Choose $\widetilde{F}$ with $\mathsf{Ind}\text{-}(p)(\widetilde{F}) = F$, and define $\epsilon_F := \mathsf{Ind}\text{-}(p)\big(\downarrow \bigvee \widetilde{F} \xrightarrow{\leq} \widetilde{F}\big)$:

$$W\,\mathrm{colim}(F) = \mathsf{Ind}\text{-}(p)\big(\downarrow\bigvee\widetilde{F}\big) \xrightarrow{\mathsf{Ind}\text{-}(p)\left(\downarrow\bigvee\widetilde{F}\xrightarrow{\leq}\widetilde{F}\right)} \mathsf{Ind}\text{-}(p)(\widetilde{F}) = F.$$

To show that this definition of ϵ_F doesn't depend on the choice of lift $\widetilde{F}$, let $\widetilde{F}'$ be another lift with $\mathsf{Ind}\text{-}(p)(\widetilde{F}') = F$. Then

$$\mathsf{Ind}\text{-}(p)\big(\downarrow\bigvee\widetilde{F}\big) = Wp\bigvee\widetilde{F} = W\,\mathrm{colim}\,\mathsf{Ind}\text{-}(p)(\widetilde{F}) = W\,\mathrm{colim}\, F$$

and similarly $\mathsf{Ind}\text{-}(p)(\downarrow \bigvee \widetilde{F}') = W\,\mathrm{colim}\, F$, hence

$$\mathsf{Ind}\text{-}(p)\big(\downarrow\bigvee\widetilde{F}' \xrightarrow{\leq} \widetilde{F}\big) = \mathsf{Ind}\text{-}(p)\big(\downarrow\bigvee\widetilde{F} \xrightarrow{\leq} \widetilde{F}\big) = \epsilon_F .$$

It easily follows from this that ϵ_F is natural in F. Finally, we see that

$$\begin{aligned}\operatorname{colim}(\epsilon_F) = \operatorname{colim} \mathsf{Ind}\text{-}(p)\big(\downarrow\bigvee\widetilde{F} \xrightarrow{\leq} \widetilde{F}\big) &= p\big(\bigvee\downarrow\bigvee\widetilde{F} \xrightarrow{\text{id}} \bigvee\widetilde{F}\big)\\ &= p(\text{id}_{\bigvee\widetilde{F}}) = \text{id}_{\operatorname{colim} F}\end{aligned}$$

and to see that $\epsilon_{W\ell} = \text{id}_{W\ell}$, choose a lift $p\tilde{\ell} = \ell$, so that $\mathsf{Ind}\text{-}(p)(\downarrow\tilde{\ell}) = Wp\tilde{\ell} = W\ell$, and

$$\epsilon_{W\ell} = \mathsf{Ind}\text{-}(p)\big(\downarrow\bigvee\downarrow\tilde{\ell} \xrightarrow{\text{id}} \downarrow\tilde{\ell}\big) = \text{id}_{W\ell} .$$

□

B.4 Two Constructions of the Topos $\mathcal{B}$

We now sketch two equivalent constructions of the topos $\mathcal{B}$. This will give a sort of analogue to Theorem 2.27 for continuous categories (see Theorem B.28). We mainly follow [JJ82]; see there for details.

B.4.1 Adjoint Endofunctors on $[\mathbb{IR}_{/\rhd}, \mathbf{Set}]$

First we construct a pair of adjoint endofunctors t^*, t_* on the functor category $[\mathbb{IR}_{/\rhd}, \mathbf{Set}]$; some intuition will be given in Remark B.22. To do so, consider $\langle\!\langle -, - \rangle\!\rangle$, from Eq. (B.7), as a functor $\mathbb{IR}^{\text{op}}_{/\rhd} \to [\mathbb{IR}_{/\rhd}, \mathbf{Set}]$. By left Kan extension along the Yoneda embedding, we get a map $t^*\colon [\mathbb{IR}_{/\rhd}, \mathbf{Set}] \to [\mathbb{IR}_{/\rhd}, \mathbf{Set}]$ as shown in the following diagram:

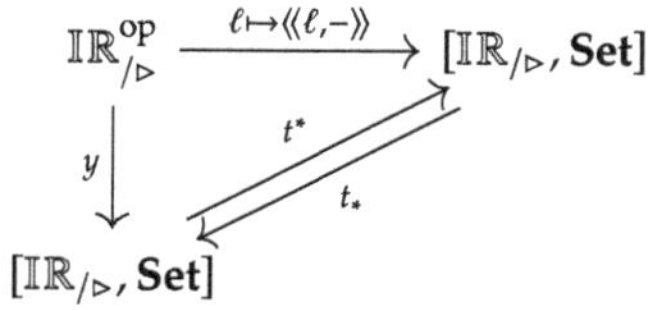

$$\begin{array}{ccc} \mathbb{IR}^{\text{op}}_{/\rhd} & \xrightarrow{\ell \mapsto \langle\!\langle \ell, - \rangle\!\rangle} & [\mathbb{IR}_{/\rhd}, \mathbf{Set}] \\ y \downarrow & \overset{t^*}{\nearrow} \ \underset{t_*}{\swarrow} & \\ [\mathbb{IR}_{/\rhd}, \mathbf{Set}] & & \end{array}$$

It can be computed for any $X\colon \mathbb{IR}_{/\rhd} \to \mathbf{Set}$ by the usual coend formula:

$$(t^*X)(\ell') = \int^{\ell} \langle\!\langle \ell, \ell' \rangle\!\rangle \times X(\ell) = \langle\!\langle -, \ell' \rangle\!\rangle \otimes X = W(\ell') \otimes X, \tag{B.10}$$

where $W: \mathbb{IR}_{/\rhd} \to \mathsf{Ind}\text{-}(\mathbb{IR}_{/\rhd})$ is as in Eq. (B.8), and can be conceived of as sending ℓ' to the diagram for the stalk at the point (closed interval) corresponding to ℓ'. Equivalently, $t^*(X)$ is given by the composite along the bottom of the following diagram, which is an expression of the usual conical colimit formula for the pointwise left Kan extension and which, by Theorem B.21, factors through the Ind-subcategories as in the top line:

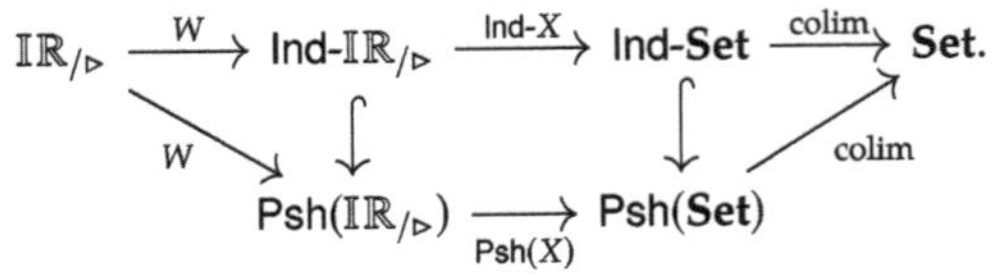

$$\begin{array}{ccccccc} \mathbb{IR}_{/\rhd} & \xrightarrow{W} & \mathsf{Ind}\text{-}\mathbb{IR}_{/\rhd} & \xrightarrow{\mathsf{Ind}\text{-}X} & \mathsf{Ind}\text{-}\mathbf{Set} & \xrightarrow{\mathrm{colim}} & \mathbf{Set}. \\ & \searrow^{W} & \downarrow & & \downarrow & \nearrow_{\mathrm{colim}} & \\ & & \mathsf{Psh}(\mathbb{IR}_{/\rhd}) & \xrightarrow[\mathsf{Psh}(X)]{} & \mathsf{Psh}(\mathbf{Set}) & & \end{array} \tag{B.11}$$

Note that t^* preserves finite limits because the functor $\ell \mapsto \langle\!\langle \ell, - \rangle\!\rangle$ is flat; see [Bor94, 6.3.8].

The functor t^* has a right adjoint $t_*: [\mathbb{IR}_{/\rhd}, \mathbf{Set}] \to [\mathbb{IR}_{/\rhd}, \mathbf{Set}]$; see [nLaba]. It can be computed by the following explicit end formula

$$(t_*X)(\ell) = [\mathbb{IR}_{/\rhd}, \mathbf{Set}](\langle\!\langle \ell, - \rangle\!\rangle, X) = \int_{\ell'} \mathbf{Set}\big(\langle\!\langle \ell, \ell' \rangle\!\rangle, X(\ell')\big). \tag{B.12}$$

We summarize the intuition for t^* and t_* in the following remark.

Remark B.22 Suppose given $X: \mathbb{IR}_{/\rhd} \to \mathbf{Set}$. For any $\ell' \leq \ell$ there are restriction maps $X(\ell) \to X(\ell')$. One such restriction map exists for each $r + \ell' + s = \ell$ with $r, s \geq 0$, but let's consider only those with $r, s > 0$, in which case let's temporarily say ℓ is way-bigger than ℓ'. The set $(t^*X)(\ell')$ is the colimit of these restriction maps over all way-bigger intervals ℓ; intuitively, this describes the stalk of X at ℓ'. On the other hand, the set $(t_*X)(\ell)$ is the limit of these restriction maps over all way-smaller intervals ℓ'; intuitively, this describes the sheafification of X evaluated at ℓ.

Proposition B.23 *The functors t^* and t_* are idempotent.*

Proof Since $w \mapsto w \times X(\ell)$ is cocontinous in w, the idempotence of t^* follows from Eqs. (B.10), (B.3) by the Fubini theorem for coends. The idempotence of t_* follows by adjointness. □

Proposition B.24 *A functor $X: \mathbb{IR}_{/\rhd} \to \mathbf{Set}$ is a fixed-point of t^* (equivalently, X is in the image of t^*) iff X preserves filtered colimits.*

Proof Using Eq. (B.11), X is a fixed-point of t^* iff $X \cong \mathrm{colim}_{\mathbf{Set}} \circ \mathsf{Ind}\text{-}(X) \circ W$. If so, then

$$\begin{aligned} \mathrm{colim}_{\mathbf{Set}} \circ \mathsf{Ind}\text{-}(X) &\cong \mathrm{colim}_{\mathbf{Set}} \circ \mathsf{Ind}\text{-}(\mathrm{colim}_{\mathbf{Set}}) \circ \mathsf{Ind}\text{-}(\mathsf{Ind}\text{-}X) \circ \mathsf{Ind}\text{-}(W) \\ &\cong \mathrm{colim}_{\mathbf{Set}} \circ \mathrm{colim}_{\mathsf{Ind}\text{-}\mathbf{Set}} \circ \mathsf{Ind}\text{-}(\mathsf{Ind}\text{-}X) \circ \mathsf{Ind}\text{-}(W) \\ & \qquad \mathbf{Set} \text{ is an } \mathsf{Ind}\text{-algebra} \end{aligned}$$

$$\cong \mathrm{colim}_{\mathbf{Set}} \circ \mathsf{Ind}\text{-}(X) \circ \mathrm{colim}_{\mathsf{Ind}\text{-}(\mathbb{IR}_{/\rhd})} \circ \mathsf{Ind}\text{-}(W)$$

Naturality, Rmk. B.7

$$\cong \mathrm{colim}_{\mathbf{Set}} \circ \mathsf{Ind}\text{-}X \circ W \circ \mathrm{colim}_{\mathbb{IR}_{/\rhd}}$$

Eq. (B.2)

$$\cong X \circ \mathrm{colim}_{\mathbb{IR}_{/\rhd}} .$$

Thus X preserves filtered colimits. Conversely, if X preserves filtered colimits, then

$$\mathrm{colim}_{\mathbf{Set}} \circ \mathsf{Ind}\text{-}X \circ W \cong X \circ \mathrm{colim}_{\mathbb{IR}_{/\rhd}} \circ W \cong X.$$

□

B.4.2 $\mathsf{Shv}(S_{\mathbb{IR}_{/\rhd}}) \cong \mathsf{Cont}(\mathbb{IR}_{/\rhd})$

In Definition 3.5, we defined $S_{\mathbb{IR}_{/\rhd}}$ analogously to our definition of $S_{\mathbb{IR}}$ in Definition 2.25. Our goal for the rest of this section is to prove that the category $\mathcal{B} = \mathsf{Shv}(S_{\mathbb{IR}_{/\rhd}})$ is not just a category of sheaves (limit-preserving functors of a given sort), but is also (isomorphic to) a category of continuous functors (colimit-preserving functors of a given sort). This is the content of Theorem B.28.

Before doing so, we can already characterize $\mathcal{B}$ as a category of fixed points. Indeed, the endofunctors t^* and t_* of $[\mathbb{IR}_{/\rhd}, \mathbf{Set}]$ are idempotent by Proposition B.23, hence each determines a subcategory, say $\mathsf{Fix}(t^*)$ and $\mathsf{Fix}(t_*)$.

Proposition B.25 *The following are the same as subcategories of* $[\mathbb{IR}_{/\rhd}, \mathbf{Set}]$*:*

$$\mathsf{Fix}(t_*) = \mathsf{Shv}(S_{\mathbb{IR}_{/\rhd}}).$$

Proof By Definition 3.5, $S_{\mathbb{IR}_{/\rhd}}$ is the site whose underlying category is $\mathbb{IR}^{\mathrm{op}}_{/\rhd}$ and whose coverage consists of one covering family $W(\ell) \subseteq \mathrm{Hom}(\ell, -)$ for each object $\ell \in \mathbb{IR}_{/\rhd}$. That is, the collection of way-below maps $\langle\!\langle \ell, \ell' \rangle\!\rangle$ covers ℓ. By Eq. (B.12), we have $(t_*X)(\ell) = [\mathbb{IR}_{/\rhd}, \mathbf{Set}](\langle\!\langle \ell, - \rangle\!\rangle, X)$. Thus the category of fixed-points (equivalently, the image) of t_* consists precisely of those functors $\mathbb{IR}_{/\rhd} \to \mathbf{Set}$ satisfying the sheaf condition for $S_{\mathbb{IR}_{/\rhd}}$. Hence this subcategory is equivalent to $\mathsf{Shv}(S_{\mathbb{IR}_{/\rhd}})$. □

On the other hand, we showed in Proposition B.24 that the fixed-points of t^* consists precisely of those functors $\mathbb{IR}_{/\rhd} \to \mathbf{Set}$ which preserve filtered colimits. We denote this category by $\mathsf{Cont}(\mathbb{IR}_{/\rhd})$. Thus we have proven the following.

Proposition B.26 *The following are the same as subcategories of* $[\mathbb{IR}_{/\rhd}, \mathbf{Set}]$*:*

$$\mathsf{Fix}(t^*) \cong \mathsf{Cont}(\mathbb{IR}_{/\rhd}).$$

The restrictions of t_* and t^* to these subcategories $\mathsf{Cont}(\mathbb{IR}_{/\rhd})$ and $\mathsf{Shv}(S_{\mathbb{IR}/\rhd})$ then provide an equivalence of categories, given by $(\mathsf{sh}' \circ U) \colon \mathsf{Shv}(S_{\mathbb{IR}/\rhd}) \leftrightarrows \mathsf{Cont}(\mathbb{IR}_{/\rhd}) : (\mathsf{sh} \circ U')$ in the following diagram of categories and functors

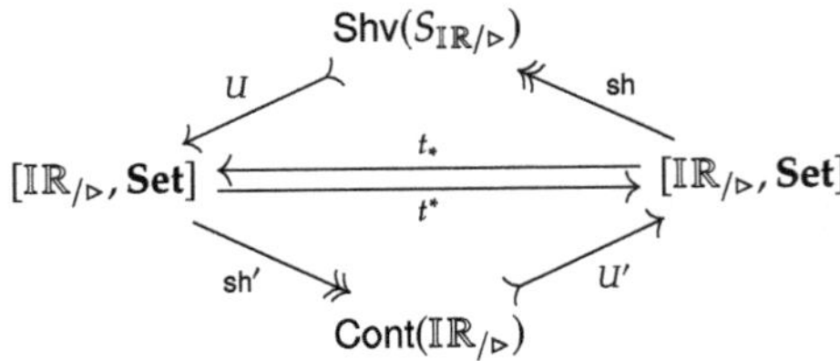

where sh is the associated sheaf functor, which is also the fixed-point functor of t_*, and sh' is the fixed point of t^*. The equivalence itself is proven in Theorem B.28.

Remark B.27 While we have been using $\mathbb{IR}_{/\rhd}$ as our primary example throughout this section, the proof of Theorem B.28 is easily generalized to any continuous category, including any domain. See also [JJ82, Prop. 3.3, Thm. 3.6, Prop. 3.8].

For any domain D, this means that there is an equivalence of categories

$$\mathsf{Shv}(D^{\mathrm{op}}) \cong \mathsf{Cont}(D),$$

where D^{op} is the posite defined in Proposition 2.20 (regarding $(D, \ll)$ as a predomain, which always has the property that $d \leqslant d'$ implies $d \lessdot d'$).

Theorem B.28 *There is an equivalence of categories between those functors* $\mathbb{IR}_{/\rhd} \to$ **Set** *that satisfy the sheaf condition and those that preserve filtered colimits:*

$$\mathsf{Shv}(S_{\mathbb{IR}/\rhd}) \cong \mathsf{Cont}(\mathbb{IR}_{/\rhd}) \tag{B.13}$$

Proof First, note that the result will follow if we can show $t^*t_* \cong t^*$ and $t_*t^* \cong t_*$. Indeed, this implies that t^* restricted to the image of t_* and t_* restricted to the image of t^* are inverses. We sketch a proof that there is an isomorphism $t^*t_* \cong t^*$; the isomorphism $t_*t^* \cong t_*$ is likewise straightforward to verify directly.

Fix any object $\ell' \in \mathbb{IR}_{/\rhd}$. By Eqs. (B.10) and (B.12) an element of $(t^*t_*X)(\ell')$ is represented by an object ℓ and a wavy arrow $\ell \rightsquigarrow \ell'$ in $\mathbb{IR}_{/\rhd}$, together with a natural transformation $\phi \colon \langle\!\langle \ell, - \rangle\!\rangle \to X$, i.e. a compatible family of sections of X, one for each interval way above ℓ. Given such data, we can choose an intermediate interval $\ell \rightsquigarrow \ell'' \rightsquigarrow \ell'$ by the isomorphism (B.2), and an element $\phi(\ell \rightsquigarrow \ell'') \in X(\ell'')$. The object $\ell'' \in \mathbb{IR}_{/\rhd}$ together with the element $\phi(\ell'' \rightsquigarrow \ell')$ represents an element of $(t^*X)(\ell')$. It is not difficult to verify, again using Eqs. (B.10), (B.12), and (B.2) that this does not depend on the choice of representative, and that it is injective and surjective. □

Glossary of Symbols

Symbol	Meaning
$[\![P]\!]$	The semantic meaning of a type or term P
im	The image of a map
$\rightsquigarrow$	A wavy-arrow, in the sense of Johnstone-Joyal
$\mathbb{N}$	The set of natural numbers
$\mathbb{Q}$	The set of rational numbers
$\mathbb{Q}^{\infty}$	The set $\mathbb{Q} \cup \{-\infty, \infty\}$ of unbounded rational numbers
$\mathbb{Z}$	The set of integers
$\mathbb{R}$	The set of real numbers
$\mathbb{R}^{\infty}$	The set $\mathbb{R} \cup \{-\infty, \infty\}$ of unbounded real numbers
$\mathbb{R}^{\infty}_{\geq 0}$	The set of nonnegative unbounded real numbers
$\mathbb{I}\mathbb{R}$	The domain of (compact real) intervals $d \leq u \in \mathbb{R}$
$\overline{\underline{\mathbb{R}}}$	The domain of extended intervals $d, u \in \mathbb{R}$
$\underline{\mathbb{R}}$	The domain of lower reals
$\overline{\mathbb{R}}$	The domain of upper reals
$\underline{\mathbb{R}}^{+}$	The domain of nonnegative lower reals
$\overline{\mathbb{R}}^{+}$	The domain of nonnegative upper reals
$\underline{\mathbb{R}}^{\infty}$	The domain of unbounded lower reals
$\overline{\mathbb{R}}^{\infty}$	The domain of unbounded upper reals
$\mathbb{I}\mathbb{R}^{\infty}$	The domain of unbounded intervals
$\overline{\underline{\mathbb{R}}}^{\infty}$	The domain of unbounded extended intervals
$\mathbb{I}\mathbb{R}_{\text{pre}}$	The predomain of intervals
$\overline{\underline{\mathbb{R}}}_{\text{pre}}$	The predomain of extended intervals
$\underline{\mathbb{R}}_{\text{pre}}$	The predomain of lower reals
$\overline{\mathbb{R}}_{\text{pre}}$	The predomain of upper intervals
$\overline{\underline{\mathbb{R}}}^{\infty}_{\text{pre}}$	The predomain of unbounded extended intervals
$\mathbb{I}\mathbb{R}^{\infty}_{\text{pre}}$	The predomain of unbounded intervals
$\underline{\mathbb{R}}^{\infty}_{\text{pre}}$	The predomain of unbounded lower reals
$\overline{\mathbb{R}}^{\infty}_{\text{pre}}$	The predomain of unbounded upper reals
$\underline{\mathbb{R}}^{+}_{\text{pre}}$	The predomain of nonnegative lower reals

P. Schultz, D. I. Spivak, *Temporal Type Theory*, Progress in Computer Science and Applied Logic 29, https://doi.org/10.1007/978-3-030-00704-1

$\bar{\mathbb{R}}^{+}_{\mathrm{pre}}$ The predomain of nonnegative upper reals
$\mathbb{N}$ The type of natural numbers (internal natural numbers object)
$\mathbb{Q}$ The type of rationals
$\mathbb{Z}$ The type of integers
$\mathbb{Q}^{\infty}$ The type of unbounded rationals
$\mathbb{R}$ The type of Dedekind reals
$\mathbb{IR}$ The type of Dedekind intervals
$\bar{\underline{\mathbb{R}}}$ The type of Dedekind extended intervals
$\underline{\mathbb{R}}$ The type of Dedekind lower reals
$\bar{\mathbb{R}}$ The type of Dedekind upper reals
$\mathbb{R}^{\infty}$ The type of Dedekind unbounded reals
$\mathbb{IR}^{\infty}$ The type of Dedekind unbounded intervals
$\underline{\mathbb{R}}^{\infty}$ The type of Dedekind unbounded lower reals
$\bar{\mathbb{R}}^{\infty}$ The type of Dedekind unbounded upper reals
$\bar{\underline{\mathbb{R}}}^{\infty}$ The type of Dedekind unbounded extended intervals
$\mathrm{c}\mathbb{R}$ The type of constant Dedekind reals
$\mathrm{c}\mathbb{IR}$ The type of constant Dedekind intervals
$\mathrm{c}\bar{\underline{\mathbb{R}}}$ The type of constant Dedekind extended intervals
$\mathrm{c}\underline{\mathbb{R}}$ The type of constant Dedekind lower reals
$\mathrm{c}\bar{\mathbb{R}}$ The type of constant Dedekind upper reals
$\mathrm{c}\mathbb{R}^{\infty}$ The type of constant Dedekind unbounded reals
$\mathrm{c}\mathbb{IR}^{\infty}$ The type of constant Dedekind unbounded intervals
$\mathrm{c}\bar{\underline{\mathbb{R}}}^{\infty}$ The type of constant Dedekind unbounded extended intervals
$\mathrm{c}\underline{\mathbb{R}}^{\infty}$ The type of constant Dedekind unbounded lower reals
$\mathrm{c}\bar{\mathbb{R}}^{\infty}$ The type of constant Dedekind unbounded upper reals
$\mathbb{IR}_{\mathrm{pre}}$ The type of the predomain for intervals
$\bar{\underline{\mathbb{R}}}_{\mathrm{pre}}$ The type of the predomain for extended intervals
$\underline{\mathbb{R}}_{\mathrm{pre}}$ The type of the predomain for lower reals
$\bar{\mathbb{R}}_{\mathrm{pre}}$ The type of the predomain for upper reals
$\bar{\underline{\mathbb{R}}}^{\infty}_{\mathrm{pre}}$ The type of the predomain for unbounded extended intervals
$\mathbb{IR}^{\infty}_{\mathrm{pre}}$ The type of the predomain for unbounded intervals
$\underline{\mathbb{R}}^{\infty}_{\mathrm{pre}}$ The type of the predomain for unbounded lower reals
$\bar{\mathbb{R}}^{\infty}_{\mathrm{pre}}$ The type of the predomain for unbounded upper reals
$\underline{\mathbb{R}}^{+}_{\mathrm{pre}}$ The type of the predomain for nonnegative lower reals
$\bar{\mathbb{R}}^{+}_{\mathrm{pre}}$ The type of the predomain for nonnegative upper reals
$\mathbb{IR}_{/\rhd}$ The continuous category of translation-invariant intervals
$\mathtt{C}$ A type of constants
$\mathtt{Prop}$ The type of propositions (internal version of subobject classifier)
$\mathtt{Time}$ The type of perfect-clock behaviors in $\mathcal{B}$
Hyb The type associated to a hybrid datum
$\frac{d}{dt}$ Derivative of a variable real
π The pointwise modality
$@^{t}_{[d,u]}$ The "At"-modality corresponds to a point of the topos
$\downarrow^{t}_{[d,u]}$ The "See"-modality corresponds to a closed subtopos
$\uparrow^{t}_{[d,u]}$ The "In"-modality corresponds to an open subtopos

AD	Approximates the derivative
$\mathrm{Fn}(X)$	The sheaf of local functions to a topological space X
unit_speed	The atomic proposition defining `Time`.
$\mathsf{Ind}\text{-}C$	The category of Ind--objects in C
$\mathsf{Psh}(C)$	The topos of presheaves on C
$\mathsf{Shv}(C)$	The topos of sheaves on C
$\mathsf{Cont}(C)$	The category of continuous set-valued functors on a continuous category C
$\mathsf{Tw}(C)$	The twisted arrow category of C
B	The "delooping" of a monoid, i.e. the associated category with one object
Ω_{Filt}	The open filters in a domain
$\mathtt{1}$	A terminal object in a category
Poset	The category of posets
Top	The category of topological spaces
Predom	The category of predomains and approximable mappings
2	The poset $0 < 1$
$\mathcal{B}$	The topos of behavior types
$S_{\mathbb{IR}/\rhd}$	The base site for $\mathcal{B}$, i.e. that of translation-invariant intervals
$\mathrm{el}F$	The category of elements of a functor $F: C \to \mathbf{Set}$
$\curlywedge$	A meet in a predomain
$\curlyvee$	A join in a predomain
sh	The associated sheaf
$\langle r, s\rangle$	The map from a subinterval into an interval, with space r on the left and s on the right
$X\big\vert_{\langle r,s\rangle}$	$X's$ restriction map to a subinterval defined by $\langle r, s\rangle$.
$\rhd$	A basic cover in a posite
$\sqsubseteq$	Order relation in a domain
$\uparrow$	The up-closure operation
$\downarrow$	The down-closure operation
$\twoheaduparrow$	The way-up closure operation (in a domain)
$\twoheaddownarrow$	The way-down closure operation (in a domain)
$\leqslant$	The lower specialization order on a predomain
$\lessdot$	The upper specialization order on a predomain
$\mathsf{Shv}_0(C)$	(0, 1)-sheaves on a posite
Id	Ideals in a poset, i.e. directed and down-closed subsets
RId	Rounded ideals in a predomain
cRId	Constant rounded ideals in a predomain
Ω_{Filt}	Rounded filters in a predomain
Ω	Opens in a predomain
#	Apartness relation
$\langle\!\langle -, -\rangle\!\rangle$	Wavy-arrows profunctor
W	Currying of the wavy-arrows profunctor
$\mathcal{U}$	The until operator from linear temporal logic
$\mathcal{S}$	The since operator from linear temporal logic

Bibliography

[AFH96] Alur, R., Feder, T., Henzinger, T.A.: The benefits of relaxing punctuality. J. ACM (JACM) **43**(1), 116–146 (1996)

[AGV71] Artin, M., Grothendieck, A., Verdier, J.-L.: Theorie de Topos et Cohomologie Etale des Schemas I, II, III. Lecture Notes in Mathematics, vols. 269, 270, 305. Springer, Berlin (1971)

[Ash13] Ashby, W.: Design for a Brain: The Origin of Adaptive Behaviour. Springer, Berlin (2013)

[ÅW13] Åström, K.J., Wittenmark, B.: Adaptive Control. Courier Corporation, New York (2013)

[Awo16] Awodey, S.: Natural models of homotopy type theory. Math. Struct. Comput. Sci. 1–46 (2016). arXiv:1406.3219 [math.CT]

[BF00] Bunge, M., Fiore, M.P.: Unique factorisation lifting functors and categories of linearly-controlled processes. Math. Struct. Comput. Sci. **10**(2), 137–163 (2000)

[BJ81] Boileau, A., Joyal, A.: La logique des topos. J. Symb. Log. **46**(1), 6–16 (1981). ISSN:0022-4812. http://dx.doi.org/10.2307/2273251

[Bor94] Borceux, F.: Handbook of Categorical Algebra. 1. Encyclopedia of Mathematics and Its Applications. Basic Category Theory, vol. 50. Cambridge University Press, Cambridge (1994)

[BT09] Bauer, A., Taylor, P.: The Dedekind reals in abstract Stone duality. Math. Struct. Comput. Sci. **19**(4), 757–838 (2009). ISSN:0960-1295. http://dx.doi.org/10.1017/S0960129509007695

[CH88] Coquand, T., Huet, G.: The calculus of constructions. Inform. Comput. **76**(2–3), 95–120 (1988). ISSN:0890-5401. http://dx.doi.org/10.1016/0890-5401(88)90005-3

[Ded72] Dedekind, R.: Stetigkeit und irrationale Zahlen. F. Vieweg und sohn (1872)

[Fer16] Ferrari, L.: Dyck algebras, interval temporal logic, and posets of intervals. SIAM J. Discrete Math. **30**, 1918–1937 (2016)

[Fio00] Fiore, M.P.: Fibred models of processes: discrete, continuous, and hybrid systems. In: IFIP TCS, vol. 1872, pp. 457–473. Springer, Heidelberg (2000)

[Fou77] Fourman, M.P.: The logic of topoi. Stud. Logic Found. Math. **90**, 1053–1090 (1977)

[Gie+03] Gierz, G., et al.: Continuous Lattices and Domains. Encyclopedia of Mathematics and Its Applications, vol. 93, pp. xxxvi+591. Cambridge University Press, Cambridge (2003). ISBN:0-521-80338-1. http://dx.doi.org/10.1017/CBO9780511542725

[GKT15] Gálvez-Carrillo, I., Kock, J., Tonks, A.: Decomposition spaces, incidence algebras and M obius inversion I: basic theory (2015). eprint: arXiv:1512.07573

P. Schultz, D. I. Spivak, *Temporal Type Theory*, Progress in Computer Science and Applied Logic 29, https://doi.org/10.1007/978-3-030-00704-1

[Gol11] Goldsztejn, A.: Modal intervals revisited, part 1: a generalized interval natural extension. Reliab. Comput. **16**, 130–183 (2011). ISSN:1573–1340

[HOW13] Hunter, P., Ouaknine, J., Worrell, J.: Expressive completeness for metric temporal logic. In: Proceedings of the 2013 28th Annual ACM/IEEE Symposium on Logic in Computer Science, pp. 349–357. IEEE Computer Society (2013)

[HS91] Halpern, J.Y., Shoham, Y.: A propositional modal logic of time intervals. J. ACM (JACM) **38**(4), 935–962 (1991)

[HTP03] Haghverdi, E., Tabuada, P., Pappas, G.: Bisimulation relations for dynamical and control systems. Electron. Notes Theor. Comput. Sci. **69**, 120–136 (2003)

[Jac99] Jacobs, B.: Categorical Logic and Type Theory. Studies in Logic and the Foundations of Mathematics, vol. 141, pp. xviii+760. North-Holland, Amsterdam (1999). ISBN:0-444-50170-3

[JJ82] Johnstone, P., Joyal, A.: Continuous categories and exponentiable toposes. J. Pure Appl. Algebra **25**(3), 255–296 (1982). ISSN:0022-4049. http://dx.doi.org/10.1016/0022-4049(82)90083-4

[JNW96] Joyal, A., Nielsen, M., Winskel, G.: Bisimulation from open maps. Inform. Comput. **127**(2), 164–185 (1996)

[Joh02] Johnstone, P.T.: Sketches of an Elephant: A Topos Theory Compendium. Oxford Logic Guides, vol. 43, pp. xxii+468+71. New York: The Clarendon Press/Oxford University Press (2002). ISBN:0-19-853425-6

[Joh99] Johnstone, P.: A note on discrete Conduché fibrations. Theory Appl. Categ. **5**(1), 1–11 (1999). ISSN:1201-561X

[Kau80] Kaucher, E.: Interval analysis in the extended interval space IR. In: Fundamentals of Numerical Computation (Computer-Oriented Numerical Analysis), pp. 33–49. Springer, Berlin (1980)

[Koc95] Kock, A.: Monads for which structures are adjoint to units. J. Pure Appl. Algebra **104**(1), 41–59 (1995). ISSN:0022-4049. http://dx.doi.org/10.1016/0022-4049(94)00111-U

[KS06] Kashiwara, M., Schapira, P.: Categories and Sheaves. Grundlehren der Mathematischen Wissenschaften [Fundamental Principles of Mathematical Sciences], vol. 332, pp. x+497. Springer, Berlin (2006). ISBN:978-3-540-27949-5. http://dx.doi.org/10.1007/3-540-27950-4

[Law86] Lawvere F.W.: State categories and response functors. Dedicated to Walter Noll. Preprint (May 1986)

[LS88] Lambek, J., Scott, P.J.: Introduction to Higher Order Categorical Logic. Cambridge Studies in Advanced Mathematics, vol. 7, pp. x+293. Reprint of the 1986 original. Cambridge University Press, Cambridge (1988). ISBN:0-521-35653-9

[Mai05] Maietti, M.E.: Modular correspondence between dependent type theories and categories including pretopoi and topoi. Math. Struct. Comput. Sci. **15**(6), 1089–1149 (2005). ISSN:0960-1295. http://dx.doi.org/10.1017/S0960129505004962

[MM92] MacLane, S., Moerdijk, I.: Sheaves in Geometry and Logic: A First Introduction to Topos Theory. Springer, New York (1992). ISBN:0387977104

[MN04] Maler, O., Nickovic, D.: Monitoring temporal properties of continuous signals. In: FORMATS/FTRTFT, vol. 3253, pp. 152–166. Springer, Heidelberg (2004)

[Mou+15] de Moura, L., et al.: The Lean theorem prover (system description). In: International Conference on Automated Deduction, pp. 378–388. Springer, Heidelberg (2015)

[MP06] Martin, K., Panangaden, P.: A domain of spacetime intervals in general relativity. Commun. Math. Phys. **267**(3), 563–586 (2006)

[nLaba] Nerve and Realization. http://ncatlab.org/nlab/show/nerve$+$and$+$realization (visited on 01/19/2016)

[nLabb] Posite. https://ncatlab.org/nlab/show/posite (visited on 06/02/2017)

[PP12] Picado, J., Pultr, A.: Frames and Locales. Frontiers in Mathematics. Topology Without Points, pp. xx+398. Birkhäuser/Springer Basel AG, Basel (2012). ISBN:978-3-0348-0153-9. http://dx.doi.org/10.1007/978-3-0348-0154-6

[RU12] Rescher, N., Urquhart, A.: Temporal Logic, vol. 3. Springer, New York (2012)

[Sai+14] Sainz, M.A., et al.: Modal Interval Analysis. Lecture Notes in Mathematics. New Tools for Numerical Information, vol. 2091, pp. xvi+316. Springer, Cham (2014). ISBN:978-3-319-01720-4. http://dx.doi.org/10.1007/978-3-319-01721-1

[Shu12] Shulman, M. Exact completions and small sheaves. Theory Appl. Categ. **27**, 97–173 (2012). ISSN:1201-561X

[Str05] Streicher, T.: Universes in toposes. In: From Sets and Types to Topology and Analysis. Oxford Logic Guides, vol. 48, pp. 78–90. Oxford Univ. Press, Oxford (2005). http://dx.doi.org/10.1093/acprof:oso/9780198566519.003.0005

[SVS16] Spivak, D.I., Vasilakopoulou, C., Schultz, P.: Dynamical Systems and Sheaves (2016). eprint: arXiv:1609.08086

[SW73] Street, R., Walters, R.F.C.: The comprehensive factorization of a functor. Bull. Am. Math. Soc. **79**, 936–941 (1973). ISSN:0002-9904. http://dx.doi.org/10.1090/S0002-9904-1973-13268-9

[Vic93] Vickers, S.: Information systems for continuous posets. Theor. Comput. Sci. **114**(2), 201–229 (1993). ISSN:0304-3975. http://dx.doi.org/10.1016/0304-3975(93)90072-2

[Voe+13] Voevodsky, V., et al.: Homotopy type theory: Univalent foundations of mathematics. In: Institute for Advanced Study (Princeton). The Univalent Foundations Program (2013)

[Wil07] Willems, J.C.: The behavioral approach to open and interconnected systems. IEEE Control Syst. **27**(6), 46–99 (2007)

Index

P. Schultz, D. I. Spivak, *Temporal Type Theory*, Progress in Computer Science and Applied Logic 29, https://doi.org/10.1007/978-3-030-00704-1

W

Z

GPSR Compliance
The European Union's (EU) General Product Safety Regulation (GPSR) is a set of rules that requires consumer products to be safe and our obligations to ensure this.

If you have any concerns about our products, you can contact us on

ProductSafety@springernature.com

In case Publisher is established outside the EU, the EU authorized representative is:

Springer Nature Customer Service Center GmbH
Europaplatz 3
69115 Heidelberg, Germany

www.ingramcontent.com/pod-product-compliance
Ingram Content Group UK Ltd.
Pitfield, Milton Keynes, MK11 3LW, UK
UKHW021010290726
14059UKWH00001BA/64

* 9 7 8 3 0 3 0 0 0 7 0 3 4 *